RECENT ADVANCES IN...
NUCLEAR STRUCTURE

RECENT ADVANCES IN...

NUCLEAR STRUCTURE

PREDEAL, ROMANIA   28 AUG – 8 SEPT 1990

edited by

D. Bucurescu, G. Cata-Danil, N.V. Zamfir
Institute of Atomic Physics, Bucharest, Romania

World Scientific
Singapore • New Jersey • London • Hong Kong

Published by

World Scientific Publishing Co. Pte. Ltd.
P O Box 128, Farrer Road, Singapore 9128
USA office: 687 Hartwell Street, Teaneck, NJ 07666
UK office: 73 Lynton Mead, Totteridge, London N20 8DH

RECENT ADVANCES IN NUCLEAR STRUCTURE

Copyright © 1991 by World Scientific Publishing Co. Pte. Ltd.

All rights reserved. This book, or parts thereof, may not be reproduced in any form or by any means, electronic or mechanical, including photocopying, recording or any information storage and retrieval system now known or to be invented, without written permission from the Publisher.

ISBN 981-02-0533-3

Printed in Singapore by Utopia Press.

PREFACE

These Proceedings contain the lectures delivered at the 1990 Predeal International School on **"Recent Advances in Nuclear Structure"**.

This edition continued the tradition of the international nuclear physics schools organized by the Institute of Atomic Physics, Bucharest. This time, after being organized for some time in Poiana Brasov, the School was held again in Predeal, the original place where it started back in the late sixties.

The subjects of the lectures covered some of the most exciting areas of the nuclear structure domain, both from the experimental and theoretical point of view. These concern mainly the reaching of the nuclei far from the stability valley and the study of their structure and decay modes. Modern techniques such as various isotope separation devices and large detector arrays for gamma ray spectroscopy were presented, along with many new results concerning the properties of light exotic nuclei, high spin states and superdeformed nuclei, beta decay and cluster radioactivities and systematics of different nucler structure properties. New aspects of the nuclear structure theories have also been represented.

The School was attended by some 100 participants from 15 countries, who enjoyed a nice scientific and social atmosphere.

We wish to thank all the invited lecturers for their effort of presenting high level courses in the most attractive manner. We acknowledge the fact that the success of the School is also due to the active participation of the students both through short contributions and lively discussion.

We would like to thank the Institute of Atomic Physics for continuous support, and especially its Director General Dr. Gheorghe Pascovici, who offered the organizing committee generous help during the whole period of organizing this School.

We acknowledge the warm hospitality of the Predeal officials.

Finally, we are very grateful to Mrs. Aurora Anitoaiei, the technical secretary of the School, for her high competence in running most of the harsh work.

The Editors

CONTENTS

Preface v

Large arrays of γ-ray detectors for nuclear structure studies:
Present status and future prospects
    J. Gizon 1

Recent progress in the study of light nuclei far from stability
using GANIL facilities
    F. Pougheon 29

Gamow-Teller beta decay of proton-rich nuclei
    O. Klepper, K. Rykaczewski 45

Nuclear structure studies by using He-jet and isotope separator
facilities at SARA
    A. Gizon, J. Gizon 61

Recent results from the Daresbury nuclear structure facility
    D. D. Warner 79

Influence of occupied orbitals on the band strucutre in the $A \approx 80$
region: Excited states in ^{84}Mo, ^{83}Nb and ^{83}Zr
    C. J. Gross, W.Gelletly, D. Rudolph, B. J. Varley, K. P. Lieb,
    M. A. Bentley, J. L. Durell, H. G. Price, J. Simpson,
    O. Skeppstedt, S. Ranstikerdar 101

Refined transient field G factor measurements
    F. Brandolini 113

The role of the γ degree of freedom in the $A = 130$ region
    P. von Brentano, D. Lieberz, W. Lieberz, A. Gelberg,
    A. Granderath, A Dewald, U. Neuneyer, F. Seiffert,
    H. Wolters, J. Yan 129

Effects of the residual proton-neutron interaction in the
development of collectivity in nuclei
    R. F. Casten 153

Physics of superdeformed bands
  W. Nazarewicz 175

Quasiparticle-phonon nuclear model and strucutre of
nonrotational states in deformed nuclei
  V. G. Soloviev 215

Description of the low-lying isovector 1^+ states
  A. Faessler 237

Low lying dipole modes in the rare earth region
  P. von Brentano, A. Zilges, R. D. Heil,
  U. Kneissl, H. H. Pitz and C. Wesselborg 275

Direct and multiple excitations in magic ^{96}Zr from 22 MeV $(\vec{d}, d')$
  D.Hofer, M. Bisenberger, **G. Graw**, R. Hertenberger,
  H. Kader, P. Schiemenz, G. Molnar 283

Linearised collective Schrödinger equation for nuclear
quadrupole surface vibrations
  M.Greiner, D. Heumann, W. Scheid 289

Microscopic insights into the shape coexistence in the $A = 70$ region
  A. Petrovici 305

Axially deformed nuclei in the relativistic mean field theory
  V. Blum, J. Fink, B. Waldhauser, **J. A. Maruhn**,
  P. G. Reinhard, W. Greiner 333

Grand unification, nuclear structure and the double beta-decay
  A. Faessler 347

Parity nonconservation in nuclear reactions and alpha-decay
  O. Dumitrescu 359

Cluster radioactivities in various regions of parent nuclei
  D. N. Poenaru 393

Exotic radioactivity: Clusters as solitons on the nuclear surface
  A. Ludu, **A. Sandulescu** 413

Heavy fragment radioactivity
  I. Silisteanu, M. Ivascu, I. Rotter 435

Towards a beam of polarized antiprotons at LEAR
  G. Graw 459

Contemporary interest in cosmic rays.
Aspects of the KASCADE Project
    H. Rebel 479

Concluding remarks
    V. G. Soloviev 509

RECENT ADVANCES IN...
NUCLEAR STRUCTURE

LARGE ARRAYS OF γ-RAY DETECTORS FOR NUCLEAR STRUCTURE STUDIES : PRESENT STATUS AND FUTURE PROSPECTS

Jean GIZON

Institut des Sciences Nucléaires
IN2P3-CNRS / Université Joseph Fourier
Grenoble, France

Abstract : Some conditions required for detector arrays used for nuclear structure studies are discussed. Several of these arrays consisting of inner ball and anti-Compton shields surrounding germanium detectors are briefly described and examples of data obtained with these multidetectors presented. The new generation of arrays (Eurogam, Gammasphere, Euroball) is also briefly discussed.

1 - Introduction

With heavy-ion beams it is possible to excite nuclei up to very high angular momentum states and by means of 4π detector arrays with large efficiency one can study the discrete γ-ray spectra and the γ-ray continuum emitted by these nuclei. The use of powerfull arrays has permitted to extend the knowledge of collective properties and single-particle excitations, and also to discover new phenomena, as for example superdeformation. The field of cold and hot nuclei under extreme conditions is now accessible. The existing detector arrays give access to weak amplitude phenomena and the new generation of arrays is designed in such a way that much weaker ones will be observable.

The aim of this lecture is to describe the main characteristics and performances of arrays used to detect gamma rays. There are many γ-ray multidetectors in the world and all have produced interesting results in the field of nuclear structure but it would take too long to give all the details for all these arrays. Therefore I limit the description to only a few ones with typical designs and those I have personally used.

This lecture is organised in the following way : the first part consists of a brief discussion of general characteristics required for γ-ray multidetectors. Then several arrays are described with some examples of results. The last part is devoted to larger devices which are under construction or at the planning stage.

2 - Some conditions required for the detector arrays

Many factors have to be considered when designing γ-ray detector arrays, in particular, quality of detection which is fulfilled by using Ge detectors equipped with Compton suppressors, large efficiency for γ-ray detection, existence of inner ball with large number of elements. Only these three points are developed in the following sections. One has also to emphasize the important role of electronics and hardware which must accomodate high counting rates.

2.1 - Rejection of Compton events

All the arrays used to detect γ-rays consist of germanium detectors surrounded with anti-Compton shields.

Most of the counts in a γ-ray spectrum obtained with a Ge crystal are Compton events and only a weak percentage is effectively in the photopeaks. An important quantity is the peak to total ratio P/T which, for a monoenergetic γ-ray, is P/T = number of counts in the photopeak divided by total number of counts. In a γ-γ coincidence experiment involving only unsuppressed detectors with $[P/T]_u \sim 0.20$ for 1 MeV γ-ray, only ~ 4% of the counts are in the photopeaks, the rest being distributed in the background (table 1).

If one uses Ge detectors equipped with Compton suppressors having $[P/T]_s \sim 0.60$, the photoelectric events represent 36% of the total number of counts. Thus the quality of the response of the detectors has been considerably improved, i.e. by a factor of 9. This improvement is even larger for 3-fold, 4-fold...coincidences since it varies as $([P/T]_s/[P/T]_u)^n$.

Table 1

Estimated P/T peak to total ratios without and with Compton suppression and improvement of the response as a function of foldness of coincident γ-rays.

Fold	1	2	3	4	5
$[P/T]_u$	0.20	0.04	0.008	0.0016	0.0003
$[P/T]_s$	0.60	0.36	0.22	0.13	0.08
Improvement	3	9	27	81	243

The suppression of Compton events is the first condition a multidetector must satisfy. The design of an anti-Compton shield surrounding a germanium crystal is closely related to the Compton cross-section per electron given by the formula of Klein-

Nishina. High density materials are required for Compton suppression. The first suppressors were made of NaI crystals and at present, bismuth germanate scintillators ($Bi_4 Ge_3 O_{12}$), commonly called BGO, are used in most of the devices. Due to the high cost of BGO, compromises are sometimes adopted [1] by using both NaI and BGO in the same suppressor as shown in fig.1.

2.2 - Gamma-ray detection

A good energy resolution is absolutely necessary for all measurements of discrete γ-ray spectra. This implies the use of coaxial hyperpure Ge crystals. In order to reduce the effects of neutron damages, n-type Ge are usually used for this purpose.

High efficiency γ-ray detection is required for in-beam nuclear structure studies in order to reduce the beam time necessary to carry out coincidence experiments which very often are relative to weak amplitude phenomena. This is achieved mainly by improving the quality of the response of the anti-Compton spectrometers and increasing the number and the total solid angle subtented by the Ge detectors.

The number of n-fold coincidences is proportional to $(N\Omega\varepsilon_p\iota_s)^n$ where N is the number of Ge detectors, Ω the solid angle and ε_p the total photopeak efficiency of each of them, ι_s, the isolated hit probability. ι_s is defined by $\iota_s \approx (1 - \varepsilon'_p)^{M_\gamma-1}$ where M_γ is the γ-ray multiplicity and ε'_p the efficiency for an individual detector after correcting for Compton suppression. Characteristics of some arrays are given in table 2. For those wich are in operation, the product $N\Omega\varepsilon_p\iota_s$ showing the performances of the device never exceeds ~.0 2. One expects a big improvement only from the new generation of arrays (Eurogam, Gammasphere, Euroball).

Table 2

Characteristics of some arrays which are in operation or planned.

	N	$\Omega \times 10^{-3}$	$N\Omega$	ε_p	ι_s ($M_\gamma = 30$)	$N\Omega \varepsilon_p \iota_s$	P/T*
TESSA 3	16	2.3	0.037	0.13	0.96	0.005	0.46
HERA	21	6.0	0.13	0.12	0.90	0.014	0.41
CHATEAU II	12	4.0	0.054	0.20	0.94	0.007	0.41
NORDBALL	20	6.6	0.13	0.15	0.90	0.018	0.42
8π	20	2.5	0.050	0.13	0.96	0.006	0.46
EUROGAM	70	5.1	0.36	0.20	0.94	0.067	0.49
GAMMASPHERE	110	4.2	0.46	0.20	0.93	0.085	0.49
EUROBALL (cluster array)	492	1.4	0.68	0.31	–	~ 0.20	0.70

* P/T = P/T (intrinsic) x ι_s x ι_n. See section 4.1.1.

2.3 - Inner ball

Many detector arrays are equipped with an inner ball which measures the total γ-ray energy emitted by the compound nucleus and the γ-ray multiplicity. These two quantities are directly related to the total energy E of the compound nucleus and its angular momentum I, respectively. Therefore it is possible with an inner ball to set gates on these two quantities and select a portion of the (E, I) plane and also measure γ-ray multiplicity distributions and entry regions. This equipment is particularly useful to enhance a given exit channel in an heavy ion induced reaction.

The energy selection of an exit channel is satisfactory if the inner ball covers a nearly 4π solid angle and has a large efficiency (several centimeter thickness of heavy scintillator like BGO is needed). This inner ball must be composed of a large number of elements to select open channels by dealing with γ-ray multiplicity since the resolution of the multiplicity distribution is strongly dependent on the number of detectors. Indeed for a given array with N detectors and for M emitted γ-rays, the probability that out of these N detectors p of them fire is :

$$P_{Np}(M) = \binom{N}{p} \sum_{n=0}^{p} (-1)^{n+p} \binom{p}{n} G(n).$$

G(n) is a function of multiplicity, efficiency Ω and diffusion [2]. As an example, for the Oak Ridge nuclear-spin spectrometer with 72 NaI detectors and Ω = 0.94, the calculated width of this probability (distribution in fold p) is of the order of 25% when M = 30 [3].

3 - Description of some arrays and examples of results

As said in the introduction, this chapter is limited to the description of only a few multidetectors. Others of excellent quality are in operation in many countries and the choice I have made is to a large extent arbitrary.

3.1 - TESSA3

TESSA (Total Energy Suppression Spectrometer Array) which was the first array of escape suppressed spectrometers consisted of five NaI shields surrounding Ge(Li) detectors [4]. It has been modified and improved several times and at present, TESSA3 is installed on a beam line of the Tandem accelerator at the Daresbury Laboratory (UK).

The first version of TESSA3 consisted of 12 BGO anticompton shields of the type shown in fig.1. The shields were in two rings of six at ±19° to the horizontal plane with detectors situated at 30°, 90° and 150° to the beam direction. The Ge detectors were placed at 23 cm from the target. This version was augmented with 4 additional shields in the horizontal plane, two at 60° and two at 120° to the beam (fig.2). TESSA3 is equipped with an inner ball of 50 hexagonal BGO crystals having nearly 4π solid angle and γ-ray efficiency of the order of unity [5].

Discovery of superdeformation

One of the best example to illustrate the performances of TESSA is the discovery of the superdeformation in ^{152}Dy. The first results were obtained by means of TESSA2 by Nyako et al. [6] who found a very large deformation in this nucleus by studying a γ-ray energy correlation matrix. The dynamical moment of inertia $J^{(2)} = (85 \pm 2)\hbar^2$ MeV^{-1} extracted from the width W of the valley along the main diagonal of this matrix (W = 8 $\hbar^2/J^{(2)}$) implied a very large deformation $\varepsilon_2 > 0.5$ (fig. 3).

Then, after these results deduced from continuum γ-ray spectra, experiments were performed by using TESSA3 equipped with 12 BGO suppressors and an inner ball to search for discrete γ-lines in this superdeformed band. A set of 19 equidistant discrete γ-rays ranging from 602 to 1449 keV was effectively observed [7] (fig.4). They are separated by 47 keV which is exactly half the full width of the valley in the correlation matrix. These superdeformed states are weakly fed (< 1% of the feeding of the lowest yrast states) and they have been detected only thanks to the use of a powerfull array.

3.2 - ESSA30

A special spectrometer was designed in order to put together a large number of detectors. ESSA30 (European Suppressed Shield Array) [8] which was the array with the largest number of Ge detectors was assembled in 1987 for less than a year and installed on a beam line at the NSF, Daresbury Laboratory. It was a common realisation of UK, Denmark, Federal Republic of Germany, Finland, Italy and Sweden. In this array (fig.5) which had no inner ball, six groups of five detectors were located at 37°, 63°, 79°, 101°, 117°, and 143° to the beam.

Neutron and proton $i_{13/2}$ alignments in ^{174}Os

The results presented in this section are part of a systematic study undertaken with ESSA 30 to study transitional nuclei with 82 > Z, N < 100. The aim was to evaluate the structure of the yrast region in these γ-soft nuclei and to establish the nature of the band crossings. I present here some results of a part of this study, namely the high spin structure of ^{174}Os.

High spin states in ^{174}Os were populated in the reaction ^{32}S (^{146}Nd, 4n) at 166 MeV. The level scheme obtained in this investigation [9] is presented in fig.6. Compared to previous studies [10,11] our data extends to higher spin and in addition a new side band has been established.

The structure of ^{174}Os is expected to be dominated by the $\nu i_{13/2}$, $\nu f_{7/2}$, $\pi h_{9/2}$ and $\pi i_{13/2}$ orbitals. The ground state band (gsb) shows a smooth increase of i_x with $\hbar\omega$ (fig.7). However, on the $J^{(2)}$ plot (fig.7) the irregularities can be observed with a peak at $\hbar\omega = 0.3$ MeV corresponding to a band crossing. The first crossing in the gsb is mostly due to the $i_{13/2}$ neutron alignment. The Routhians for the proton and neutron configurations calculated using the Woods-Saxon potential (fig.8) reveal a possibility of a proton crossing at $\hbar\omega \approx 0.55$ MeV which is slightly above the expected second neutron crossing. Since according to the calculations at higher spins the proton excitations are expected roughly 1 MeV above the yrast line which is of neutron character, the second band crossing (not clearly observed in the present experiment) may also be due to neutron alignment. The crossing frequencies in fig.8 may differ due to some other effects like reduction of Δ_n and shape changes. The later is of importance since ^{174}Os is expected to be a soft nucleus and similar effects are well pronounced in ^{172}Os [12]. TRS calculations indicate that for the ground state configuration the parameter γ is roughly around $0°$ except for $\hbar\omega \approx 0.25 - 0.30$ MeV, which is the region of the first band crossing. There a change to $\gamma \approx -10°$ and later back to $\gamma \approx 0°$ is predicted.

The fact that the crossing frequencies for the two negative parity bands and gsb in fig. 7 are roughly the same may suggest the same nature of the alignment, i.e. $vi_{13/2}$. A simple interpretation for the negative parity bands may involve a neutron pair coupled to the even core. One of the neutrons must thus originate from a negative parity configuration, most probably K = 5/2 component of $vf_{7/2}$. A strong argument for that is the observance of a well populated $f_{7/2}$ structure in ^{173}Os [13] where there is also an intense band sequence due to $vi_{13/2}$. Thus, a suggestion for the negative parity bands as the two signature partners of the $vf_{7/2} \otimes vi_{13/2}$ configuration would imply a crossing frequency much higher than for gsb because of blocking effect. On the other hand the negative parity structure is predicted to be associated with a nuclear shape corresponding to $\gamma \approx -10°$ within a wide frequency region up to $\hbar\omega \approx 0.4$ MeV. This may shift the crossing frequency down to lower values.

3.3 - Château de Cristal

This national equipment constructed by french laboratories (Bordeaux, Grenoble, Lyon, Orsay, Strasbourg) is built of hexagonal BaF$_2$ detectors (14 cm long, 10 cm between opposite faces) [14]. This type of crystal which has a density between that of NaI and BGO can be used as Compton suppressor. The most interesting property of BaF$_2$ comes from the existence of a very fast component in the light emission with λ = 220 nm and $\tau \sim 0.6$ ns. Then very good timing is possible which allows discrimination between neutrons and γ-rays and also particular studies by triggering on isomeric levels with very short half-live. I remind the wavelength of this fast component implies phototubes with quartz window.

Several geometries are possible by grouping the hexagonal detectors in rings of 1, 6, 12, 18... counters : an inner ball with 38 or 74 detectors is generally used (fig.9). The BaF$_2$ crystals are also used as anti-Compton shields for the 12 Ge detectors which are positioned at $\pm 30°$, $\pm 90°$, $\pm 150°$ to the beam. In the first phase the array was equiped with 20% efficiency detectors which have been replaced in 1988 by 70%-80% efficiency HpGe detectors. The performances listed in table 2 are those for the second phase. The "Château de Cristal" which is the only array employing BaF$_2$ for both Compton suppressors and inner ball was installed on a beam line of the M.P. Tandem at Strasbourg and is moved now near the Tandem at Orsay.

$(\pi h_{11/2})^2$ and $(\nu h_{11/2})^2$ excitations in ^{124}Ba and shape coexistence

Among the results obtained with the "Château de Cristal", I have chosen to present those relative to ^{124}Ba and discuss the coexistence of several shapes in ^{124}Ba and in neighbouring nuclei.

The A = 120-130 transitional region is well suitable to study the interplay between $h_{11/2}$ quasi-protons and $h_{11/2}$ quasi-neutrons since the proton Fermi energy lies at the bottom of the shell and the neutron Fermi level is in the upper part. For Ba nuclei, the $h_{11/2}$ protons are on the 1/2-[550] Nilsson orbital. Their spins are strongly aligned with the rotation axis of the nucleus. On the contrary the $h_{11/2}$ neutron shell is almost filled and the spins of the quasi-neutron holes are considerably less aligned since the neutron orbitals are characterized by large Ω values. Therefore we can imagine that these quasi-particles and quasi-holes will produce different polarization effects on the core of the nucleus.

Prior to our study [15], two S-bands were known in ^{126}Ba[16], ^{128}Ba[17,18] and ^{130}Ba[19] but their proton and neutron character was not unambiguously determined because they have very similar crossing frequencies and aligned angular momenta. By performing two fusion-evaporation reactions induced by ^{18}O and ^{32}S ions, two S-bands were identified in ^{124}Ba. We have shown that the $(\pi h_{11/2})^2$ band is the first to cross the yrast band [15]. One of the consequences of this result is the present situation illustrated in fig.10 where data on ^{132}Ba [20] and ^{122}Ba[21] are also indicated.

A theoretical analysis was made to fully interpret the experimental results in terms of shapes. Deformation parameters β_2, β_4 and γ were determined for the $(\pi h_{11/2})^2$ and $(\nu h_{11/2})^2$ configurations as a function of $\hbar\omega$ by doing TRS calculations based on a non-axially-deformed Woods-Saxon potential and a monopol pairing residual interaction [22]. Figure 11 illustrates very cleary the coexistence of three different shapes associated with three configurations [23]. The large differences in shapes of the heavier isotopes associated with the proton or neutron nature of the quasi-particle is related to the different positions of the proton and neutron Fermi energies in the $h_{11/2}$ shell. The γ-driving forces exist also in lighter nuclei but the descent of the Fermi level into the $h_{11/2}$ neutron shell reduces the difference between proton and neutron configurations. The rigidity of the core limits the effect of the γ-driving forces.

3.4 - Nordball

The Nordball [24] is an array of Compton suppressed germanium detectors built by the Nordic Countries, Denmark, Finland, Norway and Sweden with a participation of Holland, Italy, Japan and West Germany. Its frame structure (fig.12) based on a truncated icosahedron with 20 hexagons and 12 pentagons has a diameter of 66 cm. The Compton-suppressed Ge detectors are designed to fit in the hexagons while additional detectors like BaF_2 scintillators, neutron detectors can be placed into the pentagons.

An inner ball of 60 elements of BaF_2 crystals of 95 mm length occupies the central part of the Nordball. This calorimeter covers 96% of 4π.

Several additional devices are available with the Nordball. A 2π neutron multiplicity filter (fig.13) consisting of 16 liquid scintillators (BC-501) can be used for studies of neutron deficient nuclei far from stability [25] by gating the Ge detectors by the neutron multiplicity. This neutron wall can be combined with a 4π silicon ball [26] made of 17 Si detectors having a total solid angle larger than 90%. The combination of the neutron wall and 4π silicon ball is extremely useful to isolate a given channel among many others associated with the simultaneous emission of neutron and light charged particles.

Selection of charged particle channels can be made also by means of Hystrix [27], another charge particle detector constituted of a maximum of 11 plastic phoswich scintillators (fig.14). The scintillators are composed of NE102A ΔE-elements (100 µm thick) and NE 115 E-elements (10 mm thick). This device which has its own vacuum chamber in the centre of the Ge-array has a solid angle of $\approx$ 94% of 4π. An example of data obtained with Hystrix is shown in the next section.

First identification of collective bands in ^{121}Ba

In order to extend the knowledge of the structure of neutron-deficient transitional nuclei in the 50<Z, N<82 region, we have performed an experiment to produce and identify ^{121}Ba by using the Nordball coupled to Hystrix. Several collective excitations are known in odd-A Ba isotopes with A⩾123. It would be interesting to identify them in ^{121}Ba which is a nucleus situated midway the N = 50 and 82 shell closures where a large β_2 deformation and structural changes are expected.

A ^{106}Cd target of a thickness of 1.2 mg.cm^{-2} was bombarded with 95 MeV ^{19}F ions from the tandem accelerator at the Niels Bohr Institute in Risø. For such a fusion-evaporation reaction, calculations made with the ALICE Code show that about 15 channels with cross-sections larger than 10 mb are open. The residual nuclei which range from Ba (Z = 56) to Te(Z = 52) are produced via emission of 1p, 2p, 3p, α, αp... These predictions are verified on the experimental spectra. Then the need of channel selection by charged particle-γ coincidences appears clearly.

Due to the limited detection efficiency of the particle detector which is of the order of 50%, the one-proton-γ-γ data was contaminated with events where more particles were evaporated (but only one proton detected). By comparing to the higher particle-multiplicity matrices (2p-, 3p-, αp-$\gamma\gamma$...) these impurities could be subtracted away, leaving a γ-γ matrix consisting of γ-rays from Ba only (fig.15).

Two main collective bands have been found in ^{121}Ba [28] and preliminary results are shown in figure 16. One band is based on the 5/2$^+$ ground state which has a d$_{5/2}$ [413] 5/2 neutron configuration in contrast to heavier odd-A Ba isotopes where the positive parity band is generated from the g$_{7/2}$ neutron shell [29]. The crossing appearing in the α = +1/2 signature branch at $\hbar\omega \approx 0.36$ MeV is associated with the alignment of two h$_{11/2}$ quasi-protons.

The second band which has a large energy signature splitting ($\approx$ 200 keV at $\hbar\omega \approx$ 0.3 MeV) is assigned to the h$_{11/2}$ quasi-neutron hole excitation as in heavier isotopes [30]. It is based on a 5/2$^-$ state originating from the [532] 5/2 orbital of the h$_{11/2}$ neutron shell. This 5/2$^-$ state is observed for the first time. The crossing at $\hbar\omega \approx 0.37$ is also due to h$_{11/2}$ quasi-proton alignment.

With these results for ^{121}Ba, it is possible to follow the evolution of properties of odd-A Ba isotopes from A = 135 to A = 121 i.e. over a wide range of variations of the Fermi energy in the neutron shells, in particular the νh$_{11/2}$ one.

3.5 - HERA

HERA (High Energy Resolution Array), the first array to use BGO shields is made of 21 Compton-suppressed Ge detectors surrounding an inner ball [31]. It is installed on a beam line of the 88" cyclotron at Lawrence Berkeley Laboratory (USA).

The Ge detectors are arranged in three rings, one in the horizontal plane, two at ± 26° to this plane (fig.17). Their collimated front surface is at 14 cm from the target. This multidetector which has a large efficiency (equivalent to that of Nordball) provides large rates of triple coincidences.

The bismuth germanate ball consists of 44 elements in three concentric cylinders with two identical halves relative to the horizontal plane (fig.17). With BGO detectors which are 4 cm thick and 7.5 cm long an overall efficiency of 0.85 and a full width at half maximum of 20% for the total γ-ray energy are expected.

Three superdeformed bands in ^{194}Hg

Many results have been obtained by means of HERA since its beginning of operation, specially on the superdeformations (SD). Among these results I will briefly report on the observation of 3 SD bands in ^{194}Hg.

^{194}Hg was populated in the reaction ^{48}Ca + ^{150}Nd. All three - and higher-fold Ge detector coincidences with multipolarity and sum energy information from the inner ball were recorded. Two-fold Ge coincidences were also recorded when at least four inner ball elements were fired. With a total of 720 million events (20% of three- or higher - fold), three SD bands have been found by doing double- and triple - coincidence analysis requiring selection of high fold and high total energy.

First, 2 SD bands were observed [32] : one consisting of 18 transitions is depopulated around spin 10, another less strongly fed starting at spin 8 is made of 16 transitions (fig.18). The energy difference between transitions for these two SD bands decreases from 40 keV at low rotational frequencies to 30 keV at high rotational frequency, the $J^{(2)}$ moment of inertia ranging from $\approx$ 95 to $\approx$ 140 $\hbar^2$MeV^{-1}.

Very recently, a third SD band has been identified in the same nucleus [33]. It consists of 15 transitions with energy differences close to those in the two first bands already known (fig.18) and a weaker population ($\approx$ 1/3 of the strongest SD band).

^{194}Hg is the first nucleus in which 3 SD bands have been found.

3.6 - The Chalk River "8π" Spectrometer

This device (fig.19) is composed of a BGO inner ball surrounded by 20 Compton-suppressed germanium detectors [34]. The Ge detectors are arranged in four rings at polar angles of 37°, 79°, (180° − 79°) and (180° − 37°) with respect to the beam direction and situated at 21 cm from the target.

The design of the inner ball is based on the geometry of a 72-faced polyhedron made of 12 regular pentagons and 60 hexagons.All its BGO elements see the target with the same solid angle. The multiplicity resolution is about 25 % FWHM for M = 30 and for a probability of Compton scattering into another detector of 0.2.

With a resolving power Rγ = 4.0 (table 3), the Chalk River "8π" spectrometer is at present one of the best array in operation. A detector for particle identification is under construction.

Superdeformed band identified up to spin $\sim \frac{127}{2}$ in ^{149}Gd

These results are selected because they illustrate very well the performances of the Chalk River multidetector. Indeed in the reaction ^{124}Sn(^{30}Si, 5n) at 150 MeV weak lines of a SD band in ^{149}Gd have been observed [35].

From a total of 470 x 10^6 Ge-Ge-multiplicity M-sum energy E_s coincidence events recorded when at least 10 elements in the ball fired, 183 x 10^6 have been selected in the analysis by setting gates on the multiplicity (26 < M < 47) and sum energy (22 < E_s < 39 MeV). With such a selection of high multiplicity, high energy events, a SD band was identified in ^{149}Gd (fig.20). Its intensity is weak, only 2 % of the 5n channel for the given M,E_s conditions and 0.5 % of the total cross-section. This band of 19 transitions ranging from 618 to 1559 keV whose the bottom member has a probable spin of 51/2 $\hbar$ extends to spin 127/2 $\hbar$. Its quadrupole moment of 17 ± 2 eb corresponds to a deformation β_2 = 0.6 and its moment of inertia agrees with a structure originating from one neutron K = 1/2,$\nu j_{15/2}$ and two protons K = 1/2, $\pi i_{13/2}$.

4 - Future prospects

We have seen in the preceding section that it is possible to isolate weak γ-lines from a complex spectrum by means of the existing detectors arrays. For instance the intensity of the lines observed at the top and bottom of the SD band in ^{152}Dy is about 0.2% of the one of the strongest transitions in this nucleus. This is the limit which can be achieved with these miltidetectors. New instruments with larger number of detectors and larger efficiency are needed to go beyond this limit (to 10^{-4} or 10^{-5}). In this paragraph, I present the new generation of arrays. I describe EUROGAM, the Franco-British detector under construction, and more briefly the american project GAMMASPHERE and the new concept EUROBALL.

4.1 - EUROGAM

EUROGAM is the new array of suppressed germanium detectors proposed jointly by physicists from France and the United Kingdom [36]. The main idea was to built an instrument having an efficiency larger than the existing ones in order to search for weak phenomena (below 10^{-4} in terms of relative population intensity) using high-fold coincidence techniques.

Before presenting the design, some words have to be said about important factors like resolving power and efficiency which determine the quality of the array.

4.1.1 - Resolving power

For a rotational band with average separation SE_γ between γ-rays, a resolution of γ-rays ΔE_γ and a peak to total ratio P/T, the resolving power for n-fold coincidence events is defined as : $R_{\gamma n} = \left\{ \dfrac{SE_\gamma}{\Delta E_\gamma} \times P/T \right\}^n$.

High resolving powers are required to observe very weak intensity γ-rays. Therefore the design of new germanium arrays must satisfy the conditions of large full energy peak efficiency, good γ-ray energy resolution and good peak to total fraction.

The γ-ray energy resolution is strongly influenced by the Doppler shift and is about 7 keV for a 1 MeV γ-ray for most of the arrays (table 3).

The intrinsic P/T ratio which is of the order of 0.5-0.6 is affected by simultaneous detection of two γ-rays in the same crystal (factor $\iota_n \sim 0.85 - 0.90$). The effective value of P/T is in the range 0.45 ± 0.04. As seen in table 3, collimation of the detectors and use of the recoil mass spectrometer (RMS) improve strongly the resolving power of EUROGAM

The limit of observation of γ-rays is inversely proportional to the resolving power.

4.1.2 - Efficiency

The efficiency depends mainly on the solid angle of the germanium detectors and on the peak efficiency which is a function of the size of the germanium crystal.

Data are available on peak efficiencies as well as simulation calculations. As an example, intrinsic peak efficiencies are shown in fig.21 as a function of the diameter and length of the detectors for a 1.3 MeV γ-ray.

Table 3

Resolving powers R_γ of some arrays for each additional coincident γ-ray. From ref. [36]

	P/T	ΔE_γ (keV)	$SE_\gamma/\Delta E_\gamma^{(*)}$	R_γ
TESSA3	0.46	6.8	8.8	4.0
HERA	0.41	8.0	7.5	3.1
CHATEAU II	0.41	6.6	9.1	3.7
NORDBALL	0.42	8.0	7.5	3.2
8π	0.46	6.8	8.8	4.0
GAMMASPHERE	0.49	7.4	8.1	4.0
EUROGAM	0.49	7.7	7.8	3.8
EUROGAM (Collimated)	0.49	6.7	8.9	4.4
EUROGAM (Collimated + RMS)	0.49	4.4	13.6	6.9

(*) Calculated for $SE_\gamma = 60$ keV.

4.1.3 - The design

The design is based on an array of 12 regular pentagons (dodecahedron shown in fig.22). Each pentagon contains 5 Ge detectors surrounding a sixth one at the center of the pentagon (fig.22). The full array consists of 70 detectors, two detectors having to be removed to let the beam through.

The germanium crystals are surrounded by BGO anti-Compton shields (fig.22). The shields situated at the corners of the pentagons are made of 10 BGO crystals while the central ones consist of only 5 pieces. The BGO detectors can be operated in a shared suppression mode i.e. a BGO element will be used to reject events appearing in the neighbouring 2 or 3 Ge detectors but the modules can also work as individual suppression shields.

As said in section 2.2, the γ-ray detectors will be n-type. Two kinds are being considered : detectors with a tapered single crystal (≈ 65-75% relative efficiency) and detectors made of stacked planar detectors. The tapered ones will be large cylindrical germanium crystals (90 mm diameter) with a cone angle of 5.7° in order to fill completely the available solid angle. Detectors of this type are available so far and have been tested. The stacked detectors consist of pairs of planar Ge detectors assembled within one cryostat. They should provide a larger efficiency and the characteristics of planar detectors (good energy resolution and good timing) should be conserved. Prototypes are under developpments. Simulations have been made for these two types [36] and comparisons are given in fig.23.

The assembling of a Ge detector and its BGO shield on the support frame is shown in fig.24. The front face of the Ge detectors will be at 29 cm from the target. They will cover ≈ 40% of 4π.

There is no inner ball under construction but this possibility will be considered in a very near future.

To run the EUROGAM array which consists of up to 70 Ge detectors associated with anti-Compton shields made of 650 crystals, the electronics and data acquisition system have been designed to have a high degree of integration for the electronic components, software control of all parameters, data transfert buses, modularity of hardware and software...

EUROGAM has been funded (≈ 40 million FF) and it will be located for the first year of operation (starting summer 1991) at the target position of the recoil mass spectrometer at Daresbury. For this first phase, 45 detectors will be assembled. Then EUROGAM will be completed to 70 detectors and moved to France to be installed on a beam line of the Vivitron at Strasbourg.

4.2 - GAMMASPHERE

A project has been presented in the United States for a new array made of an inner ball and 110 Ge detectors with BGO suppression shields [37]. A Honeycomb design based on a polyhedron with 122 faces has been adopted. Each individual module (Ge detector + shield) occupies an hexagon as shown in fig.25. The Compton suppressors consist of 6 pieces each and work in shared suppression mode. Their front end is made of BaF_2.

The total efficiency ($N\Omega\varepsilon_p \iota_s = 0.85$) is about 30 % larger than for EUROGAM (table 2) but its resolution power slightly lower (table 3). These two arrays will cover similar domains of physics. If funded in 1990, GAMMASPHERE could start during summer 1992 in a reduced version and would be completed for summer 1993. The estimated cost is 17 million $.

4.3 - EUROBALL

Compared to existing arrays, EUROGAM and GAMMASPHERE will bring a large improvement in γ-ray detection efficiency (see table 2). It seems difficult now to increase their efficiencies in a significant way only using Ge detectors with Compton suppressors. Then a new concept has to be introduced to improve these performances : it consists of a full sphere of Ge detectors.

The Ge detectors would be closely packed in units of 7 detectors. For such a unit, the central detector could be operated in anticoincidence with the six surrounding ones or in add-back mode. A geometry has been suggested for this Ge ball [38]. It is based on a large number of pentagons and at least 12 pentagons. Such a "cluster" array could be realized by means of conical hexagonal Ge crystals (fig.25). A prototype of an hexagonal crystal has already been made and tested [39]. For an array of 492 detectors (60 hexagons, each containing 7 detectors and 12 pentagons, each with 6 detectors) the total solid angle covered by the Ge crystals would be ≈ 75% of 4π and the $N\Omega\varepsilon_p\iota_s$ factor ≈ 2–2.5 larger than for EUROGAM and GAMMASPHERE (table 2).

Study groups are continuing working on this concept of Ge shell which could be the next challenge for the nuclear structure community.

5 - Conclusion

In this lecture, conditions required for detector arrays have been reviewed and discussed. The selected examples of results show that the existing multidedectors have produced and are still producing very interesting data but the limit of observation is reached. Weak and very weak γ-rays associated with SD bands, their feeding and deexitation, very neutron-deficient nuclei produced with very low cross-sections, hyperdeformations... are near or below this limit of detection. New instruments under construction or in project to study these weak amplitude phenomena have been presented.

Acknowledgements

This lecture reviews the work of many people in many laboratories. I would like to thank my colleagues for many discussions and specially those I have worked with, colleagues from Grenoble, Strasbourg, Athens, Bucharest, Cologne, Debrecen, Lund, Manchester, Risø, Stockholm and Warsaw. Thanks also to David Ward for providing me with photographs and transparencies of the Chalk River array.

REFERENCES

1. - P.J. Nolan et al., Nucl. Instr. Meth. A236 (1985) 95
2. - L. Westerberg et al., Nucl. Instr. Meth. 145 (1977) 295
3. - D.G. Sarantites et al., Nucl. Instr. Meth. 171 (1980) 503
4. - P.J. Twin, Workshop on Nuclear Structure at High Spin, Riso, 1981
5. - P.J. Nolan, Proc. Int. Nucl. Phys. Conf., Harrogate (1986) 155
6. - B.M. Nyako et al., Phys. Rev. Lett. 52 (1984) 507
7. - P.J. Twin et al., Phys. Rev. Lett. 52 (1986) 811
8. - J.F. Sharpey - Schafer and J. Simpson, Progress in Particle and Nuclear Physics, vol. 21,ed. by A. Faessler
9. - L. Hildingsson et al., Proc. Int. Conf. on the Spectroscopy of Heavy Nuclei, Crete (1989) in press and to be published
10 - J.L. Durell et al., Phys. Lett. 115B (1982) 367
11 - J. Gascon et al., Nucl. Phys. A470 (1987) 230
12 - J.C. Wells et al., Phys. Rev. C40 (1989) 725
13 - C. Kalfas et al., Nucl. Phys. in press
14 - F.A. Beck, Proc. Int. Conf. on Instrumentation for Heavy-Ion Nuclear Research, Oak Ridge (1984) 129
15 - J.P. Martin et al., Nucl. Phys. A489 (1988) 169
16 - K. Schiffer et al., Zeit. Phys. A327 1987) 251
17 - L. Hildingsson et al., Physica Scripta 29 (1984) 45

18 - K. Schiffer et al., Nucl. Phys. A458 (1986) 337
19 - Sun Xiangfu et al., Phys. Rev. C28 (1983) 1167
20 - J. Gizon et al., Nucl. Phys. A252 (1975) 509 and S. Harrissopulos et al., to be published
21 - R. Wyss et al., Proc. XXV Int. Winter Meeting on Nuclear Physics, Bormio (1987) 542
22 - W. Nazarewicz et al., Nucl. Phys. A435 (1985) 397
23 - R. Wyss et al., Nucl. Phys. A505 (1989) 337
24 - B. Herskind, Nucl. Phys. A447 (1985) 395c and G. Sletten et al., Int. Conf. on the Spectroscopy of Heavy Nuclei, Crete (1990) in press
25 - S.E. Arnell et al., Conf. on Nucl Structure in the Nineties, Oak Ridge (1990) 214
26 - T. Kuroyanagi et al., Development at the University of Kyusyu, Japan, 1989
27 - F. Liden, Nucl. Instr. Meth. A288 (1990) 455
28 - F. Liden et al., Nucl. Phys. in press
29 - J. Gizon et al., Nucl. Phys. A277 (1977) 464
30 - J. Gizon and A. Gizon, Zeit. Phys. A281 (1977) 99, A285 (1978) 259 and N. Yoshikawa et al., J. Phys. 39 (1979) 209
31 - R.M. Diamond and F.S. Stephens, The High Resolution Ball, unpublished and Proc.Int. Conf. on Instrumentation for Heavy-Ion Nuclear Research, Oak Ridge (1984) 259
32 - C.W. Beausang et al., Zeit. Phys. A335 (1990) 325
33 - E.A. Henry et al., Zeit. Phys. A335 (1990) 361
34 - H.R. Andrews et al., Atomic Energy of Canada Limited report AECL-8329 (1984) unpublished
35 - B. Haas et al., Phys. Rev. Lett. 60 (1988) 503
36 - EUROGAM, unpublished reports (1990) by the Eurogam Project Scientific Committee, French and UK Eurogam local management committees
37 - Gammasphere proposal, ed. M.A. Deleplanque and R.M. Diamond, Lawrence Berkeley Laboratory, March 1988
38 - R.M. Lieder, Workshop on Nuclear Structure at High Spins, Bad Honnef, March 1989
39 - J. Eberth, Conf. on Nuclear Structure in the Nineties, Oak Ridge (1990)

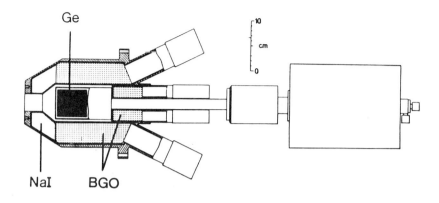

Figure 1. Bismuth germanate shielded germanium detector used in the TESSA3 array.

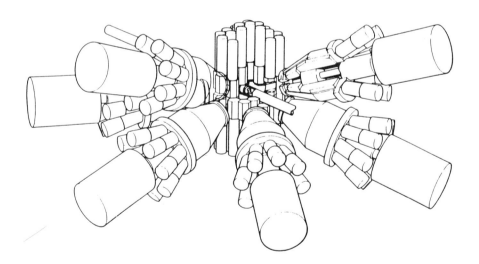

Figure 2. The TESSA3 array with a 50 BGO element inner ball and 16 BGO suppressed germanium detectors.

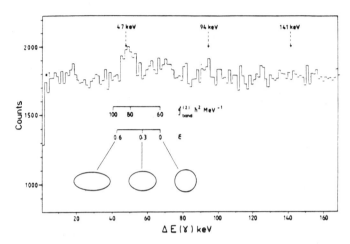

Figure 3. Spectrum perpendicular to the main diagonal of a Eγ-Eγ correlation matrix for ^{152}Dy obtained with TESSA2. The scales for $J^{(2)}$ and for the deformation ε are also indicated.

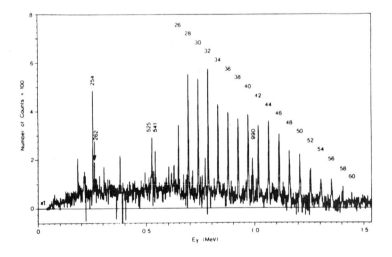

Figure 4. The γ-ray spectrum of the superdeformed band observed in ^{152}Dy by means of TESSA3. The γ-lines are labelled by spin.

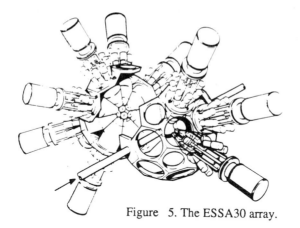

Figure 5. The ESSA30 array.

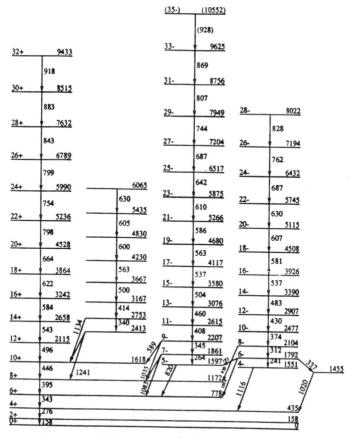

Figure 6. Level scheme of ^{174}Os deduced from experiments made with ESSA30.

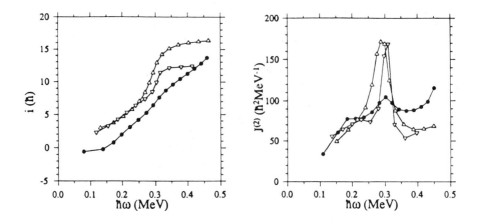

Figure 7. Experimental alignments i_x (left) and second moment of inertia $J^{(2)}$ (right) for ^{174}Os. A reference based on the Harris parameters $J_0 = 25.8 \hbar^2$ MeV^{-1} and $J_1 = 61.8$ $\hbar^4$ MeV^{-3} has been subtracted.

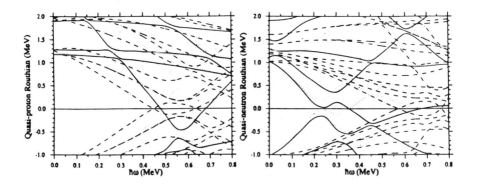

Figure 8. Calculated quasi-particle Routhians for protons (left) and neutrons (right) using a Woods-Saxon potential with $\beta_2 = 0.24$, $\beta_4 = -0.01$ and $\gamma = 1°$. Solid line $(\pi, \alpha) = (+,+ 1/2)$, dotted $(+, -1/2)$, dash-dotted $(-,+1/2)$, dashed $(-,-1/2)$.

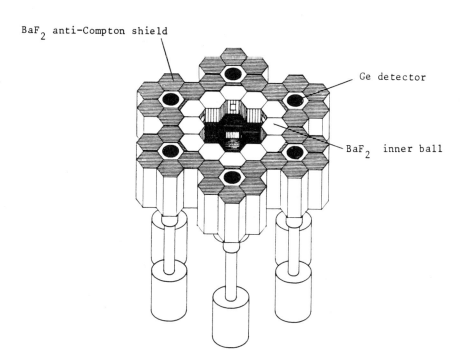

Figure 9. A cut of Château de Cristal through an horizontal plane.

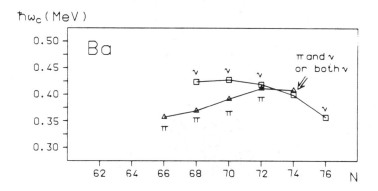

Figure 10. Observed rotational crossing frequencies between the ground state band and S-bands in Ba isotopes. The S-bands configuration is indicated by π (proton) or ν (neutron).

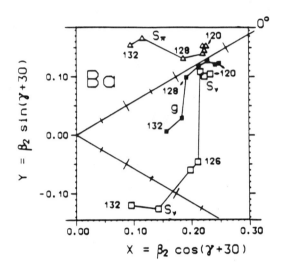

Figure 11. The positions of the first minima in the TR-Surfaces of Ba isotopes corresponding to the ground state band ∗, proton S-band (Δ) and neutron S-band (□).

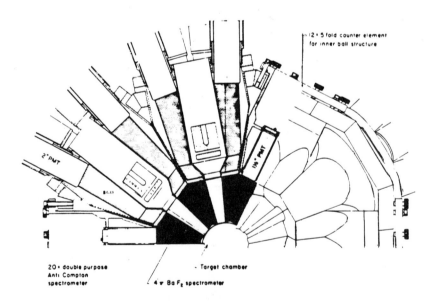

Figure 12. A cut through the Nordball showing the BGO anti-Compton spectrometers and the BaF$_2$ inner ball.

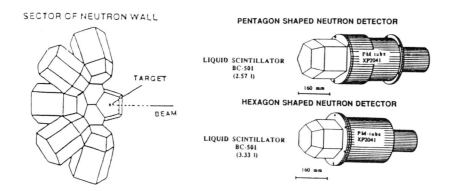

Figure 13. Sector of the 2π neutron multiplicity filter installed at the Nordball and the two types of neutron detectors.

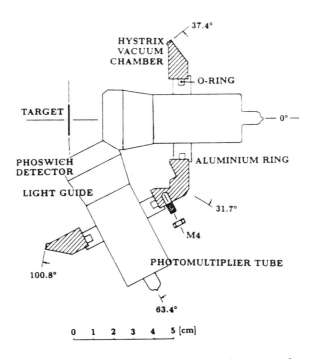

Figure 14. A cross-section through Hystrix showing the detectors and vacuum chamber. The thickness of the black plastic chamber is minimized at 37.4° and 100.8° where Ge detectors are positioned.

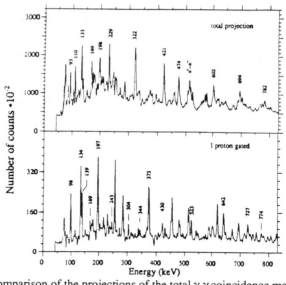

Figure 15. Comparison of the projections of the total γ-γ coincidence matrix (top) and 1proton -γ-γ matrix (bottom) for the reaction $^{19}F+^{106}Cd$ at 95 MeV. The 1 proton-channel is selected by means of the charged particle detector Hystrix.

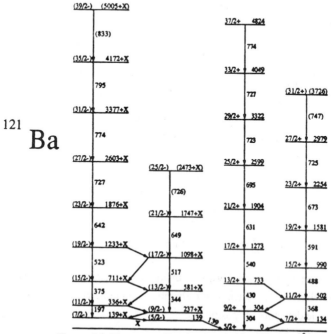

Figure 16. The two collective bands identified in ^{121}Ba.

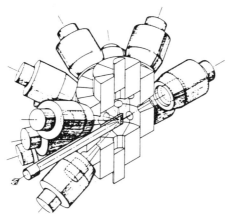

Figure 17. A view of HERA, the detector array intalled at Lawrence Berkeley Laboratory.

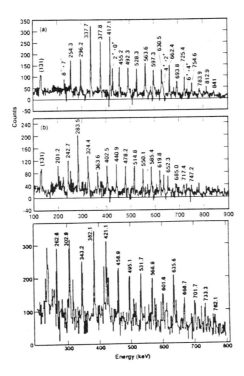

Figure 18. The spectra of γ-transitions for the 3 superdeformed bands observed in ^{194}Hg The top and middle ones are from ref.[32] and correspond to the 2 strongest SD bands. The lower one is from ref.[33] and corresponds to the third SD band.

Figure 19. The Chalk River "8π" spectrometer.

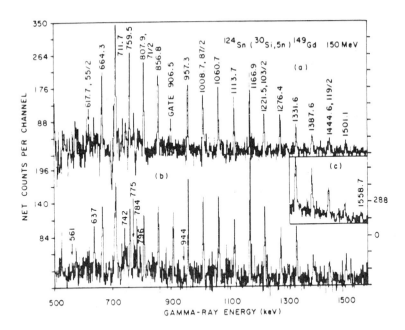

Figure 20. γ-ray spectra of the SD band in ^{149}Gd. (a) Spectrum gated by the 906 5 keV transition in the SD band. (b) Triple Ge-Ge-Ge spectrum between seven members of the SD band. (c) Part of a summed spectrum showing the highest member of the SD band.

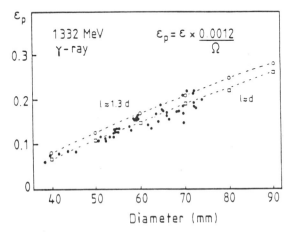

Figure 21. Measured (points) and calculated (dashed curves) intrinsic peak efficiencies for different sizes of detectors (d=diameter, l=length, Eγ=1.33 MeV).

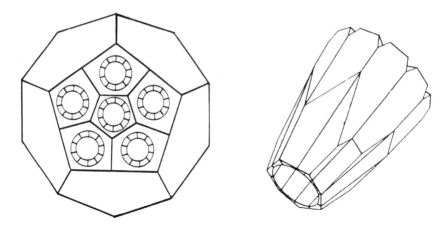

Figure 22. EUROGAM : design (left) and BGO anti-Compton shield (right).

- **Single tapered Ge detector**

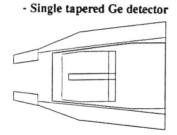

Eγ(MeV)	ε_u	ε_p	$\left(\dfrac{P}{T}\right)^u_{100}$	$\left(\dfrac{P}{T}\right)^s_{100}$
0,2	0,91	0,72	0,98	0,98
1,33	0,88	0,23	0,27	0,58
5,0	0,77	0,08	0,11	0,29
10,0	0,79	0,04	0,05	0,20

- **Stacked Ge detectors**

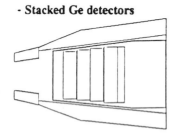

Eγ(MeV)	ε_u	ε_p	$\left(\dfrac{P}{T}\right)^u_{100}$	$\left(\dfrac{P}{T}\right)^s_{100}$
0,2	0,97	0,75	0,97	0,97
1,33	0,91	0,24	0,28	0,60
5,0	0,80	0,08	0,10	0,29
10,0	0,83	0,04	0,05	0,19

Figure 23. Simulations of two types of Ge detectors with BGO shield for EUROGAM. ε_u = efficiency for detection of γ-ray per unit solid angle. ε_p = efficiency for detection of γ-ray full energy peak per unit solid angle. The solid angle is fixed at $\Omega = 5.1 \times 10^{-3}$. The threshold is set at 100 keV on the γ-spectra.

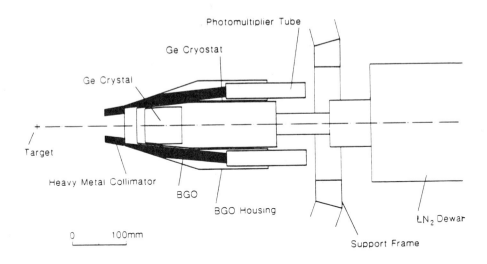

Figure 24. EUROGAM : The Ge detector and BGO anti-Compton shield module.

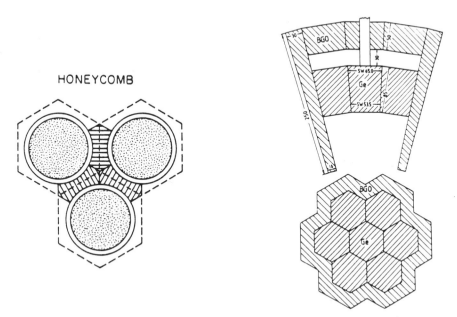

Figure 25. GAMMASPHERE (left) : three unit cells for the honeycomb design.
EUROBALL (right) : A possible design of the cluster detector.

RECENT PROGRESS IN THE STUDY OF LIGHT NUCLEI FAR FROM STABILITY USING GANIL FACILITIES

F. Pougheon

Institut de Physique Nucléaire
91406 ORSAY Cedex, France

I - Proton-rich nuclei

   1) Introduction

   2) Experimental method

   3) Results

      - towards the drip line : identification of new isotopes

      - decay of proton-rich nuclei

II - Neutron-rich nuclei

   1) Light isotopes : reaching the drip line

   2) New isotopes

   3) Decay of neutron-rich nuclei

I - Proton-rich nuclei

1) Introduction

The study of nuclei far from stability opens the nuclear frontier and provides a valuable insight into the properties of nuclei with extreme neutron to proton composition.

These exotic nuclei allow physicists to use the nucleus as a microscopic laboratory in which interesting structure effects can occur, such as the validity of magic numbers far from stability or the existence of new deformation areas.

Because both strong and electroweak forces play important roles in the binding and decay of these nuclei, fundamental information about the very nature of these forces can be obtained.

The masses of nuclei increase rapidly with neutron (proton) excess. This leads to large $Q\beta$ decay energies and thus gives the opportunity to search for new phenomena and new decay modes.

2) Experimental method

The production method used at GANIL is the so called "projectile fragmentation". The fragments are concentrated in a small opening cone around zero degree and this focusing is well suited to the use of a magnetic spectrometer. The N/Z ratio of the fragment is related to the N/Z ratio of the projectile.

At intermediate energy, due to mean field memory effects, there is an influence of the N/Z of the target and there is the possibility of nucleon transfer reactions with sizeable cross sections.

The LISE spectrometer (figure 1) is a double achromatic system set at $0°$ [1]. It consists mainly of two identical dipoles. The first one is used as a dispersive element ; the dispersive intermediate focal plane can be equipped either with a stripper or with a thick degrader. The second dipole allows the compensation of the dispersion produced by the first dipole. With such a system, it is possible to get at the achromatic focal point, a double achromatism both in angle and in position.

The detection system consists typically of two ΔE Si surface barrier detectors, a thick $SiLi$ E detector and a third detector which serves as a veto. The time of flight of the fragments is measured by deriving a start signal for a time-to-amplitude converter from the first ΔE detector and a stop signal from the radio frequency signal of the final cyclotron of the GANIL machine.

3) Results

- Towards the drip-line : identification of new isotopes

Due to the strong Coulomb repulsion between protons in the nucleus, the proton drip line is much closer to the valley of stability than the neutron dripline. Experimentally, it is easier to reach the drip-line on this side of the chart of nuclides. Nevertheless, by 1985, the drip-line had only been reached up to Aluminium. The figure 2 shows the results obtained in the ^{40}Ca (77 MeV/u) + Ni experiment [2]. On this two dimensional plot, Z over time of flight $(\alpha A/Z)$ the nuclei with a same isospin projection (Tz) are on the same lines.

The most outstanding result of this experiment is the clear evidence for the existence of the whole series $Tz = -5/2$ nuclei, namely the ^{23}Si, ^{27}S, ^{31}Ar and ^{35}Ca isotopes.

In another experiment to produce proton-rich nuclei with higher Z, a ^{58}Ni beam (55 MeV/u) was used on a Ni target [3]. Twelve new isotopes were observed for the first time ; they are indicated by an arrow on the figure 3. The predicted drip-line is reached for the odd elements but not for the even ones.

In these two experiments the limitation was clearly due to the high counting rate in the silicon telescope. To go further it was necessary to add a further degree of selection given by a thick energy degrader placed in the intermediate focal plane. In that configuration, LISE acts as an isotope separator [1]. The first experiment using LISE in that manner was devoted to the search for the ^{22}Si nucleus. About 160 counts of ^{22}Si were observed. It is the first and remains the only Tz=-3 nucleus ever observed [4].

- Decay of proton-rich nuclei

A second aspect of these experiments is the measurement of the decay properties of these exotic nuclei. These very-proton rich nuclei decay mainly by beta-delayed one or two-proton emission.

Figure 4 shows the details of the telescope placed at the focal point. The isotope of interest is slowed down by a thick aluminium foil placed in front of the telescope so that it is implanted in one of the thin detectors. The identification is performed by the ΔE and time of flight method.

On line, two single channel analysers examined the ΔE and ToF pulses. Each time that an ion of interest is identified, the beam is stopped for a duration from three to four times the expected half-life, and the heavy-ion data acquisition is set off line and a clock and a second data acquisition is started.

Up to now, the decay of the following isotopes have been studied by our group : ^{22}Si [5], ^{27}S [8], ^{28}S [6], ^{31}Ar [7,8,9], $^{39,40}Ti$ [10].

For details, look at the references. In this lecture I'll give only the example of the ^{28}S isotope. One of the reasons for studying the β-delayed proton decay is that the β^+ transition to the Isobaric Analog State (I.A.S.) is a super allowed transition with a large branching ratio. This is probably one of the best way to reach and to locate this I.A.S. in the daughter nucleus. The known location of the I.A.S. permits the use of the isobaric mass multiplet equation. This equation claims that all the masses of an isospin multiplet are linked by the very simple quadratic form :

$$M(A,T,Tz) = a(A,T) + b(A,T)Tz + c(A,T)Tz^2$$

The hypothesis of this equation is that the members of an isospin multiplet have identical wave function and that only two body forces are responsible for charge dependent effects in nuclei. To study deviations from this quadratic form isospin multiplets with T ≥ 3/2 are needed and one comes back to the interest of exotic nuclei in the general field of nuclear structure.

Figure 5 shows the energy spectrum of the β-delayed protons associated with the ^{28}S decay.

The deduced half-life of ^{28}S is 125 ms (fig. 6)

The three main peaks of the spectrum in Fig. 5 are attributed to the decay of the I.A.S. towards the three first levels of ^{27}Si. From the energies of these peaks, it is possible to deduce the excitation energy and the mass excess of the I.A.S. in ^{28}P and now the isospin quintet A = 28 is complete.

The test of the IMME shows [6] that there is no need to introduce cubic and quartic coefficients and that the quadratic form of this equation is valid for the quintet A = 28.

A partial decay scheme (fig. 7) is proposed and compared to shell model calculations performed by A. BROWN [6].

The agreement between experiment and calculations is very good for the half-life and for the location of the I.A.S. For the branching ratios, experiment and theory are compatible.

II - Neutron-rich nuclei

1) Light isotopes : reaching the drip-line

In the very first experiment performed at GANIL with the LISE spectrometer [11] the bound character of ^{23}N, ^{29}Ne and ^{30}Ne was proved. These nuclei were produced in the ^{40}Ar (44 MeV/u) + Ta reaction. Of these, only the ^{23}N is predicted to be at the drip-line. In a second experiment, using the same reaction, and after four days of beam on target, the ^{22}C and ^{19}B were observed [12]. The $^{22}C(Tz = +5)$ holds the record exoticity on this side with a neutron excess of 10 for such a light nucleus.

It was clear at that time that the experimental limit was reached with such a beam and that to go further away from the valley of stability, a more neutron rich projectile such as the ^{48}Ca was necessary.

In this aim our group started a collaboration with phycisists from the DUBNA laboratory. A very good beam of ^{48}Ca is now available at GANIL.

The results obtained in the ^{48}Ca (44 MeV/u) + Ta collision [13] are shown on the two dimensional plot (fig. 8). On this plot the ^{29}F and ^{32}Ne isotopes are clearly observed. But one of the aim was to observe the ^{26}O isotope, predicted to be slightly bound. No events corresponding to ^{26}O are seen. The plot shown on figure 9 makes this negative result more convincing. It shows the isotopic production along lines which proceed parallel to the drip line with neutron numbers N = 2 Z - 1, N = 2 Z and N = 2 Z + 2. A smooth drop in the yield is observed along these lines. From this trend, one may expect a production rate of about 30 counts for ^{26}O. Since it is almost impossible to explain the absence of 30 counts as a statistical fluctuations, this experiment gives strong evidence for concluding that ^{26}O is particle unstable. Indeed, it is a stimulating result for theory.

2) New isotopes

The view of figure 10 shows the impressive progress obtained at GANIL with this ^{48}Ca beam. All the isotopes behind the full line are nuclei observed for the first time (18 new isotopes). The spectroscopy of all the isotopes behind the dashed line is completely unknown and they are produced with a high counting rate [14].

It is interesting to note from a reaction mechanism point of view, that some of the new fragments contains more neutrons than the projectile does. This signals the presence of strong transfer reaction channels at these intermediate energies.

Another major gain has been obtained by the use of a very neutron rich target : the ^{64}Ni [15]. This target was found to be the most efficient for the production of these medium light nuclei. The magnetic rigidity was then optimized for the A/Z = 3 nuclei and seven new isotopes were observed : ^{51}Cl, $^{48,47}S$, $^{46,45}P$, ^{42}Si

and ^{39}Al.

3) Decay of neutron-rich nuclei

These very neutron-rich nuclei decay mainly by beta-delayed neutrons.

The experimental technique is about the same as for the proton rich nuclei. The fragments are implanted in a solid state detector. The telescope is surrounded by a plastic scintillator for β-ray detection and placed inside a four π neutron detector. This neutron detector is composed of six vessels containing a total of 30 litres of NE213 liquid scintillator. The decay curves are directly obtained from the time distribution of the β-neutron coincidences with respect to the implantation signal.

Up to now the half-lives and the neutron emission probabilities have been measured for the following isotopes : ^{20}C, $^{21,22}N$, $^{23,24}O$, $^{40,41,42}P$ and $^{43,44}S$ (ref. 16 and 17).

A sample set of decay curves is shown on figure 11.

The solid lines through the experimental data points correspond to a least square fitting procedure assuming a single exponential decay component and constant background.

Table 1 summarizes the experimental half-lives and neutron emission probabilities obtained in this work. All the data are calibrated on ^{15}B. These results have to be compared with the available predictions. They include the old "gross theory" of β decay by Takahashi recently improved by Tachibana [18], and the microscopic model of Klapdor et al including a new calculation made by Staudt and Klapdor [17] and based on a proton-neutron Quasi particle Random Phase Approximation (QRPA).

The relative merit of the different theoretical predictions for half lives is shown in figure 12. The ratio of theoretical to experimental results is plotted versus the isotopic mass A. $\bar{x}$ is the mean value of the logarithm of that ratio and σ is its standard deviation. The best $\bar{x}$ value is obtained for the QRPA calculation of Staudt et al but these calculations show an important dependence on the mass formula used, demonstrating the importance of having very good mass measurements.

References

[1] J.P. Dufour et al, Nucl. Instr. Meth. A248 267 (1986)
   R. Anne et al, Nucl. Instr. Meth. A257 215 (1987)

[2] M. Langevin et al, Nucl. Phys. A455 149 (1986)

[3] F. Pougheon et al, Z. für Physik A327 17 (1987)

[4] M.G. Saint-Laurent et al, Phys. Rev. Lett. 59 33 (1987)

[5] J.P. Dufour et al, to be published

[6] F. Pougheon et al, Nucl. Phys. A500 287 (1989)

[7] V. Borrel et al, Nucl. Phys. A473 331 (1987)

[8] V. Borrel et al, submitted to Nucl. Phys.

[9] D. Bazin et al, report CENBG 8938 to the Obernai workshop, November 1989

[10] C. Détraz et al, preprint GANIL 9013, to be published in Nucl. Phys.

[11] M. Langevin et al, Phys. Lett. 150B 71 (1985)

[12] F. Pougheon et al, Europhys. Lett. 2 505 (1986)

[13] D. Guillemaud-Mueller et al, Phys. Rev. C 41 937 (1990)

[14] D. Guillemaud-Mueller et al, Z. Phys. A332 189 (1989)

[15] M. Lewitowicz et al, Z. Phys. A 335 117 (1990)

[16] M. Lewitowicz et al, Nucl. Phys. A496 477 (1989)

[17] A.C. Mueller et al, Nucl. Phys. A513 1 (1990)

[18] T. Tachibana et al, Proc. 5th Int. Conf. on Nuclei far from Stability, Rosseau Lake, sept. 1987, p. 614

Table 1

A_Z	$T_{1/2}$(ms) exp.values this work	$T_{1/2}$(ms) exp.values others	P_n (%) exp.values this work	P_n (%) exp.values others
^{15}B	$10.3^{+0.6}_{-0.5}$	10.4 ± 0.3	(100)	100
^{17}B	6 ± 2	5.0 ± 0.5	70 ± 30	85 ± 15
^{18}C	78^{+20}_{-15}	66^{+25}_{-15}	50 ± 10	25 ± 4.5
^{20}C	16^{+14}_{-7}	-	50 ± 30	-
^{35}Al	170^{+90}_{-50}	130^{+100}_{-50}	40 ± 10	87^{-37}_{-25}
^{40}P	260^{+100}_{-60}	-	30 ± 10	-
^{41}P	120 ± 20	-	30 ± 10	-
^{42}P	110^{+40}_{-20}	-	50 ± 20	-
^{43}S	220^{+80}_{-50}	-	40 ± 10	-
^{44}S	200^{+50}_{-30}	-	30 ± 10	-

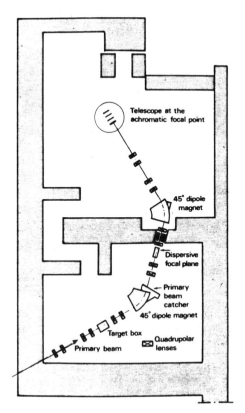

Fig. 1

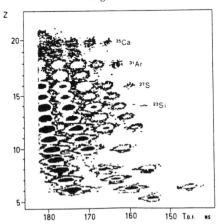

Fig. 2

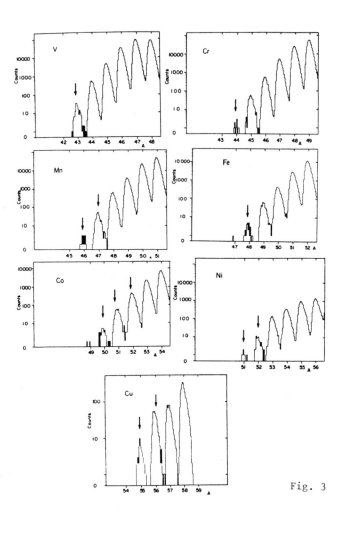

Fig. 3

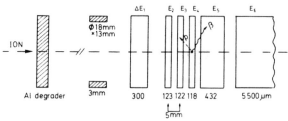

Fig. 4

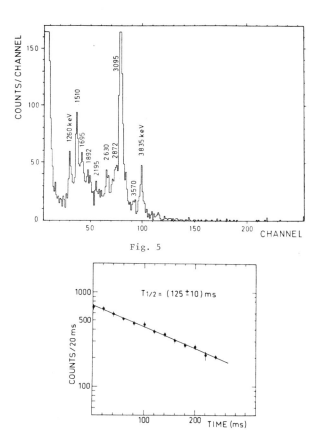

Fig. 5

Fig. 6

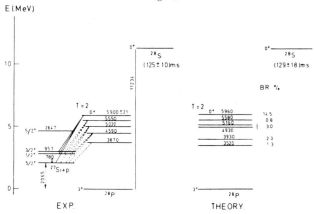

Fig. 7

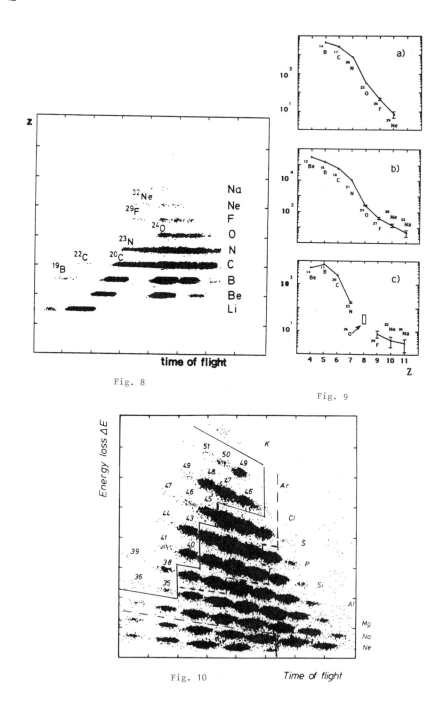

Fig. 8

Fig. 9

Fig. 10

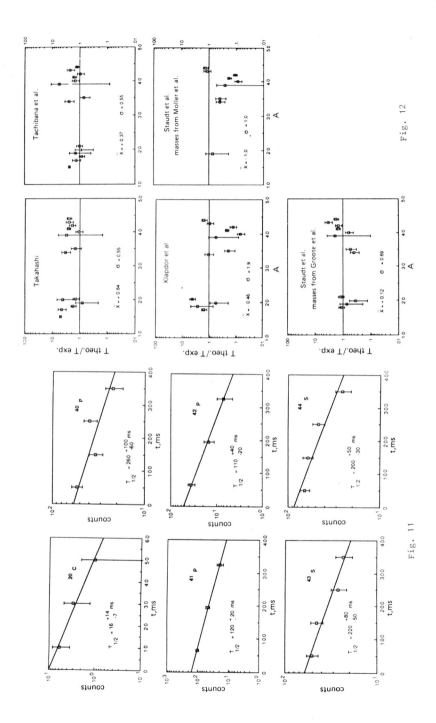

Fig. 11

Fig. 12

GAMOW-TELLER BETA DECAY OF PROTON-RICH NUCLEI

O. Klepper and K. Rykaczewski †

GSI Darmstadt, Postfach 110552, D-6100 Darmstadt 11,
Federal Republic of Germany

ABSTRACT

The beta decays of ^{48}Mn and of even-even nuclei near the double shell-closures at ^{100}Sn and ^{146}Gd are currently investigated at the GSI on-line mass separator. Their Gamow-Teller strength are surveyed in their present experimental status, together with related results from the ISOLDE (CERN) and ISOCELE (Orsay) separators, and are compared with predictions from different nuclear models. The strength of the $0^+ \rightarrow 1^+$ Gamow-Teller transitions is compiled in tables and graphs.

1. INTRODUCTION

Beta decays of very proton-rich nuclei permit to test our understanding of nuclear beta decay and to investigate, as a complementary tool to charge-exchange reactions, the phenomenon of missing strength in observed Gamow-Teller (GT) transitions compared to shell-model predictions (Quenching of GT strength). Due to the large decay energies of these nuclei a wider part of the beta strength distribution than for those close to the β-stability line can be explored. For instance, the observed GT strength averaged over nuclei in the middle of the sd shell is only 58(5)% compared to calculations with a full-basis shell-model code [BRO85]. The question arises, how nuclei in higher shells behave in this aspect.

In the $f_{7/2}$ shell, where now also large-basis shell-model codes become available, originally only the GT quenching of mirror nuclei was investigated [MIY87]. The high separation yield obtained recently at GSI for the new isotope ^{48}Mn, however, established now a proton-rich test nucleus with $T_z = -1$ in the middle of the $f_{7/2}$ shell for studying the Fermi and GT decay strength. For higher shells no similar large-basis shell-model calculations are yet available. The general interest concerns therefore more those nuclei close to the double shell-closures at ^{100}Sn and ^{146}Gd, as their relatively simple structure allows to apply and test simple shell model calculations and various approximations. Since the proton-rich nuclei, studied so far in these regions, have $T_z \geq 1$, GT transitions are the only allowed β-decay mode.

The study of β^+/EC decay of these nuclei is particularly attractive: The Fermi energies for protons and neutrons are expected to lie between the spin-orbit partner states and the β^+/EC decays around ^{100}Sn and ^{146}Gd are thus dominated by the fast spin-flip transitions $\pi g9/2 \to \nu g7/2$ and $\pi h11/2 \to \nu h9/2$, respectively.

When deducing the ft value from an experimental transition strength, the accuracy of Q_{EC} value (determining the Fermi function f), total beta-decay half-life $T_{1/2}$, and individual beta-decay branching ratio $I(EC+\beta^+)$ (determining the partial life-time t) are important. There is now a steady progress in accumulating new GT-decay data by investigating, e.g., new isotopes closer to the not-yet reached nucleus ^{100}Sn or, in the case of known decays, by improving the accuracy of the relevant parameters Q_{EC}, $T_{1/2}$, and $I(EC+\beta^+)$. As the new information is only gradually released in individual publications, conference contributions or theses, this report focuses on surveying the present experimental status, presenting it in tables and graphs, and comparing these data with model predictions.

2. EXPERIMENTAL TECHNIQUES

At the GSI on-line mass separator the proton-rich nuclei under study were produced in fusion-evaporation reactions. ^{48}Mn was synthesized by means of the reaction ^{12}C(^{40}Ca,p3n) with a graphite target/catcher inside the separator ion-source, while nuclei around ^{100}Sn and ^{146}Gd were mainly produced in ^{58}Ni induced reactions on targets such as ^{50}Cr and ^{96}Ru, respectively. These targets were mounted in front of the ion source. The method of on-line mass separation consists in principle of stopping the reaction products in a catcher inside the ion source, reionizating and accelerating them, and performing mass separation of the singly-charged ions of typically 55 keV energy in a magnetic sector field. In the case of ^{48}Mn and the nuclei around ^{100}Sn discharge ion-sources of FEBIAD-type and, in the case of ^{150}Er and ^{152}Yb, a high-temperature cavity thermoionizer [KIR90] have been applied, which achieve overall separation efficiencies in the percent range [KIR87a]. For the investigation of ^{102}Cd and ^{104}Sn the FEBIAD ion-source was operated in the bunched-beam release mode [KIR87b] to suppress the more abundantly produced isobaric contaminants. From the mass-separated beams point-like thin sources were collected on moveable tapes and transported to counting stations equipped with particle (β^+, conversion electrons, protons) and photon (γ,X) counters. For more details the reader is referred to the references given in the tables for each individual nucleus.

Recent experiments at GSI used in particular a summation-free β^+-endpoint spectrometer, which consists of a hyperpure germanium and a Si(Li) detector surrounded by an eight-segmented BGO ring [KEL90a]. The activity is collected on a thin tape and transported in stepwise operation through a three-stage differential-pumping section to the counting position. The latter is located between the two inner detectors outside the separator vacuum-chamber. Positrons are measured in the Si(Li) detector in coincidence with a characteristic γ transition observed in the germanium detector; summing of positron annihilation radiation in the Si(Li) detector is suppressed by detecting both 511-keV quanta in opposite segments of the BGO ring. With the new instrument so far seven Q_{EC} values have been remeasured with much improved accuracy [KEL90b,c].

3. THE DECAY OF ^{48}Mn

For the $T_Z = -1$ nucleus ^{48}Mn the decay proceeds dominantly via a superallowed Fermi transition with a branching ratio of 57.7% to the isobaric analogue state (IAS) at 5792 keV in ^{48}Cr. As the IAS in ^{48}Cr is bound against proton emission, Fermi and GT decays to ^{48}Cr levels up to the proton separation energy of 8.1 MeV can only be studied by γ-ray spectroscopy. GT transitions to ^{48}Cr states above the particle threshold, on the other hand, can be investigated by charged-particle measurements.

In a recent study [SZE89] of the ^{48}Mn decay at the GSI On-Line Mass Separator, a more accurate $T_{1/2}$ value of 158.1(22) ms was obtained than given in [SEK87], the γ singles, $\gamma\gamma$ coincidence, and proton single data were improved, and proton-γ coincidences were measured for the first time. Fig.1a displays the obtained GT-strength distribution: The remeasured values below 6 MeV excitation energy are derived from γ-ray spectroscopy and the ones above this energy from β-delayed protons. The analysis is based on an estimated Q_{EC} value of 13600 (100) keV [WAP85].

For comparison Fig.1b presents the strength from shell-model calculations [SEK87] including 1p-1h excitations. The resulting GT quenching factors averaged over ^{48}Cr excitation-energy intervals of 0-5.6 and 8.9-11.6 MeV, are 0.53(6) and 0.53(20), respectively [SZE89]. Calculations with updated empirical interactions for the fp shell are in progress.

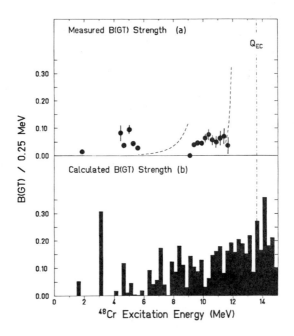

Fig.1: Comparison of the measured and calculated B(GT) strength for the ^{48}Mn decay. The measured strength is shown in part (a). The dashed lines give the experimental sensitivity limits, below which decay was not observed in this experiment. Part (b) shows the calculated strength. The B(GT) strength is integrated over 250 keV excitation-energy intervals, even though level energies up to 5.6 MeV were determined with 0.2-0.7 keV accuracies. The dash-dot line marked Q_{EC} represents the upper energy limit of the ^{48}Mn β^+/EC decay. From [SZE89]

4. DECAYS OF EVEN-EVEN NUCLEI NEAR ^{100}Sn

The synopsis of Fig.2 presents the so far observed $0^+ \rightarrow 1^+$ transition strength for all even-even nuclei with $Z \leq 50$, $N \geq 50$ that decay by β^+ emission or electron capture. The underlying parameters, as e.g. half-lives and Q_{EC} values, and the references are compiled in Table 1. While the nuclei $^{96-100}$Pd, ^{102}Cd, $^{104-108}$Sn were investigated at GSI as described in chapter 2, ^{98,100}Cd and partly ^{102}Cd were studied at CERN with chemically-pure mass-separated beams of cadmium isotopes from the ISOLDE II facility. The data for ^{94}Ru and ^{110}Sn are from the literature. By remeasuring the Q_{EC} values of ^{102}Cd and ^{104}Sn with the above-mentioned new β^+-endpoint spectrometer the errors were reduced by a factor of 7 and 5, respectively. The decay energies for ^{96}Pd, ^{100}Cd and ^{106,108}Sn were derived from the known Q_{EC} dependence of $\beta^+/(EC+\beta^+)$ ratios which were obtained from γ-ray spectroscopy. For ^{98}Cd a Q_{EC} measurement is in preparation.

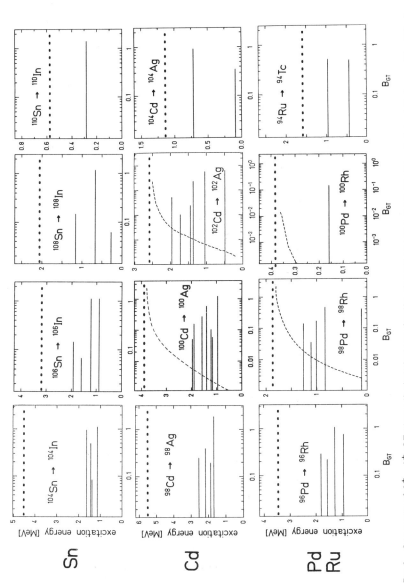

Fig.2 Synopsis of $0^+ \rightarrow 1^+$ GT-strength distributions for the decays of even-even neutron-deficient isotopes. The GT strength of transitions to 1^+ states which was considered for the summed strength B_{Σ} (GT) is plotted on a logarithmic scale against the excitation energy of these states in the β-decay daughter. The dotted lines indicate the Q_{EC}-window for the β decay and the dashed lines the estimated sensitivity limit resulting from the γ-ray spectroscopy.

For the β decay of an even-even nuclide near ^{100}Sn, the extreme single-particle shell model (ESPSM) predicts only one $\pi g_{9/2} \to \nu g_{7/2}$ transition from the 0^+ ground-state of the parent nuclide to an excited 1^+ state of the odd-odd daughter. As can be seen from Fig.2, however, experimentally one finds transitions to several 1^+ states, which is interpreted as spreading of the $(\pi g_{9/2}^{-1}, \nu g_{7/2})1^+$ configuration over several 1^+ states in the final odd-odd nucleus due to particle-core coupling and residual proton-neutron interaction [DOB88,KAL88]. For a quantitative comparison between measured GT strength and model predictions, we therefore take from experiment the reduced transition probability $B_\Sigma(GT)$, summed over all $n(1^+)$ identified 1^+ levels in the β-decay daughter according to

$$B_\Sigma(GT) = \sum_{i=1}^{n(1^+)} (B(GT))_i = \frac{6160}{(g_A/g_V)^2} \sum_i \left(\frac{1}{(ft)_i}\right) = \frac{6160}{(g_A/g_V)^2} \frac{1}{(ft)_\Sigma} \qquad (1)$$

$B_\Sigma(GT)$ as well as the calculated values of $B(GT)$, to be discussed below, are given in units of $g_A^2/4\pi$. The constant in the numerator and the free-neutron ratio of the weak coupling constants $|g_A/g_V| = 1.263$ are taken from [TOW85]. $(ft)_i$ is the comparative half-life of the i-th $0^+ \to 1^+$ transition, while $(ft)_\Sigma$ would be the corresponding value of a transition with the total strength $B_\Sigma(GT)$. Beside the values for $B_\Sigma(GT)$ and $\log(ft)_\Sigma$, Table 1 presents also the approximate excitation energy $E(1^+)^{av}$ of the $(\pi g_{9/2}^{-1}, \nu g_{7/2})1^+$ configuration in the daughter nucleus, which is calculated according to [DOB88]:

$$E(1^+)^{av} = \sum_{i=1}^{n(1^+)} (E(1^+) \times B(GT))_i \bigg/ \sum_{i=1}^{n(1^+)} (B(GT))_i \qquad (2)$$

The strength splitting of the $\pi g_{9/2} \to \nu g_{7/2}$ transition into many $0^+ \to 1^+$ transitions naturally leads to the question, how much GT strength has been missed in the experiment. Some of the strength may be shifted outside the Q_{EC} window, especially the small ones, and weaker transitions may have escaped observation. Particularly transitions to 1^+ states at the top of the Q_{EC} window are difficult to detect because of small branchings; nevertheless they may present large GT strength. The quality of the high-resolution γ-ray spectroscopy and the resulting decay scheme and branching ratios $I(EC+\beta^+)$ are therefore of major concern. For the most neutron-deficient isotopes investigated so far, the low production yield hampered very detailed spectroscopy. E.g. the fact that the yield of about 80 atoms/s for ^{98}Cd is $\approx 10^3$ times lower than that for ^{100}Cd may be the reason - beside nuclear structure effects -, that only 4 instead of 7 $0^+ \to 1^+$ transitions have been identified by now. In trying

Table 1: Gamow-Teller beta decays of even-even nuclides near ^{100}Sn. The total reduced-transition-probabilities $B_\Sigma(GT)$ and the corresponding $\log(ft)_\Sigma$ were calculated by summing the strength of the $n(1^+)$ identified 1^+ levels in the β-decay daughter according to Eq. (1). For the definitions of the average excitation energy $E(1^+)^{av}$ of the β-decay daughter and of the γ intensity ratio r_γ, see Eq. (2) and Eq. (3), respectively. A $ symbol in front of a nuclide and numbers in parentheses indicate preliminary results. For the references see Table 2.

Z	N	Nuclide	Half-life [s]	Ref.	Q_{EC}[keV]	Ref.	Strongest $E(1^+)$ [keV]	fed level $I(EC+\beta^+)$ [%]	$E(1^+)^{av}$ [keV] #	$n(1^+)$ r_γ	Ref.	$\log(ft)_\Sigma$ #	$B_\Sigma(GT)$ #	Ref.
44	50	^{94}Ru	3108(36)	1	1589(14)	2	442	77	710	2 / 42	3	3.59(4)	$0.99^{+0.10}_{-0.08}$	4
46	50	^{96}Pd	125(5)	4	3450(150)	4	939	50.4(23)	1258	4 / 200	4	3.22(14)	$2.33^{+0.86}_{-0.62}$	4
46	52	^{98}Pd	1062(18)	6	1867(25)	2	112	66(4)	674	5 / 75	7,8	3.51(2)	1.21 ± 0.05	5,8
46	54	$ ^{100}Pd	$3.14(8)\times10^5$	9	369(24)	2	159	89^{+11}_{-3}	159	1 / 770	8	$(4.37^{+0.12}_{-0.18})$	$(0.17^{+0.08}_{-0.05})$	5,8
48	50	$ ^{98}Cd	9.2(3)	10 / 5	estimated: 5500(200)	11	1691	80(5)	1850	4 / 100	5	(3.16 ± 0.13)	$(2.7^{+0.9}_{-0.7})$	5
48	52	^{100}Cd	49.1(5)	13	3890(70)	13	952	69.7	1210	7 / 910	13	3.2(1)	$2.4^{+0.6}_{-0.5}$	13
48	54	^{102}Cd	345(8)	7 / 14	2587(8)	7,14 / 15	490	64.9(23)	906	6 / 430	7 / 14	3.407 ± 0.016	1.513 ± 0.055	5 / 14
48	56	^{104}Cd	3462(60)	16	1137(12)	2	91	~72	544	2 / 420	16	$3.48^{+0.03}_{-0.05}$	$1.27^{+0.17}_{-0.08}$	5
50	54	^{104}Sn	20.8(5)	17	4515 ± 60	15	1139	58	1382	4 / 70	17 / 18	3.17(6)	$2.63^{+0.38}_{-0.33}$	5
50	56	^{106}Sn	115(5)	18	3190(60)	18	892	59.3	1123	4 / 27	18	3.20(5)	$2.44^{+0.30}_{-0.27}$	18
50	58	^{108}Sn	618(5)	19	2059(25)	18	669	87.8	~715	3 / 70	18 / 19	3.45(3)	$1.37^{+0.10}_{-0.09}$	18
50	60	^{110}Sn	$1.48(4)\times10^4$	20	576(34)	2 / 21	283	100	283	1 / -	20	3.44(12)	$1.39^{+0.48}_{-0.32}$	5

51

to characterize in some way the thoroughness of the γ-ray spectroscopic work, Table 1 lists in column 11, together with the number n(1$^+$) of identified 1$^+$ levels, the ratio

$$r_{I\gamma} = I_\gamma^{Max}/I_\gamma^{Min} \tag{3}$$

I_γ^{max} and I_γ^{min} are the maximum and minimum values, respectively, of the intensities of those γ rays included into the decay scheme. E.g., for ^{100}Cd this number is nine times higher than for ^{98}Cd. For some of the nuclei in Fig.2 the estimated sensitivity limit is indicated which results from the γ-ray spectroscopy and below which GT strength would not have been detected. In column 12 of Table 1 reference is given to the spectroscopy work from which the decay scheme and the branching ratios I(β$^+$+EC) were taken. The errors given for B_Σ(GT) and log(ft)$_\Sigma$ in this table do not include the systematic uncertainties from non-observation of 1$^+$ states at higher excitation energies due to unobserved weak transitions or limiting Q_{EC} windows.

5. DECAYS OF EVEN-EVEN N=82 ISOTONES ABOVE ^{146}Gd

In the β-decays of the even-even N=82 isotones above ^{146}Gd the $h_{11/2}$ protons transform into neutrons in the empty $h_{9/2}$ neutron orbit. In contrast to the analogous case of the N=50 isotones, however, the β decays feed mainly just one 1$^+$ level in the daughter nucleus, e.g. 97% in the case of ^{148}Dy (see Fig.3 and Table 2). The very detailed investigation ($r_{I\gamma} = 10^4$; see Table 2) of this nucleus at ISOCELE/Orsay was recently complemented by a β$^+$-endpoint remeasurement with a spectrometer based on the same principle as the one described in chapter 2. Improved Q_{EC} values were obtained for this nucleus and, at GSI, also for ^{150}Er, but not for ^{152}Yb because of a too low production rate. In the latter case the much less accurate β$^+$/EC-ratio method had to be applied (see Table 2). Concerning the GT β-decay data compiled in Table 2, the same comments are applicable as given in the previous chapter for the ^{100}Sn region presented in Table 1.

6. NUCLEAR STRUCTURE AND GAMOW-TELLER DECAY

In this section, we compare the experimental strength B_Σ(GT) with model predictions. In order to calculate the $0^+ \rightarrow 1^+$ GT transition strength, we start out with an extreme single-particle shell model (ESPSM) and then add corrections for pairing and for proton-neutron residual interactions.

Table 2: Gamow-Teller beta decays of even-Z, N=82 nuclides above ^{146}Gd. For explanation see Table 1.

Z	N	Nuclide	Half-life [s]	Ref.	Q_{EC}[keV]	Ref.	Strongest E(1$^+$) [keV]	fed level I(EC+β^+) [%]	E(1$^+$)^{av} [keV] #	n(1$^+$) r_γ	Ref.	log (ft)$_\Sigma$ #	B_Σ(GT) #	Ref.
66	82	^{148}Dy	204(12)	1	2678(10)	2,3	620	96.8	646	2 10^4	1,4	3.91 $^{+0.04}_{-0.02}$	0.472 $^{+0.023}_{-0.041}$	2
68	82	^{150}Er	19.0(4)	5	4110(15)	2,3	476	95.4	476	1 330	5	3.69 $^{+0.01}_{-0.02}$	0.792 $^{+0.038}_{-0.018}$	2
70	82	^{152}Yb	3.03(6)	5	5440 $^{+220}_{-170}$	6	482	87.2	482	1 174	5	3.51 $^{+0.09}_{-0.08}$	1.199 $^{+0.243}_{-0.224}$	7

References to Table 1: 1: BRO86; 2: WAP88; 3: GRA73; 4: RYK85; 5: RYK90c; 6: MÜL83; 7: RYK90a; 8: RYK90b; 9: SIN90; 10: BAR89; 11: from systematics and theory, see ref. 12; 12: DOB88; 13: RYK89; 14: KEL90e; 15: KEL90b; 16: BLA84; 17: SZE90; 18: BAR88a; 19: HEE82; 20: GEL83; 21: The 34 keV error is not based on the uncertainty of the mass excess of the (unknown) ground-state of ^{110}In quoted in ref. 2, but on the 30 keV accuracy of the Q_{EC} value of ^{110}In (2$^+$, 69 min). See [WAP85].

References to Table 2: 1: KLE85; 2: KEL90c; 3: KEL90d; 4: PEK90; 5: KLE89a; 6: SAL90; 7: KEL90d.

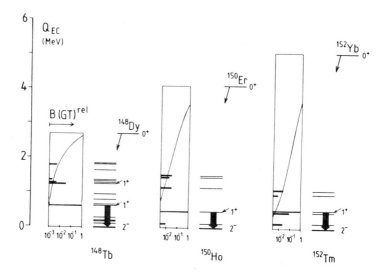

Fig.3: Simplified level scheme for the β^+/EC decays of the $N=82$ isotones ^{148}Dy, ^{150}Er, and ^{152}Yb. The excited states in the daughter nuclide and the parent state's Q_{EC} value are displayed on scale with reference to the respective daughter ground-state. To the left the relative GT strength to different levels (assuming $0^+ \to 1^+$ transitions) is plotted on a logarithmic scale and the detection sensitivity limit for 1^+ strength is included. The firmly established $0^+ \to 1^+$ transitions that were considered in the summed strength $B(GT)_\Sigma$ in Table 2 are marked by arrows and 1^+. From [KLE89b].

According to the ESPSM without residual interaction, the GT strength for the decay of even-even nuclides in the ^{100}Sn region for the one $\pi g_{9/2} \to \nu g_{7/2}$ transition (orbital angular momentum $\ell = 4$) and in the ^{146}Gd region for the one $\pi h_{11/2} \to \nu h_{9/2}$ transition ($\ell = 5$) is given by

$$B(GT) = \frac{8\ell(\ell+1)}{2\ell+1} v_\pi^2 (1 - v_\nu^2) \qquad (4)$$

v_π^2 and v_ν^2 being the occupation numbers of protons and neutrons in their orbits, respectively. For the ^{100}Sn region and within the ESPSM it is assumed, that the valence protons between niobium ($Z=41$) and tin ($Z=50$) occupy $g_{9/2}$ orbits ($v_\pi^2 = (Z-40)/10$), and that the valence neutrons above $N=50$ occupy first $d_{5/2}$ and then $g_{7/2}$ orbits ($v_\nu^2 = 0$ for $N \leq 56$; $v_\nu^2 = (N-56)/8$ for $57 \leq N \leq 64$). In order to include pairing effects, the occupation numbers in Eq.(4) were taken from the macroscopic-microscopic model calculations of [DOB88]. The formalism of Lipkin and Nogami, which was applied in this work, is more appropriate to these magic or close to magic nuclei, than the standard BCS approach. As can be seen from Fig.4, the resulting quenching factors $B_\Sigma(GT)/B(GT)$ are of the order of 0.1 to 0.2 (neglecting ^{100}Pd because of its very small Q_{EC}-value). The pairing corrections shift the calculated strength values slightly towards the experimental ones except for the decay of ^{94}Ru and ^{96}Pd, where the opposite effect is observed. Pairing calculations for ^{148}Dy, which were adjusted to experimental occupation numbers for h11/2 protons in ^{144}Sm, yielded more than two h11/2 protons in ^{148}Dy ($v_\pi^2 = 0.26$) and therefore an enhancement of the GT transition strength [KLE85]. If one extrapolates this v_π^2 value to the Z = 68 and Z = 70 isotopes in accord with the Z dependence of the N = 50 isotones, one obtaines $v_\pi^2 \approx 0.38$ and 0.5, respectively [BAR88b]. With $B_\Sigma(GT)$ from Table 2 and $B(GT)$ from Eq.(4) the following quenching factors without and with inclusion of pairing corrections result for ^{148}Dy, ^{150}Er, and ^{152}Yb: 0.13/0.08, 0.11/0.10, 0.11/0.11. For the N = 82 isotones above ^{146}Gd pairing calculations of the type performed in [DOB88] gave similar occupation factors v_π^2 as the ones quoted above [RYK90d].

An important reduction of the predicted GT strength is obtained when taking effects from proton-neutron residual interaction into account. For the N = 50 and N = 82 isotones Towner [TOW85] calculated in first-order pertubation theory the hindrance factors for various effective interactions. For the ^{100}Sn region the resulting quenching (including the above discussed pairing) is displayed in Fig.4. For ^{148}Dy, ^{150}Er, and ^{152}Yb quenching factors in the range from 0.15 to 0.3 are obtained. In a different approach to include proton-neutron residual interaction, the authors of [CHA83, KLE85, SUH88a/b, KUZ88, SOL89] applied the proton-neutron quasiparticle random-phase approximation (QRPA) to reproduce the observed suppression of strength in β^+ decay. In contrast to β^- decay, here the transition strength to low-lying states depends not only on the particle-hole interaction V_{ph}, but also quite delicately on the particle-particle interaction V_{pp}. While the strength of V_{ph} can be adjusted to the giant GT resonance via the semi-empirical energy formula, the essential problem is to obtain V_{pp}. In principle both interactions are related through angular monumentum recoupling (Pandya relationship), the truncated model space, however, leads to renormalized interaction strengths. While the particle-hole interaction V_{ph} reduces the total $B(GT)$ by about a factor of two, the particle-particle interaction V_{pp} only redistributes the strength shifting some of it from low-lying 1^+ states to higher ones. This reduces the observable strength $B_\Sigma(GT)$ in the decay Q-window by a factor of two or much larger, when varying V_{pp} between zero and the maximum value allowed by the QRPA. It was shown [SUH88a/b, KUZ88, SOL89] for the ^{100}Sn and ^{146}Gd regions that with some variation of reasonable values of V_{pp} generally the experimental strength $B_\Sigma(GT)$ could be reproduced assuming a renormalized nuclear matter value [TOW85] $|g_A/g_V| \sim 1$, corresponding to a quenching factor of 0.63 (dashed line in Fig. 4). With the exception of ^{100}Pd and ^{104,106,108}Sn [SUH88b], the experimental values were generated also for the free-nucleon value of g_A/g_V. The main difficulties arose in the cases of ^{96}Pd and some cadmium isotopes, where the models tend to predict smaller B(GT) values than the observed ones. In a recent calculation [BOR90] in the frame of the finite Fermi-system theory the quenching factors are around 0.6 or even somewhat lower. On the other hand, the QRPA approach in [KLE85] yielded a rather low quenching factor of 0.16 for the decay of ^{148}Dy.

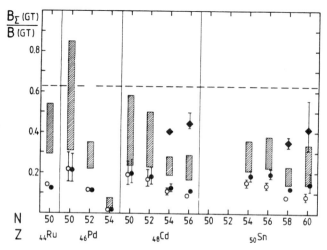

Fig. 4: GT quenching factors for $0^+ \to 1^+$ β decays of neutron-deficient ruthenium-to-tin isotopes, obtained as ratios between measured GT transition probabilities and model predictions. The experimental values $B_\Sigma(GT)$ represent summing over all observed β decay branches to 1^+ states. The predictions B(GT) result from an extreme single-particle shell-model (open circles) and from a correction of this model for pairing [DOB88] (circles). The circles are drawn slightly aside from the isotone coordinate in order to avoid overlap. The squares present results [CHA83] from QRPA calculations. All error bars in the figure indicate only experimental uncertainties. However, the range for quenching factors taking core-polarization corrections into account (hatched bars) is due to experimental uncertainties and to various choices of the effective interaction [TOW85]. These calculations were performed only for the decay of N=50 isotones between ^{94}Ru and ^{100}Sn. In this plot it is assumed, that for a given proton number these factors stay constant independent upon neutron number. The horizontal dashed line indicates how much quenching one can compensate by renormalizing $|g_A/g_V|$ from its free-nucleon value 1.263 to a nuclear matter value [TOW85] of about one.

In order to conclude if g_A has to be renormalized in nuclei, it is important to deduce the strength V_{pp} from other experimental data than β decay. As a first approach to this problem, [CHA83] adjusted the value of V_{pp} to energies of low-lying states. The resulting quenching factors, shown in Fig. 4, are about two times higher than the ones obtained for first-order core polarization.

7. CONCLUDING REMARKS

We presented results from recent measurements and from literature for ^{48}Mn and for $0^+ \rightarrow 1^+$ GT β-decays of nuclei in the ^{100}Sn region and of ^{148}Dy, ^{150}Er, and ^{152}Yb. A comparison with predictions suggests GT quenching-factors of approximately 0.5 for the decay of ^{48}Mn [SZE89], 0.2-0.5 for the ^{100}Sn region (cp. Fig.4) [CHA83,TOW85], and 0.15-0.3 for the ^{146}Gd region [KLE85,TOW85]. Recent calculations [SUH88a/b, KUZ88, SOL89, BOR90, STA90] open the chance to achieve quenching factors of 0.63 (corresponding to a renormalization of $|g_A/g_V| \approx 1$) or even unity for these two regions, if the adjusted residual particle-particle interaction strength can be confirmed by independent methods. (In the cases of ^{100}Pd and ^{110}Sn the quenching factors presently obtained may not be very significant, because strength may be missing due to Q_{EC} windows smaller than 600 keV).

These results should be compared with average quenching factors of 0.58 and 0.89 in the sd-shell (^{16}O to ^{40}Ca [BRO85] and $^{32-35}$Ar [BOR89], respectively) and a value of about 0.6 for fp-shell nuclei [MIY87, SEK87, SHI89] and from (p,n) reaction studies [GAA86]. For the β decay of the doubly-magic nucleus ^{56}Ni, which was reinvestigated recently [SUR90], a log ft value of 4.4 was obtained for the one observed $0^+ \rightarrow 1^+$ transition (Q_{EC} = 2136 keV). This means that the measured strength is only 1.1% of the value expected for the $\ell = 3$ transition $\pi f_{7/2} \rightarrow \nu f_{5/2}$ in the single particle model with no residual interaction ($v_\pi^2 = 1, v_\nu^2 = 0$ in Eq. 4). − In order to get beyond the approximate characters of quenching factors such as those shown in Fig. 4, the model calculations ought to be improved, e.g. with respect to the residual proton-neutron interaction including the particle-particle part. At least part of the missing GT strength which is up to now accounted for by renormalizing g_A/g_V should be searched for in higher-order effects such as configuration mixing due to tensor forces or mixing with Δ-isobar nucleon-hole configurations (for references see e.g. [BOR89]). Concerning emerging tasks for the experimentalist, the spectroscopy of ^{98}Cd, including a Q_{EC} measurement, and the search for ^{102}Sn (^{100}Sn) should be continued. Beside the high-resolution γ-ray spectroscopy also "more complete" spectroscopic means, e.g. total γ-ray absorption, should be applied, which allow to identify also weak GT transitions to 1^+ levels at high excitation energies and/or to derive a quantitative detection limit for such transitions. This technique would also facilitate to extend the the investigation of the GT strength-distribution to the more complicated decays of non-even-even nuclei in the ^{100}Sn and ^{146}Gd regions.

The presented results have been obtained in collaboration of the GSI on-line mass-separator group with colleagues from CEN Bordeaux, Michigan State University, Institut de Physique Nucléaire Orsay (in the case of ^{48}Mn decay), the Astronomical Observatory Frombork, the CERN ISOLDE-collaboration, KU Leuven, Manchester University, University of Warsaw (^{100}Sn region), KFA Jülich, CSNSM Orsay, RIP Stockholm, and IFIC Valencia (^{146}Gd region). The work at GSI concerning the quenching of the GT strength of proton-rich nuclei started in the ^{100}Sn region and was stimulated by fruitful discussions with J. Blomqvist from the RIP Stockholm which is gratefully acknowledged. For the support in supplying decay data for Table 2 we are obliged to H. Keller.

† On leave of absence from the University of Warsaw, Poland

References

BAR88a	R. Barden et al., Z. Phys. A329 (1988) 319
BAR88b	R. Barden "Isomerie und Betazerfall: Konversionselektronen- und γ-Spektroskopie von neutronenarmen Kernen um ^{100}Sn und ^{146}Gd", Thesis, Universität Mainz, GSI-88-18 Report, 1988
BAR89	R. Barden et al., GSI Scientific Report 1988, GSI-89-1 (1989), p. 36 and to be published
BLA84	J. Blachot et al., Nucl. Data Sheets 41 (1984) 325
BOR89	M.J.G.Borge et al., Z. Phys. A332, 413 (1989).
BOR90	I. N. Borzov et al. "Gamow-Teller Strength Functions of Stable and Neutron Deficient Nuclei", Preprint 2069, Obninsk 1990
BRO85	B.A. Brown and B.H. Wildenthal, At. Data and Nucl. Data Tables 33 (1985) 347
BRO86	E. Browne et al., Table of Radioactive Isotopes, John Wiley, New York 1986
CHA83	D. Cha, Phys. Rev. C27 (1983) 2269
DOB88	J. Dobaczewski et al., Z. Phys. A329 (1988) 267
GAA86	C.Gaarde, in Proc. Int. Conf. Weak and Electromagnetic Interactions in Nuclei, H.V. Klapdor (ed.), Springer, Berlin-Heidelberg (1986), p. 260.
GEL83	P. de Gelder et al., Nucl. Data Sheets 38 (1983) 545
GRA73	P. Graf, H. Münzel, Radiochemica Acta 20 (1973) 140
HEE82	R.L. Heese et al., Nucl. Data Sheets 37 (1982) 289
KAL88	Ł. Kalinowski, in "Nucl. Structure of the Zirconium Region", J. Eberth, R.A. Meyer, K. Sistemich (eds.), Research Reports in Physics, Springer, Berlin, (1988), p. 397
KEL90a	H. Keller et al., "A Summation-Free β^+ Endpoint Spectrometer", accepted for publication in Nucl. Instr. and Meth. in Phys. Res.
KEL90b	H. Keller et al., GSI Sci. Rep. 1989, GSI-90-1 (1990), p. 19
KEL90c	H. Keller et al., "β^+-Endpoint Measurements Near ^{100}Sn and ^{146}Gd", contribution to the 8th Conf. on Atomic Masses and Fundamental Constants (AMCO-8), Jerusalem (originally planned for Sept. 1990, new date is pending)

KEL90d	H. Keller, "β^+-Endpunktsmessungen von neutronenarmen Kernen um ^{100}Sn und oberhalb ^{146}Gd", Thesis, Technische Hochschule Darmstadt, in preparation
KEL90e	H. Keller et al., "Q_{EC}-value and Gamow-Teller Strength of the ^{102}Cd EC/β^+-Decay", in preparation
KIR87a	R. Kirchner, Nucl. Instr. and Meth. B26 (1987) 204
KIR87b	R. Kirchner et al., Nucl. Instr. and Meth. B26 (1987) 235
KIR90	R. Kirchner, Nucl. Instr. and Meth. A292 (1990) 203
KLE85	P. Kleinheinz et al., Phys. Rev. Lett. 55 (1985) 2664
KLE89a	P. Kleinheinz et al., KFA-Jülich, Institut für Kernphysik, Ann. Report 1988, Jül-Spez-499 (1989) p. 34
KLE89b	P. Kleinheinz et al., GSI Sci. Rep. 1988, GSI-89-1 (1989), p. 35
KUZ88	V.A. Kuz'min et al., Nucl. Phys. A486 (1988) 118 and Mod. Phys. Lett A3 (1988) 1533
MIY87	H. Miyatake, K. Ogawa, T. Shinozuka, and M. Fujika, Nucl. Phys. A470 (1987) 328
MÜL83	H.-W. Müller, Nucl. Data Sheets 39 (1983) 467
PEK90	L.K. Peker, Nucl. Data Sheets 59 (1990) 393
RYK85	K. Rykaczewski et al., Z. Phys. A322 (1985) 263
RYK89	K. Rykaczewski et al., Z. Phys. A332 (1989) 275
RYK90a	K. Rykaczewski et al., GSI Scientific Report 1989, GSI-90-1 (1990), p.11
RYK90b	K. Rykaczewski et al., "Gamow-Teller Beta Decay of Even-even Nuclei Close to the Doubly Magic ^{100}Sn", contribution to the XXI Summer School on Nuclear Physics, Mikolajki, Poland, Sept. 1990
RYK90c	K. Rykaczewski et al., "Gamow-Teller Beta Decay of Nuclei near the Double Shell-Closure at ^{100}Sn", contribution to the 8th Conf. on Atomic Masses and Fundamental Constants (AMCO-8), Jerusalem (originally planned for Sept. 1990, new date is pending)
RYK90d	K. Rykaczewski, private communication and to be published
SAL90	H. Salewski, "Q_{EC}-Werte aus gammaspektroskopischen Untersuchungen neutronenarmer Isotope im Bereich um ^{146}Gd ", Thesis, Universität Göttingen, 1990 and to be published
SEK87	T. Sekine et al., Nucl. Phys. A467 (1987) 93
SHI89	T. Shinozuka et al., in Proc. of Yamada Conf. XXIII on Nuclear Weak Process and Nuclear Structure, Osaka 1989, M. Morita et al. (eds), World Scientific, Singapore (1989), p. 342
SIN90	B. Singh, J.A. Szucs, Nucl. Data Sheets 60 (1990) 1
SOL89	V.G. Soloviev, in Proc. Int. Conf. on Selected Problems of Nuclear Structure, JINR publ. D4, 6, 15-89-638, Dubna 1989 (in Russ.) and in Proc. IX Int. School on Nuclear Physics, Neutron Physics and Nuclear Energy, Varna, Bulgaria, 1989, W. Andrejtscheff and D. Elenkov (eds.), World Scientific, Singapore, p. 137 (in Engl.)
STA90	A. Staudt et al., "Calculation of β-Delayed Fission of ^{180}Tl and Application of the QRPA to the Prediction of β^+-Decay Half-Lives of Neutron-Deficient Isotopes". Submitted to Phys. Rev. Lett.
SUH88a	J. Suhonen et al., Phys. Lett. B202 (1988) 174
SUH88b	J. Suhonen et al., Nucl. Phys. A486 (1988) 91
SUR90	Bhaskar Sur et al., Phys. Rev. C42 (1990) 573
SZE89	J. Szerypo et al., Contribution to the Workshop on Nuclear Structure of Light Nuclei far from Stability Experiment and Theory, Obernai, France, Nov. 1989 and to be published
SZE90	J. Szerypo et al., Nucl. Phys. A507 (1990) 357
TOW85	I.S. Towner, Nucl. Phys. A444 (1985) 402
WAP85	A.H. Wapstra, G. Audi, Nucl. Phys. A432 (1985) 1
WAP88	A.H. Wapstra, G. Audi, and R. Hoekstra, At. Data and Nucl. Data Tables 39 (1988) 281

Nuclear Structure Studies by using He-Jet and Isotope Separator Facilities at SARA (*)

Andrée GIZON and Jean GIZON

Institut des Sciences Nucléaires
IN2P3 - CNRS/Université Joseph Fourier, Grenoble, France

Abstract : Techniques used to produce neutron-deficient nuclei by means of He-jet and isotope separator facilities placed on a beam line of the Grenoble cyclotrons are presented. Data obtained on odd-A Ba nuclei, ^{124}Ba and ^{130}Ce are discussed.

1 - Introduction

Neutron-deficient A = 120 -130 isotopes from Xe (Z = 54) to Ce (Z = 58) have attracted considerable interest for more than two decades because they offer very interesting features typical of transitional nuclei. From systematic in-beam studies of these nuclei via heavy-ion fusion evaporation reactions, collective behaviour as well as particle excitations have been observed, leading to various deformations dominated by triaxial prolate shapes.

However, the low-energy low-spin level structure of the more neutron-deficient of these A = 120 -130 transitional nuclides remains poorly know. It is clear that more detailed information about these nuclei far from stability is essential. Especially important are the identification and the excitation energy of the individual low-energy intrinsic states which are needed to interpret experimental data and to improve theoretical models. In addition, these experimental informations for protons and neutrons are decisive to treat correctly the odd-odd isotopes.

Interest of precise low-energy nuclear structure studies is increased by considering the rapid development of experimental techniques in gamma-ray and low energy electron spectroscopy and the important and rich set of atomic properties available for the ground state or/and long-lived isomeric state of these isotopes. Indeed, spin, moments and isotope change of charge distribution of nuclear ground states have been progressively extracted from the hyperfine structure and the isotope shift of optical spectra. For example, a wealth of systematic results on Cs and Ba was obtained at the on-line isotope separator Isolde at CERN, using the Rabi's atomic beam magnetic resonance method (ABMR) [1]. Then, a large extension of the data has been reached with the development of the collinear fast beam laser spectroscopy applicable to a variety of

---

(*) Talk presented by J. Gizon

elements. From all these experiments, spin assignments, nuclear magnetic moments, spectroscopic quadrupole moments and finally the changes in mean square charge radii, $\delta \langle r^2 \rangle$, are now available in this transitional region for $^{120-148}$Ba [2,3], $^{118-146}$Cs [4, 5] and $^{116-146}$Xe [6, 7]. As an example, a plot of the change in mean square charge radii with respect to ^{138}Ba versus the neutron number N is reproduced in fig.1 for Xe and Ba isotopes. One observes an abrupt change in the slope at N = 82 which reflects the influence in the shell structure and a monotonic evolution for N < 82 with quite large fluctuations below N = 66. The most spectacular case was previously underlined for the coexistence of different mean square charge radii for ground and isomeric states in light odd-Z (55)-even-N (64, 66) cesium isotopes [4]. The spin of the less deformed states is 3/2 in ^{119m}Cs and ^{121}Cs while the spin of the more deformed ones is 9/2 in ^{119}Cs and ^{121m}Cs. The 9/2 spin measured in these light deformed Cs can be associated with the presence of an "intruder" proton hole in the 9/2+[404] orbital. This assignment is well supported by in-beam collective systematics in odd-A cesium isotopes [8].

As it can be observed in a proton Nilsson diagram [9], the 9/2+[404] orbital exhibits a steep upsloping. The energy required to lift a proton out of it decreases rapidly if the deformation increases. The difference in deformation observed between the ground state and the isomeric state in ^{119}Cs ($\langle \beta^2 \rangle^{1/2}$ from ~ 0.28 to ~ 0.32) has been qualitatively reproduced by using Hartree-Fock + BCS calculations [10] with blocking of the 55th proton in the relevant states [11]. The spin I = 2 and the $\langle \beta^2 \rangle^{1/2}$ deformation parameters of 0.31 - 0.32 deduced for $^{118, 120}$Cs, respectively, can be accounted for if a Ω = 5/2 orbital as [413] 5/2 or [532] 5/2, naturally filled by the last neutron, is coupled to the 9/2+[404] proton orbital.

The more recent measurements on $^{120, 121}$Ba [3] add an important experimental information to the barium isotope systematics. From isotope shift experiments relative to ^{124}Ba, $\langle \beta^2 \rangle^{1/2}$ of 0.316 and 0.320 have been estimated for ^{120}Ba and ^{121}Ba, respectively. Then, at N = 65, ^{120}Cs and ^{121}Ba exhibit a similar jump in their isotope shift. The sign change in experimental values of the ground-state magnetic moments in ^{123}Ba and ^{121}Ba is reasonably well reproduced if a 5/2+[402] orbital is associated to ^{123}Ba and a 5/2+[413] to ^{121}Ba, respectively. The similar behaviour observed in the isotope shifts of both ^{120}Cs and ^{121}Ba at N = 65 can be related to the occupation of the ν 5/2+ [413] orbital ; the neutron-proton interaction strength between the ν 5/2+ [413] and the π 9/2+ [404] orbitals is in agreement with the increase of deformation.

As shown in fig.1, the situation is quite different for Xe isotopes. After an almost linear variation from N = 82 to N = 68, the deformation increases beyond midshell (N = 66) and keeps a stable value (β ~ 0.3). Obviously, precise nuclear studies are needed to understand the differences between Xe and Ba ground states and to interpret in more details the spectacular set of atomic informations around N = 64- 66.

In the field of this puzzle, several experiments on the radioactive (β^+ + EC) decay of nuclei far from stability of this transitional region have been progressively achieved in our group, by using on-line isotope separators. When such studies include precise transition multipolarity measurements they are able to specify the characteristics of the low-spin low-energy states excited by β-decay. Moreover, through a few gamma rays, these measurements identify new nuclei studied by in-beam spectroscopy and/or establish the connections between the collective structures and with the bottom of the level schemes.

In this lecture, examples of radioactive decay results are presented to show their contribution in nuclear structure studies. As the more recent experiments have been mainly performed at the Grenoble He-je coupled mass separator on line to the heavy-ion accelerator SARA, a brief technical description of this set-up is first reviewed. Experimental results in the A = 120 - 130 mass region, reported in chapter 3 concern odd-A Ba low-lying state systematics and a comparison with the IBFM-1 model calculations. Interests of spin and parity measurements of low-spin levels in even-even ^{124}Ba and ^{130}Ce are discussed in connection with the predictions of theoretical approaches.

2 - Experimental procedure at SARA

2.1 - Production

One of the best ways for producing neutron-deficient nuclei consists of heavy-ion fusion evaporation reactions involving neutron-deficient species for both the target and the projectile.

Thin (1-3 mg/cm^2) self supporting isotopically enriched targets of ^{92,94,96}Mo, prepared by rolling, have been bombarded with various projectiles as ^{35}Cl, ^{36}Ar or ^{40}Ca, to produce light A = 120-130 rare earth nuclei. The beams were delivered by the first cyclotron of the SARA facility, with typical intensities of 200 to 800 nAe (9$^+$ or 10$^+$ ions) and energies of 5 to 7 MeV/nucleon.

2.2 - Extraction and Separation

In principle, the mass separation of new nuclei can be achieved with classical on-line isotope separator (ISOL), the target being part of the ion-source. The reaction products recoil in a thick target or are collected in a catcher placed behind a thin target. In all cases the products have to be extracted by thermal diffusion, ionized and accelerated towards the magnetic separator. Several years ago, such a technique was successfully used on-line at the Grenoble variable energy cyclotron to extract neutron-deficient indium isotopes [12] but this method is limited :
i) the delay time for release of the products from the catcher depends on their chemical nature and hinders the extraction of isotopes with half-lives shorter than a few seconds.
ii) as the ion-source was working at high temperature, only refractory targets were allowed.

The delay time can be strongly reduced with an He-jet system coupled to an ion-source. Moltz and co-workers at RAMA [13] have successfully applied this technique. The recoiling products were thermalized in a high-pressure (1-2 bars) helium atmosphere and transported along a capillary into the ion-source. Several advantages can be underlined :
- all the recoiling elements produced in nuclear reactions can be introduced in the ion-source, including the nonvolatile species
- the delay time becomes of the order of the mean transit time in the system (typically of a few tens of milliseconds per meter of capillary)
- as the target is far from the hot ion-source, in principle, any kind of material can be used.

The isotope separator on-line with the SARA accelerator in Grenoble has been rebuilt thanks to a Lyon-Grenoble collaboration. The layout of the experimental area is shown in figure 2. Characteristics of the new line have been calculated to transport all available heavy-ion beams with energies ranging from a few to 38 MeV/nucleon and to focalize the particles at two different positions, either at the recoil chamber or at the ion-source.

The first He-jet coupling which has been made with a medium current Bernas-Nier ion-source is schematized in figure 3. The radioactive products recoiling from the target are thermalized in a stopping chamber, in the middle of a pressurized (1-2 bars) cube. A 2mg/cm^2 Havar window separates the system from the vacuum. Thanks to an oven, the temperature is optimized for a given aerosol (T ~ 400°C with PbCl$_2$ aerosols). The thermalized atoms are carried via the capillary either directly to the counting station or to the separator facility. The capillary is directed in front of the skimmer (∅ ~ 1.2 mm) and their relative distance has to be adjusted to keep the ion-source in operation. A high-flow pumping system [3 Roots : 8000 m^3/h - 3000 m^3/h - 400 m^3/h and one 120m^3/h primary] is required to maintain the pressure below ~ 2.10^{-5} torr at the source for 10^{-1} - 10^{-2} torr at the injection. Tests and technical details have been already published at the 11th International EMIS Conference [14]. Through the rare earth region, the He-jet alone has an efficiency of 50 to 65 % while the coupled set-up yield (skimmer + ion source + magnet + transport) is of the order of 1-2 %. Taking into account the heavy ion cross section productions, collections of 100 to 1000 atoms/s have been currently reached.

Important technical efforts have been focussed on the single-charged ion production by means of a new ECR ion-source (electron cyclotronic resonance). The design of one of the developed prototypes is reproduced in figure 4. Technical details and characteristics of this two-stages 2.45 GHz ECR ion-source have been already reported [15]. It is important to underline that a current of 1 to 1.5mA single-charged ions (as N$^+$, Ne$^+$, Ar$^+$, Kr$^+$ or Xe$^+$) is easily extracted. Combining electric lenses and electric quadrupoles at the entrance of the 120° magnet of the isotope separator, a suitable optics has been achieved. A complete separation of the Xe isotopes is easily reached with a resolution greater than 1000.

As the properties of the ECR ion-sources for the single-charged ion production are more or less unknown, several series of tests have been undertaken to improve the coupling of the He-jet with the ion-source. With the injection of the agregats along the plasma axis, no significant ionization of the radioactive products was reached. After several essays, looking carefully at the radioactive deposits, it was observed that the ionization efficiency of the new ECR ion-source is located in the vicinity of the resonance. Consequently, a new series of tests has been undertaken with an oven or a catcher displayed inside the source to stop the agregats and to re-evaporate them towards the ionization zone.

With the ECR ion-source prototype reproduced in figure 5, which contains a graphite oven, Cs$^+$, Ba$^+$ and La$^+$ ions have been collected with the A = 127 isotopic chain produced in the ^{116}Sn + ^{14}N reaction at 80 MeV. The isotopic separation efficiencies reached were typically of 6 to 7. 10^{-2} for Cs, 5.10^{-3} for Ba and only 10^{-5} for La ions. Using the ^{112}Cd + ^{14}N nuclear reaction at 78MeV and the 3n exit channel which contains ^{123m}Cs (T$_{1/2}$ = 1.6 s), ^{123}Cs → ^{123}Xe (T$_{1/2}$ = 6 mn) and ^{123}Xe → ^{123}I (T$_{1/2}$ = 6 h), it was also observed that the ionization efficiency can reach more than 15 % for the Xe ions if the system is maintained in operation several minutes after the injection of the agregats. These observations indicate the existence of a delay in the evaporation phase in strong connection with the temperature and the surface properties of the catcher.

By using a tantalum catcher to stop the agregats and the same A = 123 radioactive isotopic chain, the ^{123m}Cs$^+$ ions (T$_{1/2}$= 1.6 s) have been observed with approximately 1 % efficiency after the isotope separator, when the temperature of the catcher reached about 1000°C. This detection of a short half-life isotope shows the important role of parameters as the material and the temperature of the catcher.

Very recently, after a careful choice of several refractory materials, a more compact 2.45 GHz metallic body ECR ion-source equipped with an internal tantalum catcher to intercept the agregat jet, has been developed to improve the coupling with the He-jet system. Off-line tests of this source are presently in progress.Its properties to ionize gas as He, Ne, Ar, Kr and Xe are quite promising and its performances for the coupling will be tested before the end of this year. This metallic ion-source will be suitable for the matching of parameters as temperatures or/and additional gas injection as CCl_4 or CF_4.

2.3 - Detection - Identification

From the collecting point of the magnet, the mass-separated ions are transported to a low background site, through a 6 m long double Einzel lens and focalized on a mylar tape. Transportation of the activity in front of the detectors, collection time and counting time are realized with a flexible automatic tape driver system.

The detection site was designed to measure simultaneously γ-ray multianalysis spectra, γ-X-t and γ-γ-t coincidences. In practice, the complete set up is used for the mass identification while the coincidence measurements are performed with the He-jet system alone.

In order to determine the multipolarity of the stronger transitions of some nuclei produced, the Isocele electron magnetic spectrometer (Orsay) has been coupled to the He-jet system on-line at SARA.The activity collected on an aluminized mylar tape was transported, at regular intervals in time, inside the magnetic selector. In this set-up, the gas pressure dynamics was a serious problem : the 12 m long capillary was connected to the 1.6 bar He atmosphere of the recoil chamber on one side, while, at the other side, the pressure inside the selector had to be kept below 10^{-6} bar. A diagram of the set-up is reproduced in figure 6. The stability of the magnetic field was controlled by Hall effect. The Si detector was protected against direct radioactivity of the source by a thick tungsten screen. The undispersive trajectories of the electrons from the sample (object) to the detector (image) are helicoïdal. Positive or negative electrons correspond to opposite curvatures ; so, the chamber geometry and the magnetic field sense select conversion electrons or β^+ ; moreover, X-rays, γ-rays or α-particles are suppressed. A special area was managed to place a Ge detector at approximately 4 cm from the radioactive sample to enable γ - e^- and/or X - e^- coincidence measurements.

For very low-energy transitions (< 100 keV) complementary experiments have been performed at Isocele (Orsay), using the same electron selector. From experiments carried out at SARA with ^{35}Cl, ^{36}Ar or ^{40}Ca projectiles on $^{92, 94, 96}Mo$ targets, a large number of isotopes have been studied and several new ones identified, as shown in figure 7. In these heavy ion reactions at 5-7 MeV/nucleon energies, medium-spin states and a few members of rotational sequences can be fed by (β^+ +EC) decay.

3 - Odd-A Neutron-deficient Barium Nuclei

3.1 - Presentation of typical β-decay measurements

- $^{127}_{56}Ba_{71}$

The spin of ^{127}Ba has been measured as $1/2^+$[1] and in the $^{118}Sn(^{12}C,3n)$ reaction, two collective structures have been observed in this nucleus : a negative parity band developed on a $9/2^-$ state and a positive parity band based upon a $7/2^+$ state [16]. Both structures were explained in the frame of the triaxial rotor-plus-particle model ; the negative band corresponds to the coupling of an $h_{11/2}$ neutron hole while the positive one is described by the coupling of a $g_{7/2}$ neutron hole.

As shown in figure 8, a relatively complex level scheme has been progressively established including the beginning of several collective structures, extracted mainly from multianalysis conversion electron spectra, γ-X and γ-e^- coincidences produced in the 1.9 s ^{127m}Ba decay [17]. In these measurements, a 80.2 keV, E3 transition appears as the cross-over of the 24.2 keV, M2 and 56.2 keV, (M1 + E2) transitions. Then, the $h_{11/2}$ negative parity band, previously identified from in beam experiments [16] and built on a 9/2⁻ state, is extended down to its 7/2⁻ basic state ($T_{1/2}$ = 1.9 s). Three other positive parity bands have been observed. The band based upon a 7/2+ level at 196 keV corresponds to the $g_{7/2}$ band identified in the (^{12}C, 3n) reaction. This band is strongly connected with a second one based upon a 5/2+ state at 81 keV. The total intensity of these two bands is deexcited by a 25 keV M1 transition to the 3/2+ state of a third collective structure starting at the 1/2+ ground state. The strong connexion observed between the two bands suggests that the two basic states 7/2+ and 5/2+ are both originating from the same $g_{7/2}$ shell. The third positive parity structure is very likely related to both the very close $s_{1/2}$ and $d_{3/2}$ configurations.

Many other levels have been observed at higher energy. Due to the high energies of their deexciting γ-transitions or to the weak feeding of some of these states, their characteristics have not been completely established. Nevertheless, the multi-quasi-particle nature of the state at 1765 keV is evident, considering its γ-decay modes to the low lying positive and negative parity states.

- $^{125}_{56}Ba_{69}$

The $^{125}La \rightarrow ^{125}Ba$ ($T_{1/2}$ = 76 s) decay has been studied at SARA in both the $^{94}Mo + ^{35}Cl$ and $^{96}Mo + ^{35}Cl$ reactions. From in beam experiments [18] $h_{11/2}$ and $g_{7/2}$ neutron-hole collective bands were identified. The decay scheme established in the present work (figure 9) is very similar to the $^{127}La \rightarrow ^{127}Ba$ one. A new rotational sequence built on the measured 1/2+ ground-state [2] has been observed with a new excited state 3/2+ at 43.7 keV. From low-energy conversion electron spectra measured at Isocèle (Orsay), an E1 multipolarity has been established for the strong 67.3 keV transition which deexcites the 7/2⁻ state and supports the 7/2⁻ → 5/2+ placement proposed. From both in beam [18] and decay measurements this 5/2+ state seems to be fed by the $g_{7/2}$ band via the 169 keV M1 transition. A careful analysis of low-energy electron and gamma spectra has been made to extract the 7/2⁻ → 3/2+ and/or 5/2+ → 1/2+ g.s. expected transitions. Up to now, no clear evidence can be retained and the excitation energy of the 5/2+ level is proposed to be 20 keV at maximum (fig.9).

3.2 - Low lying state systematics

The systematics of the known low-lying energy states in odd-A neutron-deficient barium nuclei is shown in figure 10. Spins and parities experimentally established have been indicated. The barium systematics has been extended to the ^{121}Ba case [19] for which, except the 5/2+ ground state, spin and parity assignments are unknown. As shown in figure 10, the energy of the first excited state varies rapidly between ^{123}Ba and ^{121}Ba which forbids any clear extension of the systematics. This situation has to be related to the drastic change of atomic ground-state properties of these two barium isotopes [3] and points out the urgent need of precise nuclear structure informations around the neutron number N = 65.

3.3 - Discussion and comparison with IBFM-1 calculations

It is well known that the structures of odd-A nuclei offer the unique opportunity to test the coupling of particles or holes to neighbouring even-even cores in a transitional region.

By comparison with the heavy nuclei [20], the isotopes of the transitional region from Xe to Ce are poorly known at low-spin from radioactive decays. A large number of in-beam experiments have been carried out in Grenoble to establish the collective features of odd-A barium and xenon isotopes. From (^{12}C, xn) reactions, the structures of the strongly excited negative parity $h_{11/2}$ band and of a series of positive parity bands (mainly $g_{7/2}$) have been systematically studied. The triaxial rotor-plus-particle model [21] has been successfully applied to reproduce the data [22]. The treatment has also been decisive to describe the β, γ parameter evolution of the unique parity bands observed in odd-A barium (A = 125 to 133) when the position of the Fermi level moves towards the middle of the $h_{11/2}$ neutron shell [18]. The experimental positive parity bands in odd-A barium and xenon have also been reproduced using the same theoretical approach by coupling a $g_{7/2}$ neutron hole to the even-even cores. These regular positive sequences have a less pronounced triaxiality than for the negative parity bands.

One has to underline that the triaxial rotor-particle model is unable to reproduce the relative position of the 7/2⁻, 9/2⁻ and 11/2⁻ states of the $h_{11/2}$ neutron hole band observed in the more deficient xenon isotopes (A = 119-121). In collaboration with M.A. Cunningham, a better fit was obtained with the Interacting Boson Fermion Model [23] by including additional $f_{7/2}$ and $h_{9/2}$ configurations [24].

New experimental informations are available now on low-energy low-spin states in odd-A Ba nuclei which can be compared to predictions of the Interacting Boson Fermion Model. Calculations have been made to describe unitarily these states.In these multi-shell calculations the odd fermion coupled to the boson core occupies all valence shell model orbitals relevant for this region : $2d_{5/2}$, $1g_{7/2}$, $3s_{1/2}$, $2d_{3/2}$, $1h_{11/2}$ and $2f_{7/2}$, $1h_{9/2}$ from the next shell. Starting from a realistic simple-particle level scheme (Nilsson or Woods-Saxon potential), a BCS calculation for the 7 shells gives the quasi particle energies and the shell occupancies. Three parameters which specify the strengths of the monopole, quadrupole and exchange terms are given by parametrization of the IBFM Hamiltonian. A comparison for ^{127}Ba between experimental data and calculations is shown in figure 11 for negative parity states and in figure 12 for the positive parity states. The main decay modes are well reproduced by the calculations. Large mixings ($g_{7/2}$ - $d_{5/2}$) are observed in the results.

A satisfactory description of the low-lying collective structures has been obtained in $^{123-131}$Ba, in the frame of the IBFM-1 model [25]. Calculations are also in progress for the odd-A Xe isotopes.

4 - Even-Even Nuclei

It is well known that precise radioactive decay investigations in even-even nuclei are powerful to establish low-spin states such as 0_2^+ or 0_3^+ and to identify members of vibrational sequences or 1⁻, 3⁻, 5⁻... states which could support the existence of octupole deformation. In the A = 120-130 mass region, many experimental results are well known for several even-even Xe isotopes. In the heavier ones where protons are filling orbitals that lie near the beginning of a major shell while neutrons are filling orbitals near the end of a major shell, the structures are reproduced by the O(6) limiting dynamical symmetry of the Interacting Boson Model [26]. Nevertheless the triaxiality of the lighter isotopes has been obtained by including some SU(6) structure [27].

The structures at low-spin in even-even Ba and Ce, not the same as the Xe ones, are not so well known. Near mass 130, the rotational structures resemble more the SU(3) limiting symmetry [27]. Consequently it appears interesting to add informations in these isotopes, specially towards the lighter ones with neutron mid-shell occupancy.

Among the studies undertaken at the Grenoble He-jet system, two typical examples have been reported in order to show the large variety of observables which can be deduced from β-decay measurements.

4.1 - Spins and parities in ^{124}Ba

In order to establish the basic state of collective structures observed from in-beam spectroscopy, measurements have been performed in the ^{124}La → ^{124}Ba β-decay chain. As the experiment was made with the He-jet system coupled to the electron selector (fig.6), 30 transition probabilities have been obtained in ^{124}Ba. The states fed in this study are shown in figure 13. In addition to the first members of the ground-state band, the gamma-band is quite well fed up to the 8⁺. Observed states are in agreement with previous results [28, 29, 30]. Spin and parity of the first members of well established collective structures are unambiguously assigned : the 5⁻ level at 1912.6 keV, the 7⁻ level at 2261 keV, the 5⁻ level at 2266 keV, the 6⁻ level at 2359 keV. These I^π assignments support the configurations proposed previously [28]. A new 7⁻ state at 2721 keV, well fed (8.6%) by β-decay, was not observed by in-beam spectroscopy.

The most important point is the identification of the β-band for the first time. Its assignment is mainly based on the conversion electrons of the 898 keV, $0_2^+ \to 0_1^+$ E0 transition.

As many similar excited states have been observed in the ^{126}La → ^{126}Ba decay, one thus begins to extend the systematics of the band structures of even-even Ba isotopes. Nevertheless, at the present time, due to the data scarceness and the absence of experimental E2 transition probabilities, it is difficult to test in details the applicability and the limitations of the Interacting Boson Model.

4.2 - Anharmonic vibrational state in ^{130}Ce

In a systematic search to identify new Pr isotopes, level schemes of even-even Ce fed by β-decays have been deduced from A = 126 to 132 [31]. The level scheme of ^{130}Ce obtained in the present work is reproduced in figure 14. Several excited states have been observed in addition to those recently reported [32, 33].

Contrary to the 5⁻ (1955 keV) and 7⁻ (2454 keV) basic states of collective sequences observed by in-beam experiments [32] which are weakly fed by β-decay (~ 2% of the total feeding), the 1672, 2115 and 2624 keV states are strongly populated (~ 5% each). Because of the β-feedings and γ-ray decay modes, these three last states have very likely positive parity and spin ≥ 4. Among these states, those decaying mainly to the γ-band could correspond to the anharmonic vibrational levels predicted by the multi-phonon-method calculations [34]. Using a realistic deformation ($\beta_2 \sim 0.25$) and different sets of gap energy parameters, the excitation energy of an anharmonic vibrational state in ^{130}Ce is estimated near 2340 keV ($E_{4^+}/E_{2_\gamma^+} \sim 2.8$). To prove that the level observed at 2115 keV is the good candidate, supplementary conversion electron measurements are planed to sign its spin and parity ($I^\pi = 4^+$). As the ^{130}Pr parent ground-state has very likely a spin 6 or 7 (see fig. 14), one has also to search for the existence of the first members of a possible collective structure based upon this anharmonic vibrational state.

5 - Conclusions

The He-jet system and its coupling with the on-line isotope separator at SARA appears as an efficient device to identify new neutron-deficient isotopes. In case of large productions by heavy-ion fusion evaporation reactions, very precise nuclear structure details can be reached and used in order to test the new sophisticated theoretical descriptions of nuclei at low-energy excitation.

Acknowledgements

We would like to thank the physicists from Grenoble, Bucharest, Lyon and Orsay who have participated to the experimental works and discussed the results presented in this lecture.

REFERENCES

[1] C. Ekström, Proc 4th Int. Conf. on Nuclei far from Stability, Helsingör, (1981) p.12
[2] A.C. Mueller, F. Buchinger, N. Klempt, E.W. Otten, R Neugart, C. Ekstrôm, J. Heinemeier and the Isolde Collaboration, Nucl. Phys. A403 (1983) 234
[3] S.A. Wells, D.E. Evans, J.A.R. Griffith, D.A. Eastham, J. Groves, J.R.H. Smith, D.W.L. Tolfree, D.D. Warner, J. Billowes, I.S. Grant and P.M. Walker, Phys. Lett. 211 (1988) 272
[4] C. Thibault, F. Touchard, S. Bûttgenbach, R. Klapisch, M. de Saint-Simon, H.T. Duong, P. Jacquinot, P. Juncar, S. Liberman, P. Pillet, J. Pinard, J.L. Vialle, A. Pesnelle and G. Huber, Nucl.Phys. A367 (1981) 1
[5] A. Coc, C. Thibault, F. Touchard, H.T. Duong, P. Juncar, S. Liberman, J. Pinard, M. Carre, J. Lerme, J.L. Vialle, S. Buttenbach, A.C. Mueller, A. Pesnelle and the Isolde Collaboration, Nucl. Phys. A468 (1987) 1
[6] W. Borchers, E. Arnold, W. Neu, R. Neugart, K. Wendt, G. Ulm and Isolde Collaboration, Phys. Lett. 216 (1989) 7
[7] R. Neugart, private communication
[8] U. Garg, T.P. Sjoreen and D.B. Fossan, Phys. Rev. Lett. 40 (1978) 831
[9] T. Bengtsson, I. Ragnarsson, Nucl.Phys. A436 (1985) 14
[10] P. Bonche, H. Flocard, P.H. Heenen, S.J. Krieger and M.S. Weiss, Nucl. Phys. A443 (1985) 39
[11] A. Coc, P. Bonche, H. Flocard, and P.H. Heenen, Phys. Lett. B192 (1987) 263
[12] J. Treherne, J. Genevey, A. Gizon, J. Gizon, R. Béraud, A. Charvet, R. Duffait, A. Emsallem and M. Meyer, Zeit. Phys. A309 (1982) 135
[13] D.M. Moltz, J.M. Wouters, J. Aÿsto, M.D. Cable, R.F. Parry, R.D. Von Dinclage, J. Cerny, Nucl. Instr. and Meth. 172 (1980) 519
[14] A. Plantier, R. Beraud, T. Ollivier, A. Charvet, R. Duffait, A. Emsallem, N. Redon, V. Boninchi, S. Vanzetto, A. Gizon, J. Treherne, N. Idrissi, J. Genevey and J.L. Vieux-Rochaz, Int. Conf. EMIS 11 (1986), Nucl. Intr. Meth. and Meth. in Phys. B26 (1987) 314
[15] G. Gimond, J.L. Belmont and A. Gizon, Proc. Int. Workshop of ECR ion-sources, J. de Phys., Colloque C1, 50 (1989), C1-791
[16] J. Gizon and A. Gizon, Zeit. Phys. A281 (1977) 99
[17] C.F. Liang, P. Paris, A. Peghaire, B. Weiss and A. Gizon, Proc.4th Int.Conf. on Nuclei far from Stability, Helsingör (1981) p. 487
[18] J. Gizon and A. Gizon, Z. Phys. A285 (1978) 259

[19] T. Sekine, S. Ichikawa, M. Oshima, M. Limura, Y. Nagame, K. Hata, N.Takahashi and A. Yokoyama, Z. Phys. A331 (1988) 105
[20] E.F. Zganjar in Future Directions in Studies of Nuclei Far From Stability (J.H.Hamilton et al., eds), North-Holland, Amsterdam (1980) p. 49
[21] J. Meyer-ter-Vehn, Nucl.Phys. A249 (1975) 111, 141
[22] J. Gizon, A. Gizon and J. Meyer-ter-Vehn, Nucl. Phys. A277 (1977) 464
[23] F. Iachello and O. Scholten, Phys. Rev. Lett. 43 (1979) 679
[24] V. Barci, J. Gizon, A. Gizon,J. Crawford, A. Plochocki and M.A. Cunningham,Nucl. Phys. A383 (1982) 309
[25] D.Bucurescu, G.Cata, A. Gizon, J. Gizon and N.V. Zamfir, Contr. to the Oak Ridge Conference "Nuclear Structure in the Nineties" April 1990
[26] R.F Casten and P. Von Brentano, Phys. Lett. 152B (1985) 22
[27] A. Sevrin, K. Heyde and J. Jolie, Phys. Rev. C36 (1987) 2631
[28] J.Ph. Martin, V. Barci, H. El-Samman, A. Gizon, J. Gizon, B. Nyako, W. Klamra, F.A. Beck, T. Byrski and J.C. Merdinger, Zeit. Phys. A326 (1987) 337 and Nucl. Phys. A489 (1988) 169
[29] T. Komatsubara, T. Hosoda, H. Sakamoto, T. Aoki and K. Furuno, Nucl. Phys. A496 (1989) 605
[30] S. Pilotte et al., Nucl. Phys. A514 (1990) 545
[31] D. Barnéoud, J. Blachot, J. Genevey, A. Gizon, R. Béraud, R. Duffait, A. Emsallem, M. Meyer, N. Redon and D. Rolando-Eugio, Zeit. Phys. A330 (1988) 341
[32] D.M. Todd, R. Aryaeinejad, D.J.G. Love, A.H. Nelson, P.J. Nolan, P.J. Smith and P.J. Twin, J. Phys. G : Nucl. Phys. 10 (1984) 1407
[33] M. Kortelahti, E.F. Zganjar, R.M. Mlekodaj, B.D. Kern, R.A. Braga, R.W. Fink and C.P. Perez, Z. Phys. A327 (1987) 231
[34] M.K. Jammari and R. Piepenbring, Nucl. Phys. A487 (1988) 77

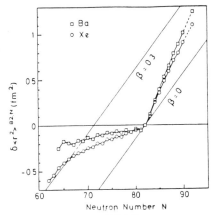

Fig. 1 - Variation of the charge radii in $^{116-146}$Xe (Z = 54) and $^{120-148}$Ba (Z = 56). Figure taken from ref. 6.

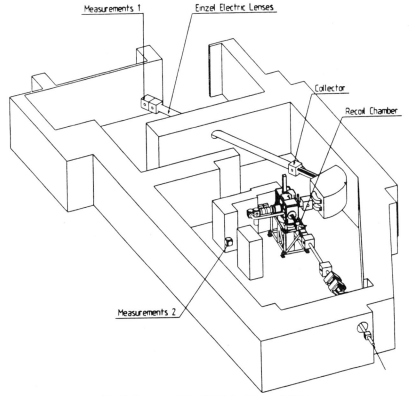

Fig.2- Layout of the ISOL facility at SARA.

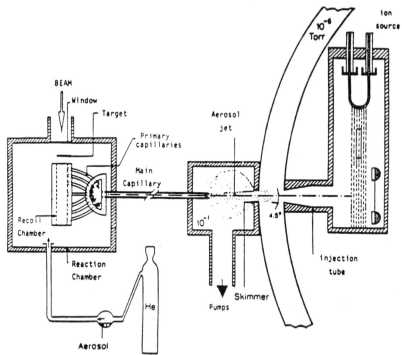

Fig. 3 - Schematic view of an He-jet ion-source coupling

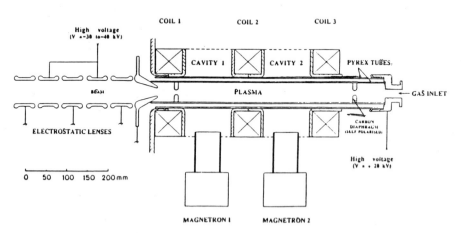

Fig. 4 - Schematic view of the two-stages ECR ion-source system

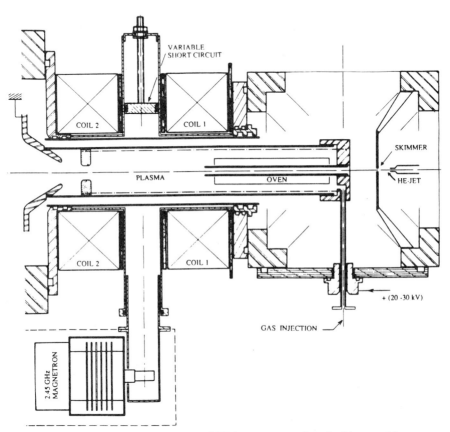

Fig.5 - Schematic view of a compact ECR ion-source equipped with a graphite oven

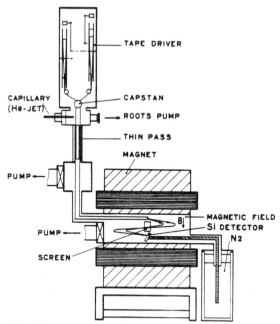

Fig. 6 - Coupling of an electron selector with the He-jet system at SARA.

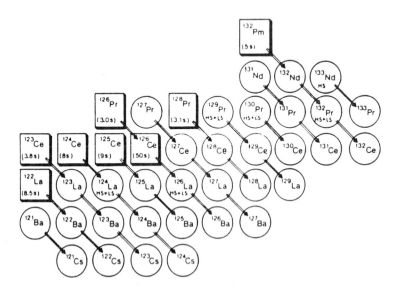

Fig. 7 - Isotopes of A = 120-130 mass region produced and investigated at SARA via heavy ion fusion reactions. New identified isotopes are indicated by solid squares.

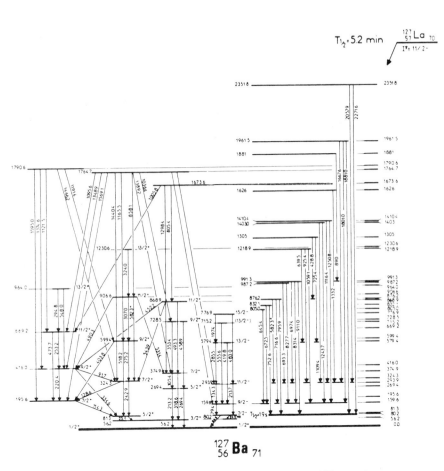

Fig.8 - ^{127}Ba level scheme fed in the β-decay of ^{127}La

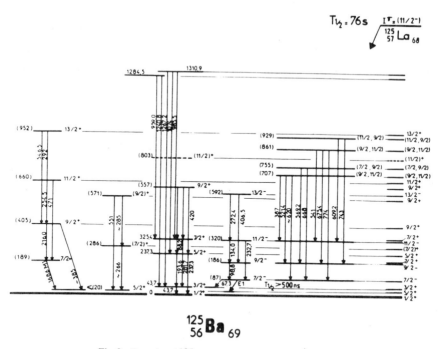

Fig.9 - Levels of ^{125}Ba fed by the β-decay of ^{125}La.

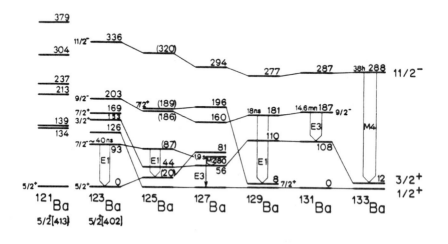

Fig.10 - Low-energy state systematics in odd-A barium isotopes

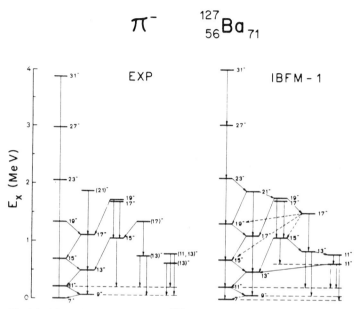

Fig.11 - Negative parity states in ^{127}Ba : comparison between experimental energies (left) and IBFM-1 calculations (right). The levels are labelled by 2I.

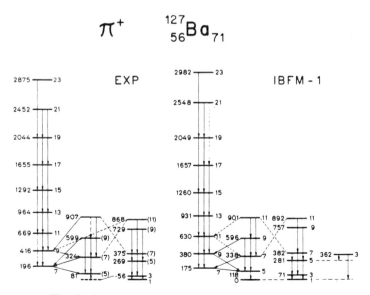

Fig.12 - Same as fig. 11 but for positive parity states.

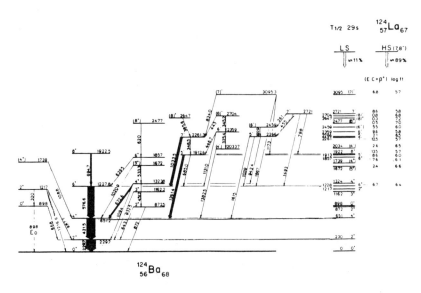

Fig.13 - ^{124}Ba level scheme fed by the ^{124}La β-decay

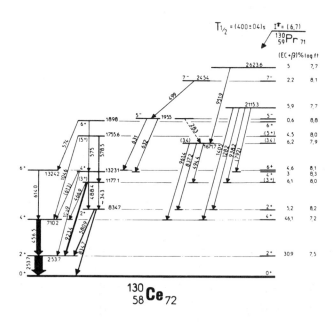

Fig.14 - Low-energy levels fed in ^{130}Ce by the ^{130}Pr β-decay

RECENT RESULTS FROM THE DARESBURY NUCLEAR STRUCTURE FACILITY

D.D. Warner,
SERC Daresbury Laboratory, Warrington WA7 4AD, UK

ABSTRACT

This lecture presents a brief overview of the facilities available at the Daresbury Nuclear Structure Facility and of some of the recent nuclear structure results obtained there.

1. INTRODUCTION

The Nuclear Structure Facility (NSF) at Daresbury is centred on a large vertical tandem accelerator capable of operating at terminal voltages of up to 20 MV. The NSF operates as a user facility, mainly for the UK Universities, but also with a significant degree of foreign participation. It shares the site at Daresbury with one other major facility, the Synchrotron Radiation Source (SRS) which is based on a 2 GeV electron storage ring.

The accelerator is housed in a 70 m high concrete tower and consists of a 42 m long vertical insulating stack which supports a 4.5 m high terminal at its centre. The stack is contained within a pressure vessel which holds a charge of 8 atmospheres of SF_6 weighing 150 tonnes. Two ion source enclosures are located at the top of the tower, one of which is dedicated to the production of polarised beams of ^{23}Na, ^7Li and ^6Li. At the base of the tower a 90° analysing magnet can be rotated to direct beam into any one of seven beam lines or into the newly-installed linear accelerator. A total of 79 beam species have been successfully accelerated to date.

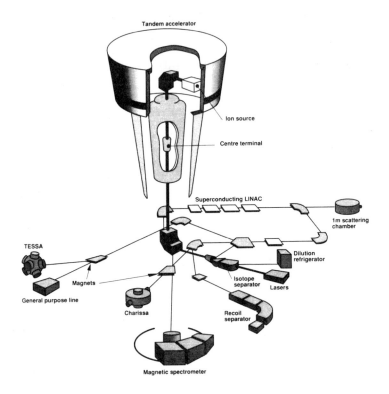

Fig. 1. Layout of accelerator and beam lines.

The layout of tandem and beam lines is illustrated schematically in fig. 1. The experimental equipment is located in four experimental areas which surround the base of the tower, all of which can receive tandem beams. In addition, the tandem beam can be directed into the superconducting linac. This device, which has just been commissioned, consists of 3 modules, each containing 3 split-loop resonators with a resonant frequency of 150 MHz. The option exists to add a further 3 modules. The buncher at the top of the tandem is phase-locked to a superbuncher in front of the linac to produce 100 ps beam bunches. The first modules give an energy gain of ~5 MeV per charge unit and their addition extends the mass range of beams with energies above the Coulomb barrier on the heaviest targets, from ~80 to ~130. The installation of an additional resonator as a rebuncher/debuncher module will allow the experimenter freedom in optimising the timing or energy characteristics of the beam. The first linac beam into the 1 metre scattering chamber is expected very shortly, and the lines to the other experimental areas are planned or under construction.

The wide-ranging capabilities of the NSF stem largely from the variety and versatility of the instruments available, and from their inherent precision and sensitivity. This latter aspect is best reflected in the results of recent studies undertaken at the NSF. The remainder of this lecture will therefore be devoted to a brief discussion of the characteristics of some of the instruments in the context of some of the nuclear structure results which have emerged from them in the last few years.

2. SPECTROSCOPY OF SUPERDEFORMED BANDS WITH TESSA3

The discovery[1] of discrete states in nuclei with a 'superdeformed' ellipsoidal shape (i.e. with a 2:1 major-to-minor axis ratio) represents one of the most significant results of nuclear spectroscopy in recent years. The discovery was made possible by the extreme sensitivity provided by the new generation of gamma-ray arrays which are in use at the NSF and which fall under the generic title of TESSA arrays. The basic configuration of TESSA3 consists of 16 escape-suppressed Ge detectors surrounding a 50 element 'ball' of BGO. The inner ball permits the determination of multiplicity and sum energy of γ-rays emitted in a fusion evaporation reaction and thus provides a degree of channel selection. In general, for a given excitation energy above the yrast line of the compound system, the greater the number of particles evaporated, the lower will be the final multiplicity and total energy detected.

The study of superdeformed structures has taken a dramatic step forward in the last year or so, by virtue of the observation of excited bands within the second (superdeformed) minimum of the potential. Moreover, a remarkable feature of these results[2] was the observation of near-identical γ-ray transition energies in superdeformed bands in two different nuclei, as illustrated in fig. 2. Part (a) of the figure shows the difference in γ-ray energies between the lowest band in ^{152}Dy and an excited band in ^{151}Tb, while part (b) repeats this process for the lowest band in ^{151}Tb and an excited band in ^{150}Gd (all 'bands' are superdeformed). The degree of similarity exhibited in the two cases extends to a level of about 0.1%; the expected $A^{5/3}$ mass dependence of the moment of inertia would lead to a change in the γ-ray energies of about ten times the observed amount.

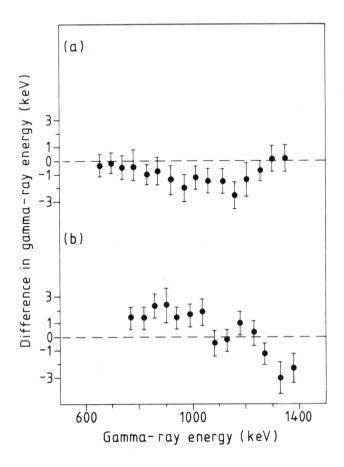

Fig. 2. Differences in γ-ray energies in superdeformed bands (see text).

The underlying origin of these remarkable degeneracies has yet to be understood. The starting point in the discussion is represented schematically in fig. 3 which shows the nature of the single particle orbitals near the Fermi surface in the superdeformed minimum. The point to note here is that, in both the examples of fig. 2, the excited superdeformed band in the nucleus with Z protons is formed by removing a proton from orbit (b) in its Z+1 partner. Orbit (b) is, in fact, the 1/2[301] Nilsson orbit and the relevance of the above observation is discussed in detail by Nazarewicz in his lectures[3].

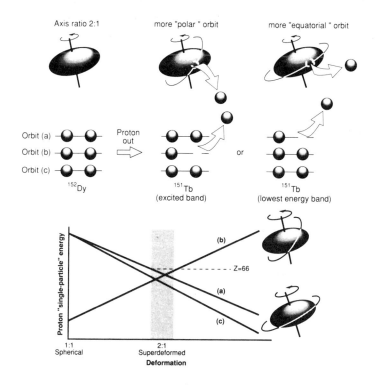

Fig. 3. Schematic representation of the proton orbitals near Z=64 in the superdeformed minimum.

3. STUDIES WITH THE RECOIL SEPARATOR

3a. Spectroscopy of N=Z nuclei

The recoil separator at the NSF has proved to be an invaluable tool in the identification of the weakest channels occurring in fusion evaporation reactions. It makes use of the forward focusing of recoil products, characteristic of such reactions, and uses a pair of Wein filters with crossed electric and magnetic fields to reject the primary beam and produce a velocity dispersion matching the momentum dispersion of a 50° sector dipole magnet. The result is a velocity acceptance of ±1% and a dispersion in A/q in the final focal plane which is sufficient to resolve 1 mass unit if q is known.

Total energy measurement and Z selection is achieved with a split anode ion chamber providing ΔE and E signals. The resolution in Z improves with increasing v/c of the recoils and with decreasing mass. A full description of the facility is given elsewhere[4].

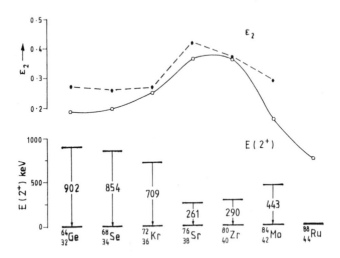

Fig. 4. $E(2^+_1)$ for even-even N=Z nuclei from A=64-84[5] (see text for details).

One of the best examples of the power of the recoil separator has been provided by the study of the N=Z nuclei in the N=Z=28-50 shell. These experiments essentially represent the current limit of sensitivity of the combination of the POLYTESSA gamma-ray array and the recoil separator. For ^{84}Mo, the production cross section is ~10 μb, and only 1 in 10^3 of the recoils reaching the ion chamber are ^{84}Mo. The current status of this work is summarised in fig. 4 which shows the results for all the even-even, N=Z nuclei from Ge to Mo. These results offer conclusive evidence that there is indeed something to be learned in nuclear structure by going to new regions of the N-Z chart. The 'naive' expectation that the doubly magic nucleus ^{80}Zr should be spherical has to be replaced by the observation that it represents the peak of deformation in the shell. The upper part of the figure shows the deformation extracted from the 2^+ energies via Grodzins formula[6], compared with the PES calculations of Moller and Nix[7] which are based on a folded Yukawa potential. The Woods-Saxon calculations of Nazarewicz et al[8] predict a larger ϵ_2 for ^{84}Mo, in better agreement with the empirical value. However, it is interesting to note that both calculations predict oblate minima for this nucleus.

3b. Mirror Nuclei at High Spin

Coulomb energy differences have been studied in the past for low-lying low-spin states. Recently, however, an experiment[9] at the Daresbury recoil separator provided new information on the behaviour of these quantities at higher spin for the $f_{7/2}$-shell mirror nuclei ^{49}Mn and ^{49}Cr. The resulting level schemes are shown in fig. 5. Before this study, essentially nothing was known about the excited states of ^{49}Mn, while states in ^{49}Cr were known only up to J=15/2$^-$. The A=49 isobars were produced in the 3

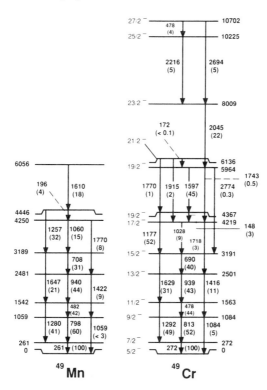

Fig. 5. Level schemes of ^{49}Mn and ^{49}Cr.

nucleon exit channels from the reaction ^{40}Ca+^{12}C at c.m. energies near 40 MeV. This highly inverse reaction results in a high v/c (~7%) allowing good Z discrimination in the gas detector. Gamma rays were detected using the POLYTESSA array of 16 escape suppressed Ge detectors surrounding the target position of the recoil separator. Recoil-γ, γ-γ and recoil γ-γ coincidence spectra were selected.

Inspection of fig. 5 reveals the presence of couplets of levels with spins J, J+1 which exhibit a monotonic increase of energy with spin, reminiscent of signature-split rotational bands. Moreover, a $f_{f/2}$ shell model calculation for ^{49}Cr fails to reproduce these characteristics. Both aspects point to the applicability of a collective approach. In the lower part of fig. 6 the data are displayed in terms of the rotational frequency ω versus J. It can be seen that the low spin states indeed show a smooth positive slope consistent with collective rotation. Moreover, the discontinuity which appears at J=17/2 can, in principle, be understood in terms of quasiparticle alignment. In fact, a cranked

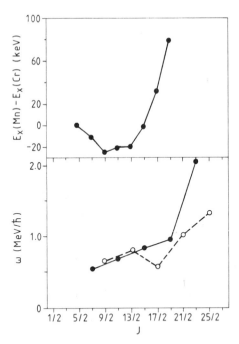

Fig. 6. Coulomb energy differences (upper) and rotational frequency (lower) vs J for ^{49}Mn, ^{49}Cr.

shell model calculation[10] in the full f,p space indicates that the first two-quasiproton crossing in ^{49}Cr should occur at ω=0.8 MeV/$\hbar$ and involves an alignment of 6$\hbar$, as expected for the $f_{7/2}$ orbital.

The Coulomb energy differences are displayed as a function of spin in the upper part of fig. 6, and show that the apparent alignment phenomenon is accompanied by a dramatic increase in the Coulomb energy difference. This effect can be understood by recalling that the first alignment involves two protons in the odd neutron nucleus ^{49}Cr, and two neutrons in ^{49}Mn. The consequent reduction in spatial overlap of the aligned nucleons produces a concomitant reduction of the Coulomb energy in ^{49}Cr. An explicit calculation of the Coulomb energies in the context of a $f_{7/2}$ cranked shell model calculation[10] supports this interpretation.

Thus, in summary, these results have identified the onset of collective behaviour in the middle of the $f_{7/2}$ shell, and have demonstrated the correlation between alignment and Coulomb energy differences in mirror nuclei. Note, however, that effects stemming from a breakdown in collectivity due to the smallness of the available configuration space could become important at higher spins, as signalled by the sudden increase in frequency evident at J=23/2 in fig. 6.

4. COLLINEAR LASER SPECTROSCOPY WITH THE ISOTOPE SEPARATOR

We turn now to use of collinear laser spectroscopy to study the charge radii of exotic nuclei. The nuclei of interest are produced in a fusion evaporation reaction and then ionized and mass separated with the Daresbury Online Isotope Separator (DOLIS)[11]. The aim is to measure the small shift in the frequency of resonantly scattered photons which can be ascribed to the finite size of the nucleus in its interaction with the atomic electrons. To this end, the ions from DOLIS are accelerated to energies in the range 20-60 keV and made to interact with a counter-propagating laser beam in collinear geometry. A few years ago, a coincidence technique was developed[12] to improve the sensitivity of detection in this type of study. This method reduces the background of non-resonantly scattered light by recording only photons which are in coincidence with the ions of the beam from the Isotope Separator. The resulting sensitivity, in favourable cases, can allow measurements to be made with fewer than 100 atoms/sec.

This technique has been used recently[13] to measure the charge radii and moments of the extremely neutron deficient isotopes $^{120\text{-}124}$Ba. The isotopes were produced by bombarding ^{92,94}Mo with 160-180 MeV ^{32}S beams from the NSF. The resonance condition was obtained by varying an accelerating/decelerating voltage in the region of the light collection whilst keeping the laser frequency fixed at a wavelength close to the barium transition. Then the velocity of the ions could be scanned across the region where the Doppler-shifted resonance frequency matched the laser frequency. The results for isotope shifts (relative to ^{124}Ba) and moments are given in Table 1.

Table 1.

A	Δr^2_{A-124} (fm^2)	$\langle\beta^2\rangle^{1/2}$ (from IS)	μ (n.m.)	Q (b)
120	- 0.087(2)	0.316(1)		
121	- 0.008(4)	0.320(1)	+ 0.660(1)	+ 1.79(12)
122	- 0.030(3)	0.305(1)		
123	- 0.046(4)	0.289(1)	- 0.680(1)	+ 1.49(12)
124	0	0.287(1)		

The spin J=5/2 can definitely be assigned to ^{121}Ba, while the spin and moments for ^{123}Ba are consistent with a previous measurement[14]. Moreover, for ^{121}Ba, the combination of J=5/2 and positive magnetic moment unambiguously identifies the 5/2$^+$[413] Nilsson orbital as the ground state. This coincides with the expectation from an axially symmetric Nilsson potential. The spin and negative sign of μ for ^{123}Ba picks out the 5/2$^+$[402] orbital, but in this case, the axially symmetric Nilsson scheme cannot give this orbital as the ground state for N=67. Mueller et al[14] have explained this anomaly by introducing a small degree of γ-distortion for ^{123}Ba. It is, of course, well established that the heavier Ba isotopes around A=130 exhibit triaxial features. The current results thus suggest that this triaxiality dissipates with increasing quadrupole distortion near mid-shell, until axial symmetry is achieved at A=121.

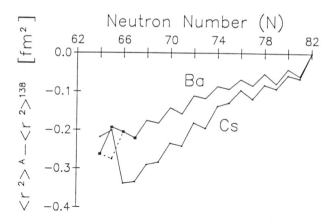

Fig. 7. Changes in mean square charge radii vs. neutron number for Ba and Cs isotopes.

The isotope shift measured for ^{121}Ba also attracts attention, in that it departs from the trend set by the higher mass isotopes, as shown in fig. 7. In fact, ^{121}Ba exhibits a change in charge radius, relative to ^{124}Ba, of -0.008 fm^2 compared with -0.1 fm^2 which would be expected if the staggered trend were continued. The neutron number N=65 of ^{121}Ba is identical to that of ^{120}Cs where, as can be seen, a similar discontinuity occurs. For N>65 the 9/2$^+$[404] band appears as a deformed 'intruder' configuration in the odd-A Cs isotopes resulting from the excitation of a proton across the Z=50 shell gap. The steeply up-sloping nature of the 9/2$^+$[404] orbital results in a potential energy surface for the associated proton hole configuration which minimizes at larger deformation and whose excitation energy decreases with decreasing N, until it becomes the ground state in ^{119}Cs. The sudden jump which occurs at N=65 can be attributed to the additional contribution to the overall n-p interaction which results from the $g_{7/2}$ parentage of the 5/2$^+$[413] neutron orbital and the $g_{9/2}$ parentage of the 9/2$^+$[404]. While the overlap is not maximal, due to the differing Ω-values, it is presumably still sufficient to give rise to the abrupt increase in deformation observed.

The sensitivity of the proton-ion coincidence technique, while representing a great improvement, remains limited by the coincident time resolution between photons and ions, which determines the peak-to-background ratio achievable. This time resolution is set principally by the differing path lengths of the ions from the point of emission of the photon to the final ion detector, which stems from the finite length of the interaction region. Very recently a new technique has been implemented[15] in which the position from which the photon is emitted is measured using an imaging photomultiplier, so that an appropriate time correction can be applied to compensate for the ion path length.

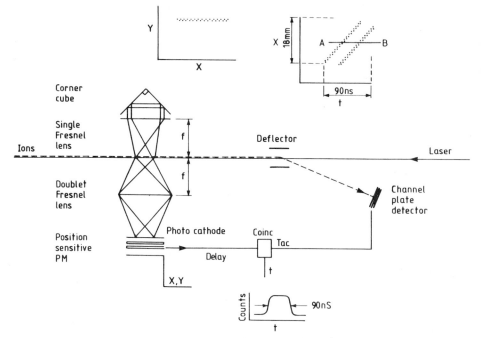

Fig. 8. Position sensitive detection of ions in collinear laser spectroscopy.

Figure 8 gives a schematic layout of the new system. An interaction region of about 18 mm in length is imaged onto the photocathode of the photomultiplier using Fresnel lenses and a retro-reflector as shown. The photon detector consists of a series of channel electron multipliers behind a photocathode. The amplified electrons are collected on a resistive anode which yields x and y signals via four conducting strips along its outside edges. Data is collected in event mode, each event consisting of the scan voltage, the TAC output and the digitised x and y information. The resulting x-t distribution, where the slope represents the ion velocity, is shown in fig. 9, along with the 'raw' and corrected time peak. The improvement is obvious, but should be considerably better. Work is currently under way to improve the quality of the fast timing pulses from the photomultiplier, to allow the expected limiting resolution of ~5 ns to be attained.

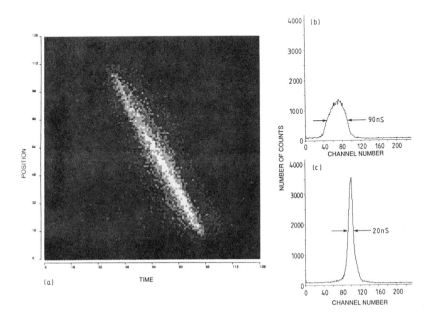

Fig. 9. a) x versus t plot; b) total projection t-axis; c) t spectrum after correction for position.

5. STUDIES WITH THE MAGNETIC SPECTROMETER

5a. Probing the Single Particle Structure of the Scissors Mode

The magnetic spectrometer at Daresbury is of the QMG/2 type, one of the Q3D family of such devices. Its characteristics include a 10 msr acceptance angle and a resolving power of 4300. The focal plane is inclined at an angle of 45° to the particle trajectories and, in most applications, the focal plane detector used is 550 cm in length, giving a momentum range of 5%. This instrument is described in detail elsewhere[16].

A recent study with the QMG/2 utilised the (t,α) reaction to probe, for the first time, the microscopic structure of states which were thought to represent candidates for the collective M1, or 'scissors', mode. An illustration of the type of collective modes under discussion is given in fig. 10. The first evidence for such states came from electron scattering studies[17] in ^{156}Gd which have subsequently led to an extensive series of measurements with both electrons and photons to characterise the location and fragmentation of the scissors mode in deformed nuclei. This mode corresponds to an oscillation in the angle between the deformed neutron and proton mass distributions, and gives rise to a collective $K^\pi = 1^+$ band at an excitation energy of about 3 MeV in the rare earth region with an M1 excitation strength of 1-3 μ_N^2.

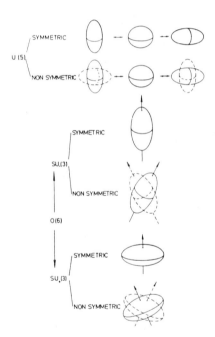

Fig. 10. Shapes associated with symmetric and non-symmetric n-p modes.

The wealth of data which has emerged over the past few years confirming the existence of mixed symmetry states has led to a concomitant theoretical effort to understand the microscopic origins of such modes. It is thus of particular interest that the first experimental results probing this question have recently been published.

The study[18] centred on the ^{165}Ho(t,α)^{164}Dy reaction. The ground state of ^{165}Ho is the 7/2⁻[523] Nilsson proton orbital, which has dominantly $h_{11/2}$ parentage. The aim of the experiment was to identify $l=5$ transfer into one or more of the $K^\pi=1^+$ bands in ^{164}Dy, whose bandheads had previously been identified via their enhanced M1 excitation probabilities from electron[19] and photon[20] scattering work. Clearly, in this region, contributions from $(h_{11/2})^2$ two-quasiproton configurations might be expected to constitute a major part of the total observed M1 strength.

The results are shown in fig. 11 and are surprising for a number of reasons. A $K^\pi=1^+$ band is clearly identified with the expected $l=5$ angular distribution and with the 'fingerprint pattern' of the 7/2⁻[523]×5/2⁻[532] configuration. The bandhead energy is 2.557(15) MeV, while the square of the amplitude of the $(h_{11/2})^2$ component in the intrinsic wave function is estimated to be ~0.9, indicating very little fragmentation. In the (e,e′) and (γ,γ′) experiments, two groups of 1⁺ states were identified in ^{164}Dy, with excitation energies centred around 2.5 MeV and 3.1 MeV, carrying total M1 strengths of 1.67 μ_N^2 and 3.15 μ_N^2, respectively. The upper group is less fragmented and corresponds in energy and M1 strength to the expected behaviour of the scissors mode state; the transfer data show no significant population in this energy region, consistent

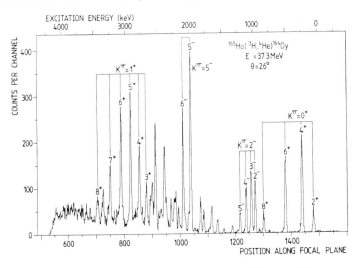

Fig. 11. Position spectrum of α-particles from the ^{165}Ho(t,α)^{164}Dy reaction.

with a collective interpretation. In the lower group, there are two states identified in (γ,γ') at 2.539 and 2.578 MeV which could correspond to the two-quasiproton $K^\pi=1^+$ band of fig. 11. However the corresponding M1 strengths are only 0.30 and 0.48 μ_N^2 respectively. Hence the majority of M1 strength in this nucleus remains unexplained. A recent RPA calculation[21] successfully predicts the dominance of the 7/2⁻[523]×5/2⁻[532] configuration at 2.5 MeV, but also predicts a dominant component of this configuration at 3.1 MeV. Thus the single particle structure of the collective M1 modes in deformed nuclei is still an open question which requires further experimental and theoretical study.

5b. Location of High-j Orbitals in W Nuclei

The high j unique-parity orbitals which occur in each major shell play a crucial role in determining many of the structural features observed in well deformed nuclei. The most familiar example is that of the low Ω orbitals stemming from the $i_{13/2}$ neutron state in the rare earth region where the large component of angular momentum directed along the rotational axis maximises the Coriolis effects and gives rise to the lowest lying examples of the backbending phenomenon. The deformed shell model also predicts that the low Ω orbitals of the N=7 $j_{15/2}$ state should descend into the rare earth region at higher deformation. These orbitals can also give rise to a significant degree of rotational alignment and appear to be[22] an essential ingredient in the description of superdeformed states in ^{152}Dy.

The position of high j orbitals in the deformed scheme, relative to the 'normal' states in any shell, depends critically on the strengths of the l^2 and $l.s$ interactions in the spherical shell model potential. These strengths can be found empirically in a given region by a determination of the relevant quasiparticle bandhead energies. For the $i_{13/2}$ orbits, such information for particle states is fragmentary since it relies largely on data from the (d,p) reaction which strongly favours low spin states, while no example of a state of $j_{15/2}$ parentage has yet been identified. Nevertheless, in the heavier W nuclei, the Nilsson scheme suggests that the lowest $j_{15/2}$ orbitals may occur as particle states in the region of 1-3 MeV of excitation. A study has therefore been initiated by Daresbury and Manchester groups to search for these and other high j orbitals in ^{185,187}W by exploiting the known[23] high selectivity of the (^{12}C,^{11}C) and (^{16}O,^{15}O) transfer reactions, using the QMG/2. Given the obvious limitations in resolving individual states and obtaining accurate energies from particle spectra alone, particle-gamma coincidence measurements have also been undertaken.

The selectivity of specific heavy ion induced transfer reactions stems from three features. The large negative Q-values of both reactions cited above ensure a mismatch between the incoming and outgoing grazing angular momenta so that small l-transfers are suppressed. In addition, there is a strong selectivity between $j=l+1/2$ and $l-1/2$ final states dependent on the reaction used, resulting from the different ground states of the projectiles ($p_{1/2}$ and $p_{3/2}$ for ^{16}O and ^{12}C, respectively) and the empirical fact that the spin of the transferred neutron does not flip. Thus the ^{12}C-induced reaction favours final states with $j=l-1/2$ and the ^{16}O favours those with $j=l+1/2$.

The first experiments were performed using targets of 100-200 μg/cm^2 of enriched ^{184}W and beams of ^{16}O at 140 MeV and ^{12}C at 100 MeV. Spectra for the two reactions taken at the grazing angle are shown in fig. 12. Particle-γ coincidence data were taken for the ^{16}O case using a 10 cm diameter chamber with four Ge detectors positioned at backward angles. A value of 40 nsec was obtained for the FWHM time resolution for γ-rays in coincidence with ^{15}O particles; about a third of this stems from variation in time of flight through the spectrometer.

While the analysis of the data is still in progress, and a further study of ^{187}W is planned, certain preliminary observations can be made concerning the results in fig. 12. These are summarised in Table 2. The aforementioned strong selectivity in the two reactions is immediately evident from an inspection of the relative intensities in peaks E-H, which correspond to low-lying levels whose parentage is already established. There is a clear enhancement of the $h_{9/2}$ strength in the ^{11}C spectrum and of the $i_{13/2}$ strength in the ^{15}O spectrum. Of the remaining peaks, those labelled A, B and D are consistent with $j=l+1/2$ assignments, while peak C is manifestly $j=l-1/2$ and can then only be associated with the low Ω orbit(s) of the $i_{11/2}$ state.

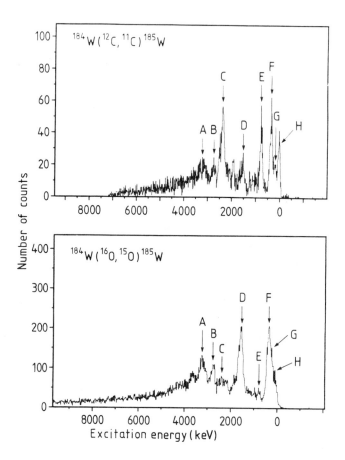

Fig. 12. ^{11}C and ^{15}O particle spectra.

Table 2.

Peak[a]	j_f[b]	Tentative assignment	Peak[a]	j_f[b]	Tentative assignment
A	$j_>$	$j_{15/2}, g_{9/2}, i_{13/2}$	E	$j_<$	$h_{9/2}[505]$
B	$j_>$	$j_{15/2}, g_{9/2}, i_{13/2}$	F	$j_>$	$i_{13/2}[615]$
C	$j_<$	$i_{11/2}$	G	$j_>$	$f_{7/2}[503]$
D	$j_>$	$j_{15/2}, g_{9/2}, i_{13/2}$	H	$j_<$	$f_{5/2}[512]$

a) See fig. 1.
b) $j_> = \ell + 1/2; j_< = \ell - 1/2$

The coincident γ-ray spectra indicate that the decays of the $13/2^+$ member of the $11/2^+$[615] band and the $9/2^-$ member of the $7/2^-$[503] and $3/2^-$[512] bands have been identified for the first time. Most importantly, in the case of the largest peak D in the ^{15}O spectrum, they show population of only the low lying negative parity structure, and not the $i_{13/2}$ band, thus ruling out $i_{13/2}$ parentage for this state. Given the large observed single particle transfer strength, it seems clear that this peak must represent the lowest lying $j_{15/2}$ strength in ^{185}W.

6. SUMMARY AND FUTURE DEVELOPMENTS

There are clearly many current areas of interest at Daresbury which have not been touched on in this lecture. The most notable omissions can be found in the domain of reaction studies where active programmes exist to study near and sub-barrier effects, break-up and the use of polarised beams. Nevertheless, this brief overview has, hopefully, given some impression of the range of nuclear structure problems which are currently being addressed at the NSF.

There are also a number of major new developments planned and/or under way for the near and long term future. The installation of the first phase of the linac booster has already been mentioned. Work is already in progress to build a new broad range spectrometer to make optimum use of the higher energies which will shortly become available. This device, named SUSAN (Spectrometer for Universal Selection of Atomic Nuclei) is projected to have a mass resolution of 1 in 200 and a solid angle of ~80 msr, with a full angular range. It will basically consist of a large superconducting solenoid followed by a dipole magnet and ion chamber, and will utilise time-of-flight and momentum measurement to determine mass. The problem of aberrations which would normally arise because of the large solid angle will be dealt with by ray-tracing using position-sensitive heavy-ion detectors and fast event-by-event data acquisition.

Another major addition to our apparatus, which should come into operation early in 1992, is a new, improved γ-ray array of up to 70 suppressed Ge detectors. This device, referred to as EUROGAM, is the next logical step in the family of TESSA arrays, and is being built jointly by groups from France and the UK. Its design aim should lead to an increase of two orders of magnitude in sensitivity, relative to existing arrays. It will be capable of being used alone, or in conjunction with either the recoil separator or SUSAN. The first experiments will be performed during the initial year at Daresbury, after which it will be used for a further year at the new VIVITRON accelerator currently under construction in Strasbourg.

Finally, the UK community sees the long term future of nuclear structure as lying in the direction of accelerated radioactive beams. The problems of nuclear structure are naturally bounded in energy. Excursions into the relativistic regime, while producing results of interest in their own right, will not address questions relevant to the nuclear many-body problem. On the contrary, in nuclear physics, the range of nuclei and phenomena which can be studied is limited by the beam and target species available. Accordingly, a Working Group has already put forward a preliminary proposal which envisages an Exotic Beam Facility (EBF) based on spallation and fission induced by high energy protons. The proposal suggests the use of the ISIS facility of the Rutherford Appleton Laboratory as a possible source of 750 MeV protons, with subsequent acceleration of the unstable nuclei being provided by an RFQ linac and synchrotron. Further acceleration can then be obtained by re-injection of the heavy ions into the primary proton ring producing maximal energies of up to 40 MeV/A for uranium.

Based on the experience provided by the ISOLDE facility at CERN, the range of both neutron-deficient and neutron-rich nuclei which can be produced in this manner with usable intensities is truly remarkable. It would not be unreasonable to compare such a development to the first appearance of heavy-ion accelerators themselves. Nevertheless the scale of such a facility mandates that it be designed and constructed in the context of a pan-European collaboration. Similar proposals have been made in the USA, Japan and Canada over the past year and an international consensus seems to have developed that the concept of the EBF represents the next major step forward for nuclear physics.

REFERENCES

1) P.J. Twin et al, Phys. Rev. Lett. *57* (1986) 811.
2) T. Byrski et al, Phys. Rev. Lett. *64* (1990) 1650.

3) W. Nazarewicz, contribution to these proceedings.
4) A.N. James et al, Nucl. Instr. and Meth. *A267* (1988) 144.
5) W. Gelletly et al, in press and references therein.
6) L. Grodzins, Phys. Lett. *2* (1962) 88.
7) P. Moller and J.R. Nix, Nucl. Phys. *A361* (1981) 117; At. and Nucl. Data Tables *26* (1989) 165.
8) W. Nazarewicz et al, Nucl. Phys. *A435* (1985) 397.
9) J.A. Cameron et al, Phys. Letts, *235B* (1990) 239.
10) J.A. Sheikh et al, Phys. Letts. (in press).
11) I.S. Grant et al, Nucl. Instr. and Meth. *B26* (1987) 95.
12) D.A. Eastham et al, Optics Commun. *60* (1986) 293.
13) S.A. Wells et al, Phys. Letts. *211B* (1988) 272.
14) A.C. Mueller et al, Nucl. Phys. *A403* (1983) 234.
15) D.A. Eastham et al, Optics Commun. (in press).
16) R.A. Cunningham et al, Nucl. Instr. and Meths. *A234* (1985) 67.
17) D. Bohle et al, Phys. Lett. *137B* (1984) 27.
18) S.J. Freeman et al, Phys. Lett. *222B* (1989) 347.
19) D. Bohle et al, Phys. Lett. *195B* (1987) 326.
20) C. Wessellborg et al, Phys. Lett. *207B* (1988) 22.
21) K. Sugawara-Tanabe and A. Arima, Phys. Lett. *229B* (1989) 327.
22) I. Ragnarsson and S. Aberg, Phys. Lett. *180B* (1986) 191.
23) P.D. Bond, Comments Nucl. Part. Phys. *11* (1983) 231.

Influence of Occupied Orbitals on the Band Structure in
the A≈80 Region: Excited States in ^{84}Mo, ^{83}Nb and ^{83}Zr

C.J. Gross[1], W. Gelletly[2], D. Rudolph[1], B.J. Varley[3],
K.P. Lieb[1], M.A. Bentley[2], J.L. Durell[3], H.G. Price[2],
J. Simpson[2], Ö. Skeppstedt[4], S. Rastikerdar[5]

[1] II. Physik. Inst., Univ. Göttingen, D-3400 Göttingen, FRG
[2] SERC Daresbury Lab., Warrington, WA44AD, UK
[3] Schuster Lab., Univ. Manchester, Manchester M139PL, UK
[4] Chalmers Univ. of Technology, S-41296 Goteborg, Sweden
[5] Isfahan Univ., Isfahan, Iran

Abstract

Excited states in the very neutron deficient isotopes ^{84}Mo, ^{83}Nb and ^{83}Zr were observed in the reaction ^{28}Si(^{58}Ni,xnyp) at 195 MeV. Identification of the γ rays in the previously unknown nuclei ^{84}Mo and ^{83}Nb was made with the Daresbury Recoil Separator and an array of 20 Ge detectors. A rich band structure in ^{83}Zr was also observed. The deformations of the N=Z=32-42 even-even isotopes are discussed. The influences of the potential energy minima corresponding to a particular particle number, as determined from the N=Z nuclei, can be observed in the population of rotational bands in other nuclei. The importance of the occupied orbitals in influencing the overall nuclear structure is also discussed.

Introduction and Experimental Details

The single particle spectrum at N≈Z≈40 contains [Na85] several pronounced energy gaps developing as a function of the deformation. These gaps correspond to minima in the potential energy and hence, determine the overall nuclear shape. Relatively low single particle densities permit large fluctuations in shape with a small change in particle number [Lü86] or particle excitations [Gr89]. The influence of the changing particle number is similar for protons and neutrons as they are filling the same orbits.

It remains an open question if a strong np interaction arises in this mass region. Competition between the large deformed shell gaps at Z=38,40 and the spherical shell gap at N=50 manifests itself quite dramatically in the Zr isotopes at N≈45: the collectivity observed [Pr83] in ^{84}Zr is reduced by almost a factor of 3 in ^{86}Zr [Wa85]. The shapes of the odd-A nuclei depend on which orbital is occupied. The lifetimes and mixing ratios of states in three rotational bands in ^{83}Zr indicate [Fi89] that each has a different deformation. In fact, when interpreted [Fi89] with an axially symmetric, particle plus rotor model, the deformation of the $K^{\pi}=5/2^+$ band is $\beta_2=-0.28$ (oblate shape) while that of the $K^{\pi}=5/2^-$ band is $\beta_2=+0.33$ (prolate shape).

The nuclei on or near the N=Z line are crucial to our understanding of the influence of the single particles on a strongly deformed core. In the absence of a strong np interaction the N=Z nuclei should define the potential energy minimum for a given particle number. Thus, their study establishes the position of the energy gaps as a function of deformation in the single particle spectrum. By adding or subtracting nucleons from these N=Z cores, the effects on the deformation by single particles may be measured. In this contribution we report on the first observation of excited states in ^{84}Mo and ^{83}Nb and on the extension of the band structure in ^{83}Zr. In particular, by examining the level structure of these nuclei, we look at the role of the occupied orbitals in determining the observed nuclear structure.

Unfortunately, the N=Z nuclei with A≈80 lie far from stability and are weakly populated in fusion-evaporation reactions. With typical cross-sections of ≤ 100 μb, sensitive particle detectors and large Ge arrays are required to identify these nuclei. We have used the Daresbury Recoil Separator (RS) and a 20 element Compton-suppressed Ge detector array [No85,Ja88]. A 195 MeV ^{58}Ni beam was focussed on a self-supporting, 99.6% enriched ^{28}Si target of 700 μg/cm^2 thickness. The prompt γ rays resulting from the fusion-evaporation products were detected in the Ge array located at the target position. The high velocity (v/c=5.4%), inverse reaction ensured good efficiency for detection of the recoils in the 0° electromagnetic RS and good Z resolution based on the energy loss signal from the ionization chamber located at the end of the RS. Four mass groups, A=84, 83, 81, and 80, were focussed and well resolved on the focal plane of the RS where a position sensitive detector provided mass identification. The tagged recoils then pass through an ionization chamber where their resulting energy loss (ΔE) is Z dependent. Thus, γ rays can be unambiguously matched to a particular recoil.

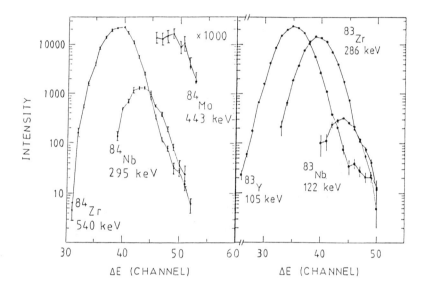

Fig. 1 - γ ray yield vs ΔE for masses 84 and 83

After identifying the particular mass groups, γ ray intensity curves as a function of ΔE can be plotted as in fig. 1 for A=84 and 83. Gates set in ΔE produce γ spectra which, when carefully subtracted from each other, result in spectra containing transitions belonging to a particular nucleus. The resulting "pure" γ ray spectra for the N=Z=42 nucleus ^{84}Mo and its Z=41 isotone ^{83}Nb are shown in figs. 2 and 3, respectively, and will be discussed below.

N=Z Nuclei

Through the evaporation of two neutrons from the ^{86}Mo compound nucleus, ^{84}Mo is populated with a cross-section of 6±3 µb. The ^{84}Mo spectrum shown in fig. 2 contains only one transition at 443 keV. It is assumed that this is the usual $2^+ \rightarrow 0^+$ transition common to all even-even nuclei. Although only one transition was identified [Ge91], we are able to draw some conclusions about the

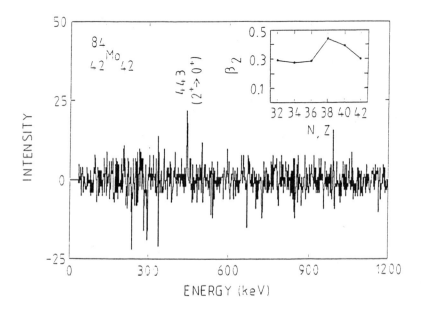

Fig. 2 - "Pure" ^{84}Mo spectrum. The inset shows the deformation β2 as a function of N=Z.

deformations of the even-even, N=Z nuclei [Oo86,Li87a, Li87b,Va87] with proton numbers between 32 and 42. A value for the quadrupole moment β_2 can be derived from Grodzins' formula [Gr62,St72]

$$\beta_2 = \left[\frac{1225}{A^{7/3} E(2^+)} \right]^{1/2} .$$

For nuclei in this region, values of β_2 derived in this way have been found [Bu90] to agree with that obtained from measured lifetimes [Li82]. We find a value of $\beta_2=0.30$ for ^{84}Mo which may be compared with the theoretical prediction of $\beta_2=-0.16$ by Möller and Nix [Mö81] using a folded Yukawa potential. Although we cannot determine the sign of β_2 from our measurement, the magnitude is a factor of 2 larger. The single particle spectrum shown in [Na85] contains minima corresponding to values of $\beta_2=+0.29$ and -0.28 for N=Z=42. Both values agree surprisingly well with experiment.

The relationship of the deformation as one proceeds along the N=Z line is shown in the inset of fig. 2. The maximum in β_2 ($\beta_2 \geq 0.40$) occurs at N=Z=38 corresponding to the large deformed shell gap [Na85] in the single particle spectrum. The reduction observed in ^{72}Kr [Va37] and ^{84}Mo is consistent with an expected decrease in deformation as one moves away from the middle of the shell and toward the shell closures at N=Z=28 and 50. These deformations correspond surprisingly well to the energy gaps in the single particle spectrum calculated in [Na85]. We conclude that in the absence of a strong np interaction, these deformations are the most favorable energetically at low angular momentum for a given particle number.

Excited States in ^{83}Nb

The nucleus ^{83}Nb has one proton less than the ^{84}Mo core, and as such, we expect rotational bands to be built on a similarly deformed core. The energy curves shown in fig. 1 clearly identify ^{83}Nb which was populated with a cross-section of 175±75 µb. A "pure" ^{83}Nb spectrum is displayed in fig. 3a. Once the γ ray transitions have been identified [Gr91] as originating from ^{83}Nb, γγ coincidences can be used to construct a level scheme. A gate set on the 122 keV transition produced the spectrum in fig. 3b and revealed that the 130, 623, and 848 keV transitions are in coincidence. Through such relationships the ^{83}Nb γ rays may be grouped into two distinct cascades which are presented in the center of fig. 4.

Both cascades are shown decaying to the same state (assumed ground state) although states with effective

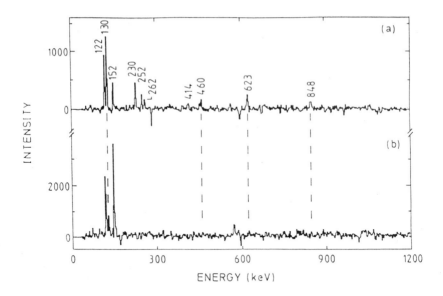

Fig. 3 - (a) "Pure" 83Nb spectrum. (b) Coincidence spectrum gated by the 122 keV transition. Contaminant peaks are from the 119 keV transition in 81Y [Li85] and an unknown 122 keV γ ray seen in coincidence with itself.

lifetimes longer than a few nanoseconds or with depopulating transitions smaller than 50 keV could not be observed in our experiment. We do note, however, that only one β^+ decay branch has been observed [Ku88] and it feeds the 77 keV, $7/2^+$ state in ^{83}Zr. On the basis of the Kurie plot shown in [Ku88] we calculate an allowed transition with log ft=4.48(13) which suggests a ground state with $I^\pi=(5/2,7/2,9/2)^+$. We favor the $5/2^+$ assignment because of the marked similarities between the energy levels of ^{83}Nb with Z=41 and the N=41 isotones, ^{77}Kr [Wö84] and ^{79}Sr [Li82,He90] also shown in fig. 4. Although the relative energies of the bandheads may be interchanged, ie., the ground state of ^{77}Kr has $I^\pi=5/2^+$ while ^{79}Sr has $I^\pi=3/2^-$, the level structure is similar to that of ^{83}Nb.

It is evident that the occupied orbitals, regardless of which type of nucleon, have the dominant influence in determining the nuclear structure. The presence of two bands in ^{83}Nb indicate that more than one potential energy

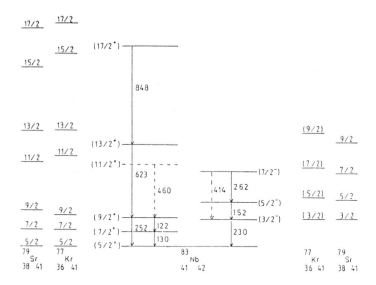

Fig. 4 - Comparison of ^{83}Nb energy levels with the N=41 isotones, ^{77}Kr and ^{79}Sr.

minimum can exist for a particular N and Z, in this case for N=42 and Z=41. In the next section we will see that these different minima can interact, favoring the occupation of particular orbitals, and thus, influencing the observed nuclear structure.

The Level Structure of ^{83}Zr

In moving away from the N=Z=42 single particle potential energy minimum the interactions between minima corresponding to different N and Z can occur. The 2pn channel leading to ^{83}Zr is populated with a cross-section of 23±5 mb. This strength allows detailed spectroscopy to reveal the effects of such interactions on the band structure of ^{83}Zr [Ru91] with N=43 and Z=40. The ^{83}Zr level scheme shown in fig. 5 contains three previously unobserved decay sequences built on the 328, 580, and 3587-3625 keV levels; the last of these indicating that the position and depth of the minima are spin dependent.

Fig. 5 - Proposed level scheme for ^{83}Zr.

The presence of many different decay sequences can be associated with the unpaired neutron. The decay schemes of the N=43 isotones ^{79}Kr [Sc90] and ^{81}Sr [Mo88,Ar83] are similar in that each has a strongly populated $g_{9/2}$ neutron band characterized by a sharp upbend-backbend at the 1qp-3qp band crossing and three bands labeled with $K^{\pi}=1/2^{-}$, $3/2^{-}$, and $5/2^{-}$. However, the influence of the minima associated with changing Z in these isotones can also be observed. The series of levels built on the 580 keV state in ^{83}Zr has an irregular pattern characteristic of single-particle excitations. Indeed, such states arising from the semi-magic nature of the Z=40 spherical shell gap were predicted [Hü88] to lie low in energy. No such decay sequence has been observed in the other N=43 isotones.

Similarly, the tentatively identified $d_{5/2}$ band in ^{81}Sr shown in fig. 6 has no counterpart in the other isotones. As was observed in the inset to fig. 2, the potential energy minimum for Z=38 occurs at the largest deformation $\beta_{2}=0.44$. This allows the odd neutron to occupy the $d_{5/2}$ orbital

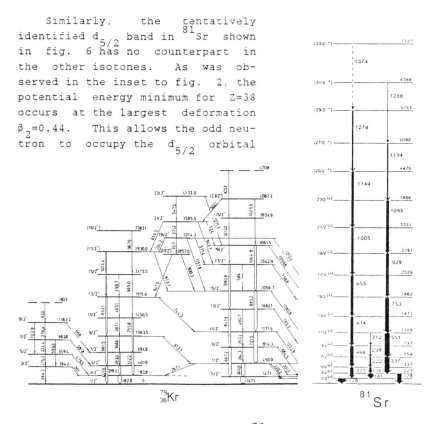

Fig. 6 - Negtive parity states in ^{79}Kr [Sc90] and the $d_{5/2}$ band in ^{81}Sr.

which comes down from above the N=50 shell gap and approaches the N=43 Fermi surface at $\beta_2 > +0.4$. A deformation of $\beta_2 \approx 0.42$ has been measured [Mo88] for this band. Although the minimum for N=Z=40 occurs at a slightly smaller deformation there is no evidence for such a structure.

The decay sequence built on the 3587-3625 keV levels occurs in the vicinity of the 1qp-3qp band crossing region. A corresponding structure also shown in fig. 6 has been observed [Sc90] in ^{79}Kr and was interpreted in terms of non-collective states coupled by strong M1 transitions. The presence of stretched E2 transitions in ^{83}Zr, however, makes a similar interpretation doubtful. A similar structure in ^{81}Sr has not been observed.

To conclude, it has been shown that the study of N=Z nuclei enhances our understanding of the effects on the nuclear core of the addition or removal of single particles. By locating the absolute potential energy minimum for a given number of nucleons at low spins the interaction between these minima may be studied in nuclei with N≠Z. These minima can be associated with the energy gaps between single particle levels. A comparison of the Z=41 isotope ^{83}Nb with the N=41 isotones ^{77}Kr and ^{79}Sr indicates that the unpaired particle exerts a large influence on the nuclear structure regardless of whether it is a proton or neutron. A comparison of the structures of the N=43 isotones ^{79}Kr, ^{81}Sr, and ^{83}Zr reveals similarities indicating the importance of the unpaired neutron in determining the overall deformation. However, the various minima associated with different values of Z also influences the nuclear shape: the large deformed shell gap at Z=38 allows the population of the $d_{5/2}$ intruder orbital by the odd neutron in ^{81}Sr while the semi-magic nature of the spherical shell gap at Z=40 allows single-particle states to be populated along with rotational bands.

These studies would be enhanced further by comparing the odd-odd and even-even nuclei with N=Z to look for the effects of any np interactions and by comparing mirror nuclei, ie., nuclei with Z±1,N and Z,N±1 with N=Z, in order to examine the effects of the Coulomb energy difference on a well deformed core. Such a study has recently been completed [Ca90] for ^{49}Mn and ^{49}Cr.

Acknowledgements and References

This work was funded by the BMFT (Germany), the SERC (United Kingdom), and Isfahan University (Iran).

[Ar83] S.E. Arnell, et al., J. Phys. G 9, 1217 (1983).
[Bu90] F. Buchinger, et al., Phys. Rev. C 41, 2883 (1990).
[Ca90] J.A. Cameron, et al., Phys. Lett B235, 239 (1990).
[Fi89] W. Fieber, et al., Z. Phys. A332, 363 (1989).
[Ge91] W. Gelletly, et al., Submitted to Phys. Lett.
[Gr62] L. Grodzins, Phys. Lett. 2, 88 (1962).
[Gr89] C.J. Gross, et al., Nucl. Phys. A501, 367 (1989).
[Gr91] C.J. Gross, et al., Submitted to Phys. Rev. C.
[He90] J. Heese, et al., Phys. Rev. C 41, 603 (1990).
[Hü88] U.J.Hüttmeier, et al., Phys. Rev. C 37, 118 (1988).
[Ja88] A.N. James, et al., Nucl. Instr. and Meth. A 267, 144 (1988).
[Ku88] T. Kuroyanagi, et al., Nucl. Phys. A484, 264 (1988).
[Li85] C.J. Lister, et al., J. Phys. G 11, 969 (1985).
[Li87a] C.J. Lister, et al., Proceedings of the Fifth International Conference on Nuclei Far from Stability, ed. I.S. Towner, p 354 (1987).
[Li87b] C.J. Lister, et al., Phys. Rev. Lett. 59, 1270 (1987).
[Lü86] L. Luhmann, et al., Europhys. Lett., 1, 623 (1986).
[Mo88] E.F. Moore, et al., Phys. Rev. C 38, 696 (1988).
[Mö81] P. Möller and J.R. Nix, At. Data Nucl. Data Tables 26, 165 (1981).
[Na85] W. Nazarewicz, et al., Nucl. Phys. A435, 397 (1985).
[No85] P.J. Nolan, et al., Nucl. Instr. and Meth. A 236, 95 (1985).
[Oo86] S.S.L. Ooi, et al., Phys. Rev. C 34, 1153 (1986).
[Pr83] H.G. Price, et al., Phys. Rev. Lett. 51, 1842 (1983).
[Ru91] D. Rudolph, et al., Submitted to Z. Phys. A.
[Sc90] R. Schwengner, et al., Nucl. Phys. A509, 550 (1990).
[St72] F.S. Stephens, et al., Phys. Rev. Lett. 29, 438 (1972).
[Su88] S. Suematsu, et al., Nucl. Phys. A485, 304 (1988).
[Va87] B.J. Varley, et al., Phys. Lett 194B, 463 (1987).
[Wa85] E.K. Warburton, et al., Phys. Rev. C 31, 1211 (1985).
[Wö84] B. Wörman, et al., Nucl. Phys. A431, 170 (1984).

REFINED TRANSIENT FIELD G FACTOR MEASUREMENTS

F. Brandolini

Dipartimento di Fisica dell' Universita' di Padova
Istituto Nazionale di Fisica Nucleare, Sezione di Padova, Italy

ABSTRACT

Recent g factor measurements on heavy even even nuclei by means of Coulomb excitation are reported and compared with theoretical expectations. The prospects for g factor measurements of short-lived states in fusion reaction are discussed.

1.Introduction

The measurement of excited state g factors is an essential piece of information for the understanding of the nuclear structure. In the present contribution some examples of this rather obvious statement will be illustrated. The most interesting aspect concerns at present the case of short lived states (0.1-100 ps), for which hyperfine interaction techniques are mostly necessary. In the last years many data have been accumulating, which have been recently reviewed [1], however much remains to be done, as many interesting cases were not measured. Moreover the precision is often poor or discording values are reported, therefore they are not useful for a detailed comparison with theoretical models.

The method mostly used is the Transient magnetic Field (TF) technique, because, in many circumstances, it is superior to the other ones. The main part of this contribution is devoted to illustrate recent examples of TF g factor measurements by means of Coulomb excitation at the Tandem of the Legnaro National Laboratories. The technique is well established in his main aspects [2], however we have introduced specific improvements which allow an higher sensitivity and precision.

As many interesting states can be only populated efficiently in fusion reaction, a comment will be given at the end about the perspective of individual g factor measurement in fusion reaction, which present great experimental difficulties.

The aim of this contribution is mainly centered on the comparison of experimental data with theory, therefore only a brief description will be given of the experimental apparatus and of the origin of the TF.

We shall moreover make a brief comparison with other hyperfine techniques, which have been also recently reviewed [3].

2. The Transient magnetic field technique

2.1. The origin of the TF

A nucleus, when moving in a saturated ferromagnetic medium, feels a hyperfine magnetic field, oriented along the polarizing external field. The details of this phenomenon are in part still not understood, however it is generally accepted [4] that the TF is due to the polarization of the moving ion ns electrons consequent to the capture of polarized electron from the ferromagnet. One may then write $B_{TF} = q_{ns}p_{ns}B_{ns}$, where B_{ns} is the hyperfine field generated at the nucleus by a ns electron, q_{ns} is the average probability that a single electron ns state exists and finally p_{ns} is the degree of polarization of that state. Values of q_{ns} have been measured in some cases for the 1s states and can be estimated on the basis of theoretical models. At low velocities (v= 1-3 v_0, $v_0=1/137$) q_{ns} may show discontinuities in function of Z of the order of 20-30%, which have been related to an anomalously low vacancy production, explained in the frame of the molecular orbital (MO) model for the atomic scattering [4]. Values of p_{ns} range roughly around 0.14 for iron and 0.20 for gadolinium, at least for 1s states. The mechanism of polarization transfer has been not well established; anyhow a direct spin flip scattering model has been put forward [5]. The mapping of the transient field as function of the velocity, the atomic number of the moving nucleus and for different ferromagnetic media (mainly Gd and Fe), is far from being complete. For the heavy ions, which we are discussing in the present work a usual parametrization is the Chalk River (CR) one [6]: $B_{TF}=aZv/v_0 exp(-bv/v_0)$, where for Fe a=17.5(15) and b=0.10(2), while for Gd a=27.5(10) and b=0.135(20). In fig. 1 the parametrization is specified for reproducing the experimental data for ^{169}Tm. TF strengths of some kT are produced.

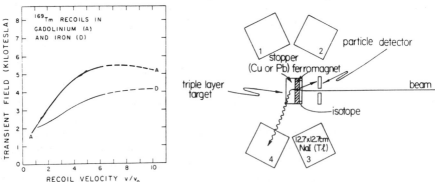

fig.1 TF for ^{169}Tm in Fe and Gd.   fig.2 Typical TF experimental set up

The given parametrization is quite reliable, however it should be mentioned that, in the case of Pt in Fe a remarkable reduction has been found, which was related to the MO effects, mentioned before [7].

2.2 The experimental set up

A sketch of the experimental set up for Coulomb excitation is shown in fig 2. A more detailed description is given elsewhere [9]. The target is essentialy composed of three layers: the proper target, the ferromagnetic foil and a metallic backing. The recoil nucleus is produced in the target, traverses the ferromagnetic foil and comes to rest in the perturbation free metallic backing. The γ rays de-exciting the nucleus are taken (at least) in two detectors, simmetrically located to the beam axis, in coincidence with backscattered projectiles and at angles, where the slope of the angular distribution is large. For lifetimes longer than the transit time (which should be shorter than the stopping time and typically 0.5-1.0 ps) the interaction time is approximately the transit time. As generally in integral angular precession techniques, the effect ε is defined as

$$\varepsilon = \frac{\sqrt{r} - 1}{\sqrt{r} + 1} \quad \text{where} \quad r = \frac{N^\uparrow_1 N^\downarrow_2}{N^\downarrow_1 N^\uparrow_2}$$

The double ratio r is obtained with the number of counts of the two detectors at field up and field down. For small rotations one may write $\varepsilon = S\Delta\theta$, where the slope S is the logarithmic derivative of the angular distributions.

The formula giving the angular precession is:

$$\Delta\theta = g\frac{\mu_N}{\hbar}\int_{t_i}^{t_f} B_{TF} e^{-\frac{t}{\tau}} dt = g\frac{\mu_N}{\hbar} A \int_{v_i}^{v_f} B_{TF} \frac{e^{-\frac{t(v)}{\tau}}}{\frac{dE}{dx}} dv$$

As dE/dx(Gd) $\cong$ 0.5dE/dx(Fe) and B_{TF} is in general bigger in Gd, the use of gadolinium is usually advantageous, even if more difficult.

Four detectors were normally used, located at ±68 and ±112 degrees, or nearby angles, since quadrupole transitions have mainly been observed. The effect for two opposite detectors should be zero and was used as a consistency test of the data. The external polarizing field of 0.02-0.03 T is periodically inverted every few minutes. The experimental slopes was normally measured by us rotating the detector assembly by ±3 degrees. A typical sensitivity in Gd was 150-200 mr/g. For detecting the backscattered projectiles a 4cm×8cm Parallel Plate Avalanche Counter has been used, at a distance of 3 cm from the target (fig. 3).

The TF technique is very precise, when measuring relative g factors. For that purpose a multi-isotope target was often used. Any parametrization should be otherwise used with prudence, as systematic errors could come into play.

A correct procedure implies several delicate points. A proper annealing of the ferromagnetic foil and the measurement of its magnetization are necessary, as well as a precise thickness measurement, which is done by weighing, α source, micrometer and, sometimes, backscattering. An often used procedure is to attach the foils to each other via a thin layer (0.2-0.3 mg/cm^2) of evaporated indium. A good target adherence between the layers is necessary. For this purpose the foil surfaces should be cleaned via glow discharge. In fact even a small detachment, may result in a remarkable attenuation of the anisotropy. To check this point it has been found essential in Coulomb excitations to make a comparison of the experimental slopes with those predicted by the code Coulex [9].

3. Short comparison with other hyperfine techniques

In the following the abbreviations used in ref. 1 will be adopted. The main characteristics of the compared methods are displayed in table 1 in a very qualitative way. The integral precession methods (TF, IMPAD, TMF) use a similar detector assembly.

table 1: Schematic comparison of hyperfine techniques

technique	lifetime range(ps)	sign	Differential Integral	Absolute Relative
TF	0.1-1000	Y	I	R
IMPAD	50-1000	Y	I	R
TMF	1-1000	Y	I	R
RIG	10-1000	N	I	R
RIV/D	1-1000	N	D	A

- **IMPAD:** Perturbed angular distribution after implantation in ferromagnetic media. In this case the recoil nucleus stops in the ferromagnet. The g factors can be deduced, after correction for the transient field effects, if one knows the static hyperfine magnetic field strength. As in the TF technique, the sign can be measured. However the hyperfine field is not always utilizable, because it varies with Z very amply [10] and the implantation site, on which the field value depends, may not be unique [11].

- **RIV/D**: recoil in vacuum/differential. This is an interesting case but limited in its applications. If the recoiling ion emerges from the target in vacuum with a velocity of the order of the Bohr velocity ($v=Zv_0$), it is for a remarkable fraction in a hydrogen-like configuration, for which the static hyperfine field can be exactly calculated. The nuclear spin I preceeds arount $F=I+J$, where J is the atomic spin. Using a plunger device one may follow differentially the perturbation caused by the static hyperfine magnetic field at the nucleus due to the 1s electron [12]. This can be considered an absolute method, however is not sensitive to the sign, as the perturbation is isotropic, and moreover can be applied only to light nuclei (Z<16) because the period of the perturbation becomes otherwise shorter than a ps. Attempts to use other ns configurations were not as successful, owing to the large amount of competing atomic configurations.

In connection with recoil in vacuum it should be noted that the integral γ angular distribution is attenuated, because it results from the integration of periodic perturbations. The attenuation is large for low nuclear spins (I≤J, J≅5).

- **RIG**: Recoil in gas. In the standard set up the plunger technique is used as before, in order to have similar interaction time for states with different lifetimes [3]. However, in this case during the flight the nucleus experiences many collisions with the molecules of the gas. The interaction in vacuum is therefore interrupted at every collision and time dependent perturbations, owing to the atomic de-excitation, could become remarkable. The resulting effect at constant pressure is a gradual decrease of the anisotropy with the plunger distance. By varying the pressure at constant distance the anisotropy varies in a rather complex way, but at sufficiently high pressure the full anisotropy should be restored. The variation of the anisotropy is less marked for high spin states; i.e. when I>J. This method is not sensitive to the sign and its conclusions depend on the model of the statistical description for time dependent and static perturbations.

- **TMF**: tilted multifoil method. Also in this case the recoil nucleus emerges in vacuum, but now the target is inclined with respect to the beam axis. It may be shown [3] that the ion is then partially polarized along the target and perpendicularly to the beam. Therefore I starts to preceed around F and then approximately around J. If the ion encounter a stack of inclined foil at each passage the angular distribution accumulates a small precession $\Delta\theta$. The total precession is in some cases comparable to that observed with transient field in iron, but during the vacuum recoil the angular distributions is sensibly attenuated. Another limitation is that the target stack preparation is rather difficult. The method has been used for g factor measurement only in some cases. It has been otherwise successfully used for polarizing nuclei beam [13].

4. Selected examples of TF g factor measurements in even even heavy nuclei using Coulomb excitation.

4.1 g factor of the first 2+ state in the stable N=82 nuclei [14]

The characteristic of the studied levels are displayed in table 3. The very short lifetime of the levels under study must be underlined. Owing to the little angular precession and the small cross section (1-2 mb) 8 NaI detectors have been used in order to get the highest efficiency in γ counting. A schematic sketch of the experimental set up is shown in fig. 3.

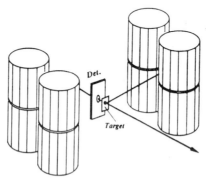

Fig 3 Experimental set up for N=82 isotopes emploing 8 NaI detectors

table 2. Relevant data for the 2+ g factor in N=82 isotones

nuclide	2+ energy	τ(fs)	Δθ (mr)	g_{sp}	g_{th}
138Ba	1436	296(12)	17.2(18)	0.72(11)	1.05
140Ce	1596	133(4)	11.4(10)	0.97(9)	1.16
142Nd	1574	156(5)	12.2(10)	0.84(7)	1.20
144Sm	1660	122(4)	8.5(8)	0.76(11)	1.24

The levels were excited with a 110-116 MeV ^{32}S beam. Typically the target consisted of few hundred micrograms of metallic isotope on 2 mg/cm² Fe backed with Ag. Some ^{148}Sm was added to the targets in order to have an internal reference, as its first 2+ state has a known g factor. In this case Fe was used because the short lifetimes rendered less advantageous the use of Gd. Moreover owing to the small cross section, a high beam current (10-20 pnA) was used, which could warm up the Gd. The experimental g factors are also reported in table 3 together with the predictions of shell model calculations [14] in the proton subspace of the $g_{7/2}$, $d_{5/2}$, $h_{11/2}$, $s_{1/2}$, $d_{3/2}$ orbitals. The sensibly lower experimental values are explained by QRPA calculations [15], which predict a contribution of the $\nu(h_{11/2}^{-1}\text{-}f_{7/2})$ excitation, increasing with Z from 15 to 25 %.

4.2 g factor of the 3⁻ state in ^{144}Sm [16]

In this case, as in the following ones, a four Ge's set up has been used. The 3⁻ state in ^{144}Sm has been populated in the Coulomb excitation induced by a 217 MeV ^{58}Ni beam with 2-3 pnA intensity. The relevant information is presented in fig. 4a.

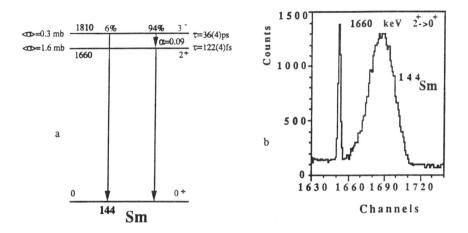

Fig. 4. a) Level scheme in ^{144}Sm. b) Lineshape of the $2^+ \to 0^+$ transition.

The level under study mainly decays to the first 2^+ state and its contribution to that population is seen an a unshifted part in the γ lineshape (fig. 4b). The cross section is only about 0.3 mb. The target consisted of a 2 mg/cm² foil of metallic samarium (85% ^{144}Sm and 15% ^{148}Sm), attached with a thin In layer to a 5.6 mg/cm² annealed Gd foil on whose backside about 6 mg/cm² Ag were evaporated. Also in this case ^{148}Sm served as internal calibration. The measured effect is ε =0.108(10), from which Δθ =76.1(71) and g =0.76(9) are deduced. This value agrees with QRPA calculations [15], which describe this level as due to a 50% of π(d5/2-h11/2) configuration, while the rest is mainly distributed among several core excitation, which can be approximately described as a vibrational excitation with g_v = 0.3.

4.3 g factor of excited states in even even Pt and Hg isotopes [17]

Also these nuclei are rather spherical, because of the proximity to the magic number Z=82.

- Concerning Pt our main goal was to establish unambiguously the variation of the lowest 2^+ state g factors along the isotopes, as this is a severe test for the proposed nuclear models. For this purposes we also measured the 4^+ and 2_2^+, for which not very precise data were available. A ^{58}Ni beam of 210 MeV and 180 MeV was used. The 2 mg/cm^2 foil of metallic Pt (consisting of equal amount of ^{192,194,196}Pt) was attached by means of a thin In layer to a 4.6 mg/cm^2 foil of Gd. The target was backed with 25 mg/cm^2 of Ag in order to ensure a good thermal contact. Measurements were further made with a similar target having 2.5 mg/cm^2 Fe instead of Gd. A typical γ spectrum is shown in fig. 5, while a preliminary summary of the results is presented in table 3

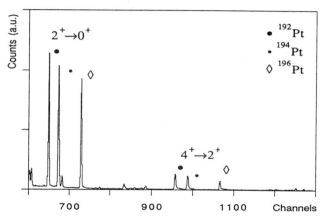

Fig.5   Coincidence γ spectrum for Pt isotopes.

All measured g factors are within the error equal to each other. In order to get absolute g factors one may multiply by 0.32, got from a average of data reported in the literature [1]. On the other side, by adopting the CR calibration in Gd and Fe we obtain for first 2^+ states 0.32 and 0.28 respectively. It seems therefore that the CR calibration in Fe underestimates the 2^+ g factor of 10-15% and this is in agreement with the MO effect already mentioned.

Measurements have been also carried out at various beam intensities up to 5 pnA. The change of the measured effect was negligeable. Owing to the great data reproducibility possible beam correlated effects [18] have not practical consequences in our case.

table 3. Experimental data for Pt and Hg isotopes

	2_1^+	4_1^+	2_2^+	g_{th}	
$g(I,^{192}Pt)$	1 (ref)	1.05(10)	0.80(15)	1.00[a]	1.00[b]
$g(I,^{194}Pt)$	0.97(5)	1.02(10)	1.02(15)	0.98	1.13
$g(I,^{196}Pt)$	0.97(5)	1.15(12)		0.96	1.30
$g(I,^{198}Hg)$	1 (ref)	1.50(15)		1.00	1.00
$g(I,^{200}Hg)$	0.82(4)	0.70(13)		1.15	1.30
$g(I,^{202}Hg)$	1.04(5)	1.42(18)			1.73

a) Dynamical Deformation Model [20] ,b) IBA/2 [21]

- For the Hg isotopes the experimental conditions were quite similar to those of the previous measurement. In this case the target material consisted of a 2 mg/cm^2 of HgS (containing comparable amounts of ^{198,200,202}Hg) on 5.5 mg/cm^2 Gd foils, backed with Ag. Very uniform and stable HgS layers was evaporated with a special procedure [19]. The experimental data are summarized in table 3. An absolute normalization is obtained using g (2+, ^{200}Hg)= 0.326(26), measured by us [17]. The very small cross section of the 4$^+$ states must be underlined. The three isotopes behave rather differently. While ^{200}Hg is similar to Pt isotopes, ^{198,202}Hg have a bigger 2$^+$ g factor , this tendency is even more marked for the 4$^+$ states. This is reflected also in the level scheme as in ^{200}Hg the level density is higher, pointing to a bigger deformation. The large 4$^+$ g factor in ^{198,202}Hg may indicate a specific contribution of the shell model configuration $\pi(h_{11/2})^{-2}$.

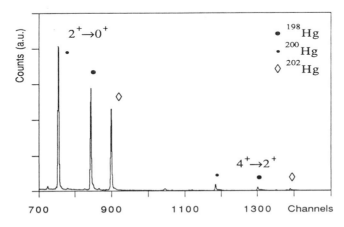

fig. 6 Coincidence γ spectrum for Hg isotopes

Referring to both Pt and Hg, the theoretical predictions for the first 2^+ states are reported in the last two columns of table 3. While the IBA/2 predictions fail, the microscopical DDM calculations give a correct trend, with the exception of ^{200}Hg. An failure of the IBA near to magic numbers was already seen for N= 84, 86 [14].

4.4 g factor in the g.s. rotational band of the isotopic doubles ^{156,158}Gd and ^{166,168}Er [22]

It has been observed that the g.s. rotational band of even even deformed rear earths may somewhat departs from the rotor I(I+1) rule. Great effort has been done in order to explain this fact as an effect of the alignment of quasiparticles pair, with cranked HFB [23] and cranked shell model [24] calculations. Concerning the g factor in general all calculations predict a small reduction at low spin states. From the experimental side a systematic work has been done some years ago [25] by using the TF technique by the CR group in Sm, Gd, Dy, Yb isotopes, which has found a g factor reduction of the order of 10% in average at the 10^+ states with respect to the 2^+ ones, in agreement with the calculations. Those g factor values were fitted with the formula $g(I)=g_0(1+\alpha I^2)$ and therefore were not individually measured.

However subsequently the 6^+ g factor in ^{166}Er was found to be about 20% smaller than that of the 2^+ state [26,27] and in ^{156}Gd a similar reduction has been reported already for the 4+ state [28]. The authors related this finding to the upbander character of the bands. No reduction was otherwise observed in ^{168}Er [29] and ^{158}Gd [28]. The found values could not be quantitatively explained. For example in ^{166}Er only an attenuation of 6% has been calculated [30] and the discrepancy is even bigger for ^{156}Gd [24]. As ^{158}Gd and ^{158}Er are expected to behave normally, we have measured simultaneously the isotopic doublets using a composite target in order to clarify this ambigous situation.

- Concerning the Gd measurement ^{58}Ni beam at the bombarding energies of 117, 130, 180 and 217 MeV has been used in order to determine with good reliability individual g factors. Several targets have been used. typically they consisted of about 1 mg/cm^2 Gd (50% of ^{156}Gd and ^{158}Gd) on a suitable foil of ^{160}Gd or Fe. ^{160}Gd was used in order to clearly distinguish between target and ferromagnetic medium. All data were simultaneously fitted with the code MagMo [31], which has been written for this purpose. A summary of data is reported in table 4. They are in agreement with the CR ones, as well as with the theoretical estimates.

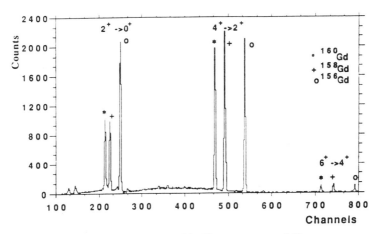

fig. 7 Coincidence γ spectrum with the ^{156,158}Gd target on ^{160}Gd at 117 MeV.

table 4. Experimental g factor in Gd and Er isotopes

nuclide	2+	4+	6+	8+
^{156}Gd	0.386(4)[a]	0.41(2)	0.37(2)	0.31(4)
^{158}Gd	0.38(3)	0.39(2)	0.36(3)	0.40(4)
^{166}Er[b]	0.316(5)[a]		0.29(2)	0.28(2)
^{168}Er[b]	0.321(6)[a]		0.33(2)	0.30(2)

a) Taken from ref 1.
b) The quoted error doesn't include the error in the CR calibration

- In the case of ^{166}Er and ^{168}Er we have done only a measurement at 210 MeV. The target consisted of a 1.5mg/cm^2 Er foil (50% ^{166}Er and ^{168}Er) on a 5.5 mg/cm^2 ^{160}Gd foil, backed with Ag. Again ^{160}Gd was used as ferromagnet in order to avoid line overlap between target and Gd. A typical γ spectrum is shown in fig.8. It is evident the upbander character of ^{166}Er as compared to the good rotor ^{168}Er. The obtained data are also reported in table 4. The relative g factors do not shows a different behaviour between the two isotopes. Furthermore, adopting the CR parametrization with an error of 10% the estimated g factor of the 6+ and 8+ don't differ much from that of the 2+. Our measurement is more sensible than a previous TF one [29]. For example we got ε(8+)=0.042, against 0.024.

In order to effectively extend the individual g factors to the 12+ state, for which the theory predict a strong difference in g factor between ^{166}Er and ^{168}Er, heavier projectiles as Kr or Br would be necessary.

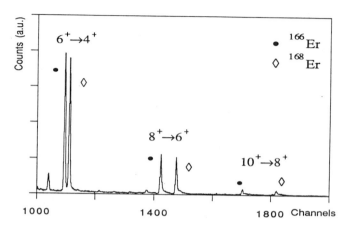

fig. 8 Coincidence γ spectrum with a mixed ^{166,168}Er on ^{160}Gd

4.5 g factor of 2_γ^+ states in ^{162}Dy and ^{164}Dy.

^{162}Dy and ^{164}Dy were studied with a similar procedure as that reported for Er isotopes. Owing to the small cross section for populating the 2_γ^+ states, monoisotopic targets have been used, in order not to increase the background. In this case the ferromagnet was a 156Gd foil, in order to avoid line overlap. The preliminary values are: $g(2_\gamma^+)/g(6^+)$ = 1.15(12) and 1.05(8) for ^{162}Dy and ^{164}Dy respectively.

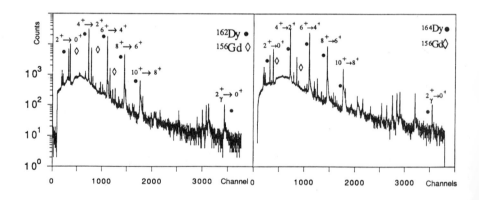

fig. 9 Coincidence γ spectrum with ^{162}Dy and ^{164}Dy targets on ^{156}Gd

The 2_γ^+ states have then a similar g factor to that of the g.s. band and possibly slightly bigger. The equality is predicted by RPA calculations (30) and, in the frame of IBA, points to a little F-spin contribution. For ^{164}Dy our value agrees with a result already published [29], but we got a precision two times better.

5. TF g factors of short lived states populated in fusion reactions

It has been shown that great progress has been made for the g factor measurements in Coulomb excitation. However many states of physical interest can only be populated in fusion reactions. The determination of individual g factors is in this case a difficult goal, mainly because the levels under study may be indirectly fed in a dominant part from upper levels in a time which may be comparable with transit time in the ferromagnet. Therefore in this case the observed precession has to be mainly related to some average g factor of the upper levels, which may also belong to the continuum. In some particular cases a limit has been estimated for the effect of feeding, as recently was done for the ^{82}Sr and ^{84}Sr isotopes [32]. Also at the LNL laboratory a measurement for the 19/2$^-$ state at 8028 kev in ^{39}K has been recently accomplished. The data analysis is still in progress, anyhow a value similar to that of the corresponding level in ^{43}Sc has been estimated.

One may circumvent the feeding problem for states with lifetimes of at least some ps and with a shorter feeding time. One can in fact use a plunger and select the time of fligth in order to prepare the recoiling nucleus in the state on study. In this case the stopper is the ferromagnetic medium. This has been done for ^{160}Yb [33], where the backbender yrast 14$^+$ state (τ= 11.9(6)ps) was studied. For this purpose the reaction ^{64}Ni(^{100}Mo,4n)^{160}Yb at 430 MeV has been employed. In order to effectively clean the complex gamma spectra, γ-γ coincidence have been performed between four Ge's and the 2$^+ \to$ 0$^+$ line, detected in the NaI Heidelberg Crystal Ball. A sketch of the exprimental conditions is shown in fig.9. After four days of effect measurement g=-0.23(31) has been obtained. In spite of the limited precision the backbending could be ascribed to a two neutron quasiparticle alignment, namely $\nu i_{13/2}$, rather than to a proton one.

In the perspective our group is considering for a possible application of the γ-multiarray GASP [34], with 40 Compton suppressed Ge's, to be installed at the LNL. In fact, when using the plunger in γ-γ coincidence with a feeder level of shorter lifetime, it is possible to clean drastically the spectrum and analize all the de-exciting cascade terms, as they should have the same accumulated angular precession. This has been recently also elsewhere undertaken [35]. The feasibility of such measurements has to be still verified in particular for the second backbending region.

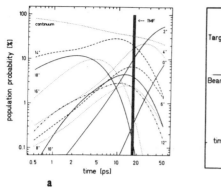 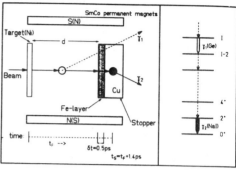

fig. 10 a) Population of the yrast states in ^{160}Yb as function of the time. The vertical bar corresponds to a recoil distance of 240 μm. b) Sketch of the experimental set up.

Acknowledgments

All the colleagues participating in the experiments and quoted in the references are warmly acknowledged, as well as A.Buscemi and R.Zanon for their skilful collaboration during the experimental work. Much gratitude is also due to G. Manente and dr. R.Pengo for their contribution in preparing the delicate targets.

References

[1]   P.Raghavan Atomic Data and Nuclear Data Tables, **42**(1989)19.
[2]   N.Benczer-Koller, M.Hass and J.Sak, Ann. Rev. Nucl. Part. Sc., **30** (1980) 53.
[3]   G.Goldring and M.Hass. Magnetic Moment of Short Lived Nuclear Levels in Treatise on Heavy Ion Science, D.Allan Bromley ed., vol.3 (1984) pg. 539.
[4]   N.Rud and K.Dybdal, Physica Scripta, **34**(1986)561.
[5]   H.-Simonis, F.Hagelberg, M.Knopp, K.-H.Speidel, W.Karle, J.Gerber, Z. Phys **D7** (1987) 223.
[6]   O.Häusser et al., Nucl Phys A **412**(1984)142.
[7]   A.E.Stuchbery, C.G.Ryan and H.H.Bolotin, Hyperf. Interact. **13**(1983)275.
[8]   D.Bazzacco, F.Brandolini, K.Löwenich, P.Pavan, C.Rossi-Alvarez, R.Zannoni and M.De Poli, Phys. Rev. C **33**(1986)1785.
[9]   A.Winther and J. De Boer, in Coulomb Excitation. ed K. Alder and A. Winter, Academic Press, New York, 1966.
[10] K.S. Krane, Hyperfine Interactions, **16**(1983)1069.

[11] F.Brandolini, M. De Poli, C.Rossi-Alvarez, C.Savelli and G.B.Vingiani, Hyperfine Interactions 4(1978)323.
[12] W.L.Randolph et al., Phys.Lett.44B(1973)36.
[13] J.Bendahan, C.Broude, M.Hass, E.Dafni, G.Goldring, D.Habs, W.Korten and D.Schwalm, Z.Phys.A331(1988)343.
[14] D.Bazzacco, F.Brandolini, K. Löwenich. P.Pavan and C. Rossi-Alvarez, Z.Phys.A 328(1987)275,and to be published.
[15] C.Conci, V.Klemt, J. Speth, Phys.Lett.148B(1984)405.
[16] D.Bazzacco, F.Brandolini, K. Löwenich. P.Pavan, C. Rossi-Alvarez, M.De Poli, A.M.I.Haque, Hyperfine Interactions, 59(1990)133.
[17] D.Bazzacco, F.Brandolini, K.Löwenich. P.Pavan, C.Rossi-Alvarez, M.De Poli, A.M.I.Haque, Hyperfine Interactions, 59(1990) 129; and to be published.
[18] K.-H.Speidel et al., Phys.Lett. 227B(1989)16.
[19] P.Pavan and F.Brandolini, LNL Annual Report 1988, 020/89.
[20] Th.J.Köppel, H.Fässler, U.Götz, Nucl.Phys. A403(1983)263.
[21] M.Sambataro, O.Scholten, A.E.L.Dieperink, Nucl.Phys.A423(1984)333.
[22] D.Bazzacco, F.Brandolini, K.Löwenich, P.Pavan, C.Rossi-Alvarez, M.De Poli, A.M.I.Haque, Hyperfine Interactions, 59(1990) 125 and to be published.
[23] M.Diebel, A.N.Mantri, V.Mosel, Nucl.Phys.A345(1980)72.
[24] Y.S.Cheng and S. Frauendorf, Nucl.Phys A393(1983)135.
[25] O.Häusser et al., Nucl. Phys.A406(1983)339, and references therein.
[26] A.Alzner, E.Bodenstedt, B.Gemünden, J. van den Hoff and H. Reif, Z. Phys. A322(1985)467.
[27] C.E.Doran, H.H.Bolotin, A.E.Stuchbery, A.D.Byrne, Z.Phys. A325 (1986) 285.
[28] A.Alzner et al., Z.Phys. A331(1988)277.
[29] C.E.Doran, A.E.Stuchbery, H.H.Bolotin, A.D.Byrne, G.J.Lampard, Phys. Rev. C40(1989)2035.
[30] K.Sugawara-Tanabe and K.Tanabe, Phys.Lett. 207B(1988)243.
[31] R.V.Ribas, to be published.
[32] A.I.Kucharska, J.Billowes and C.J.Lister, J. Phys. G 15(1989)1039.
[33] E.Lubkiewicz et al., Z.Phys.A335(1990)369.
[34] The GA.SP Project. C.Rossi-Alvarez et al. LNL Annual Report 1989, 030/90
[35] U.Birkental, A.P.Byrne, S.Heppner, D.Fallon, P.D.Forsyte, J.W.Roberts, H.Kluge, G.Goldring, E.Lubkiewicz, Daresbury Annual Report 1988/1989

The Role of the γ Degree of Freedom in the A = 130 Region

P. von Brentano, D. Lieberz, W. Lieberz, A. Gelberg, A. Granderath,
A. Dewald, U. Neuneyer, F. Seiffert, H. Wolters, J. Yan
Institut für Kernphysik der Universität zu Köln,
5 Köln 41, W. Germany

ABSTRACT

We present new data of ^{126,129}Xe and ^{127,128}Ba. All nuclei were investigated with the OSIRIS-12 detectorsystem. The Xe isotopes were measured at the FN-Tandem accelerator of Cologne, ^{127,128}Ba at the VICKSI facility of the Hahn Meitner Institute in Berlin. In ^{126,129}Xe we found new low lying collective bands. The data are compared to the predictions of the interacting boson model and to geometrical models such as the rotation vibration model, the asymmetric rotor model and the rigid triaxial rotor plus particle model. The data of ^{126}Xe are in contradiction with the rigid asymmetric rotor model. Experimental routhians, alignments and signature splitting of ^{126}Xe, ^{127,128}Ba are discussed.

I. INTRODUCTION

The nuclei in the A = 130 region constitute a typical transitional region between the vibration like character of the excitations, close to the N = 82 and Z = 50 closed shells and a well deformed rotational character far off the magic numbers.

At low spins the Xe and Ba nuclei of this region exhibit a competition between excitations built on the rotational and the different vibrational degrees of freedom. At high spins one has to consider the interplay of neutron and proton quasi particle excitations and their effect on the nuclear shape and excitation modes. Thus an interesting question is concerns the role of the γ degree of freedom.

Recently Casten et al. [1] proposed that the low spins states in this region can be described by the O(6)-limit of the interacting boson model (IBM) [2], which is related to a γ soft rotor [3]. Another approach was done by Rohoziński et al. [4]. They have interpreted the Xe and Ba nuclei using the γ soft rotor model of Wilets and Jean [3]. However, in all these studies mostly data of the ground state and the quasi γ band were used. It is well known that the asymmetric rotor model [5,6,7] is able to describe rather well the general pattern of the levels and the decay properties of the ground band and the quasi γ band.

In order to be able to distinguish the different collective models we have studied the nucleus ^{126}Xe in more detail. We used the non-selective (α,n) reaction and have populated high lying side bands. In the next chapter the spectroscopic data and the properties of this reaction type are presented [8,9]. Following, we will compare the data with the predictions of three different collective models, namely the O(6)–IBM, the rotation vibration model (RVM) [7,10,11] and the asymmetric rotor model (ARM).

The negative parity spectra of the odd Xe and Ba nuclei form rotational bands built on states, in which the valence particle in the $h_{11/2}$ single particle orbital is coupled to core states. Since the odd particle has many different possibilities to couple to the different core states, an examination of the role of triaxiality is rather intricate. For a description of the yrast and the yrare band in odd nuclei it seems that one needs only the levels of the yrast and yrare band in even nuclei, i.e. the ground band and the γ–band being well described by the rigid triaxial rotor. Thus we content ourselves with studies in which the particle is coupled to a rigid core and we used the rigid triaxial rotor plus particle (RTRP) model [12,13] for our calculations.

We shall present new results of the measurements of ^{129}Xe and compare these data with the predictions of the RTRP model [12,13,14]

At high spins the levels near yrast of the nuclei in this region are rotational bands built on quasi particle excitations. A systematic study of the influence of the excitation of neutron $h_{11/2}$ and proton $h_{11/2}$ quasi particles on the γ deformation is presented in ref. [15]. We shall present new high spin data of ^{126}Xe and ^{127,128}Ba and discuss the $h_{11/2}$ quasi particle alignment.

II. LOW SPIN STATES IN THE NUCLEUS ^{126}Xe

The nucleus ^{126}Xe was studied using the ^{123}Te(α,n) reaction [16]. A γ–γ coincidence experiment was performed by means of the OSIRIS–12 detectorsystem [17] at the FN–Tandem accelerator facility of the University of Köln. Spins and parities were determined with standard γ-ray experiments.

The analysis established 150 levels in ^{126}Xe. The ground state band was observed up to a 10^+ state. We could identify the $2^+ - 7^+$ states of the quasi γ–band and found candidates for the 8^+ (3062 keV) and 9^+ (3521 keV) states of the quasi γ band.

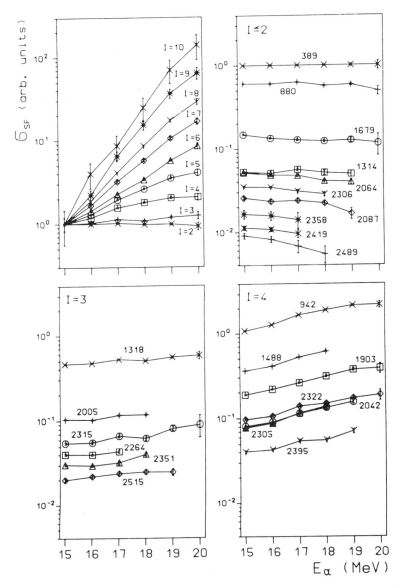

Fig. 1: Excitation functions of continuum feeding cross sections $\sigma_{CF} = \sigma_{SF}$ for states in ^{126}Xe versus beam energy E_α extracted from the (α,n) reaction between 15 and 20 MeV. The top left diagram shows excitation functions of levels with different spin. The other diagrams present excitation functions of levels with the same spin (spin determination is not unique for all states).

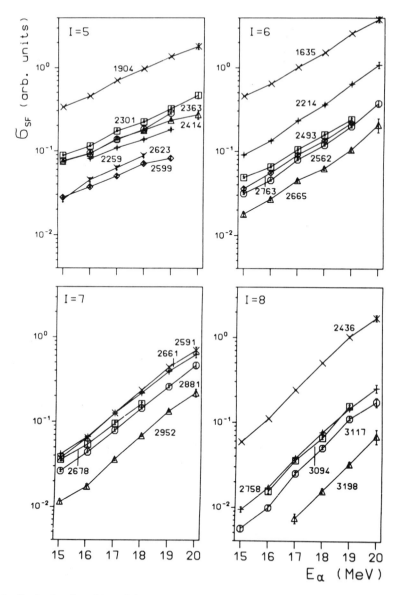

Fig. 2: Excitation functions of the continuum feeding cross section $\sigma_{CF} = \sigma_{SF}$ of levels in ^{126}Xe with the same spin. See also fig. caption 1.

We observed two new high lying side bands. A "K=0"-band starting at 1314 keV, 0^+ and a "K=4"-band starting at 1903 keV, 4^+. These bands are discussed in detail in ref. [18].

Measurements of the energy dependence of the cross sections were used to obtain continuum feeding excitation functions which were calculated with the following formula:

$$\sigma_{CF}^i(E_\alpha) = \sum(\sigma_-^i(E_\alpha) - \sigma_+^i(E_\alpha)) \quad (1)$$

where σ_-^i and σ_+^i are the cross sections for depopulating and populating a level (i). Since weak transitions have a small influence on σ_{CF} and consequently on the slope of $\ln(\sigma_{CF})$, only strong transitions were used to calculate σ_{CF}.

In fig. 1 and 2 σ_{CF} is plotted versus beam energy E_α. The following two properties can be noted [8,9]:

1. The logarithmic slope of the excitation functions $\sigma_{CF}^i(E_\alpha)$ increases with increasing spin and is equal for states with the same spin. This allows a determination of spins by comparing the continuum feeding excitation functions of levels with unknown spin with those with known spins.

2. The continuum feeding intensities of levels with the same spin decrease monotonously with the excitation energy. This leads to a criterion for completeness in a given window of spins and energies. If a level is observed at a certain excitation energy and spin all levels with the same spin and lower excitation energy should have a bigger amount of continuum feeding cross section and thus should be observed in the data. Of course there may be some exceptions due to e.g. E0 transitions or transitions with low energy, or due to bad doublets.

One of the crucial points in comparing experimental data with model predictions is the test of a one to one correspondence between the experimental data and the predictions of the theory. This requires complete spectroscopy, which can be achieved at least partially with the (α,n) and (α,2n) reactions using multidetector arrays with high sensitivity.

The data of ^{126}Xe are compared to the theoretical results of the O(6)-limit of the IBM [2], with the γ rigid asymmetric rotor model (ARM) [5,6,7] and with the rotation vibration model (RVM) [7,10,11]. The O(6) symmetry is related to a γ-soft rotor, the ARM has a fixed rigid γ deformation and RVM a variable γ deformation. A detailed comparison of the experimental data with the different models is given in ref. [18]. In table 1 the experimental branching ratios are compared with the predictions of the different models. Fig. 3 presents the experimental and predicted energies. All models are able to reproduce the energies and E2-branching ratios of the ground state and quasi γ-band rather well. The "K = 4"-band is quite well reproduced by the O(6)-limit and RVM. The ARM describes two ratios ($4_3^+ \to 2_2^+$, $5_2^+ \to 4_2^+$) stronger than the experimental ones.

Table 1: "B(E2)"-ratios in ^{126}Xe [16,18]. In the ratios are marked with $^{a)}$ M1 parts were taken into account. Any undetected M1-portion will lower the experimental ratios (exceptions: $4_3^+, 5_2^+$). All normalization lines (B(E2)-ratio=100) have e. g. in the IBM-O(6) at least a B(E2)-value of 25% of the B(E2:$2_1^+ \to 0_1^+$). The intensities are extracted from the coincidence measurement and the energies are given in keV. In the IBM-O(6) symmetry the B(E2)-ratios are calculated with a χ parameter $\chi = -0.5$ in the transition operator. $^{b)}$: The B(E2)-ratios of the ARM within the band built on the one β-phonon excited 0^+-state are calculated with the assumption that this band corresponds to the experimental "K=0"-band.

level	spin	transition	exp.		O(6)	ARM	RVM
879.9	2_2^+	491.3→ 2_1^+	100	$^{a)}$	100	100	100
		879.9→ 0_1^+	1.5 ± 0.4	$^{a)}$	1.5	1.9	8.6
1317.7	3_1^+	375.7→ 4_1^+	34^{+10}_{-34}	$^{a)}$	40	39	31
		437.8→ 2_2^+	100	$^{a)}$	100	100	100
		929.0→ 2_1^+	$2.0^{+0.6}_{-1.7}$	$^{a)}$	2.0	2.1	8.3
1488.4	4_2^+	546.4→ 4_1^+	76 ± 22	$^{a)}$	91	70	52
		608.5→ 2_2^+	100	$^{a)}$	100	100	100
		1099.8→ 2_1^+	0.4 ± 0.1	$^{a)}$	2.0	5.9	0.03
1903.5	5_1^+	268.5→ 6_1^+	75 ± 23		45	58	31
		415.1→ 4_2^+	76 ± 21		45	105	63
		585.8→ 3_1^+	100	$^{a)}$	100	100	100
		961.6→ 4_1^+	2.9 ± 0.8		2.0	0.02	2.8
2214.3	6_2^+	579.3→ 6_1^+	34^{+15}_{-25}	$^{a)}$	47	14	28
		725.9→ 4_2^+	100		100	100	100
		1272.1→ 4_1^+	0.49 ± 0.15	$^{a)}$	1.1	1.2	0.9
2661.5	7_1^+	447.4→ 6_2^+	40 ± 26		18	45	35
		758.0→ 5_1^+	100	$^{a)}$	100	100	100
1313.9	0_2^+	434.0→ 2_2^+	100	$^{a)}$	100	100 $^{b)}$	100
		925.3→ 2_1^+	7.7 ± 2.2	$^{a)}$	2.0	567 $^{b)}$	0.3
1678.5	2_3^+	360.8→ 3_1^+	67 ± 25		125	5.5 $^{b)}$	69
		364.2→ 0_2^+	100	$^{a)}$	100	100 $^{b)}$	100
		736.4→ 4_1^+	2.0 ± 0.8	$^{a)}$	1.1	20 $^{b)}$	0.4
		798.8→ 2_2^+	2.2 ± 1.0		3.5	5.6 $^{b)}$	11
		1290.0→ 2_1^+	0.14 ± 0.06		0	6.9 $^{b)}$	0.04
		1678.2→ 0_1^+	0.13 ± 0.04	$^{a)}$	0	4.4 $^{b)}$	0
2042.1	4_4^+	363.4→ 2_3^+	100	$^{a)}$	100	100 $^{b)}$	100
		1100.2→ 4_1^+	7.9 ± 3.4		0	3.4 $^{b)}$	0.08
		1653.5→ 2_1^+	0.9 ± 0.4	$^{a)}$	0	3.4 $^{b)}$	0.02
1903.1	4_3^+	414.8→ 4_2^+	100		100	100	100
		585.3→ 3_1^+	43 ± 13		115	150	159
		961.2→ 4_1^+	4.5 ± 1.4		0.4	12	24
		1023.2→ 2_2^+	2.8 ± 0.9		2.5	75	4.3
2363.0	5_2^+	459.8→ 5_1^+	100		100	100	100
		460.0→ 4_3^+	100 ± 45		286	667	394
		727.7→ 6_1^+	2.2 ± 1.8		0.03	2.7	15
		874.5→ 4_2^+	1.9 ± 0.9		1.5	77	0.5
		1045.3→ 3_1^+	4.2 ± 1.6		3.2	11	13

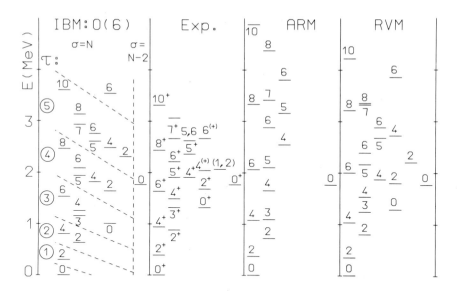

Fig. 3: Scheme of the positive parity levels of ^{126}Xe [16,18] compared with the IBM–O(6) symmetry, the asymmetric rotor and rotation vibration model. The third 0^+ state is taken from ref. [19]. N is the boson number (^{126}Xe: $N = 7$).

The branching ratios in the "K=0"-band are reproduced qualitatively by the IBM–O(6) symmetry and by the RVM model, whereas the ARM is not able to reproduce this band. In the ARM model excited 0^+ states are built on β-oscillations and the 0_2^+ state of ARM therefore mainly decays to the ground state 0^+-state, contrary to the experimental situation. Thus the experimental data rule out the ARM model with rigid triaxiality for ^{126}Xe.

Since in neighbouring nuclei the "K=0"-band is mostly unidentified, we can only compare the decay of the first excited 0^+ state. In table 2 the B(E2)-ratios of the $0_2^+ \to 2_1^+$ and $0_2^+ \to 2_2^+$ transitions are listed for several Xe and Ba nuclei. In all cases one finds that the 0_2^+ state is connected strongly to the 2_2^+ state of the quasi γ-band indicating variable γ deformation for these nuclei.

The energy staggering in the quasi γ band is another fingerprint of γ softness. A quantitative measure of the energy staggering is

$$S = \frac{R(E_I)}{R(E_I)_{rigid}} - 1 \quad \text{with} \quad R(E_I) = \frac{2 \cdot [E_I - E_{I-1}]}{[E_I - E_{I-2}]} \quad (2)$$

$R(E_I)_{rigid}$ is the value of $R(E_I)$ for a rigid axial rotor ($E \sim I(I+1)$). It is noted that S is independent of the moment of inertia. In fig. 4 experimental and predicted values of S are plotted versus spin for the quasi γ band. The experimental staggering S of the quasi γ band shows an alternating behaviour. For even spins it is found to be positive and for odd spins negative.

The IBM–O(6) symmetry and the RVM model corresponding to a γ soft behaviour predict the same sequence, whereas in the ARM model one obtains the opposite. It is important to note, that the γ soft models generally predict a smaller spacing between the 4^+ and the 3^+ states as compared to the 3^+ to 2^+ energy difference. This results can not be obtain by γ rigid rotor model.

Table 2: Experimental B(E2)-ratio of the $0_2^+ \rightarrow 2_1^+$-transitions and $0_2^+ \rightarrow 2_2^+$-transitions in Xe- and Ba-nuclei compared to the predicted values of different models (data from ref. 1 and refs. therein). [b)] see caption of table 1.

	^{124}Xe	^{126}Xe	^{128}Xe	^{134}Ba	O(6)	ARM	RVM
B(E2;$0_2^+ \rightarrow 2_1^+$)	1	8	14	4	2	567 [b)]	0.3
B(E2;$0_2^+ \rightarrow 2_2^+$)	100	100	100	100	100	100 [b)]	100

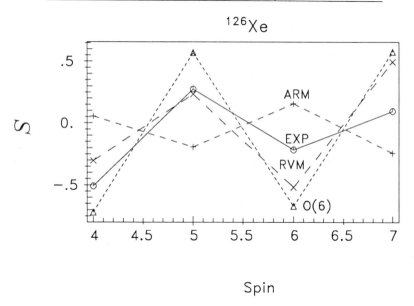

Fig. 4: Experimental and predicted values of the S (see text) versus spin of the quasi γ–band in ^{126}Xe. S measures the signature splitting of the quasi γ band.

137

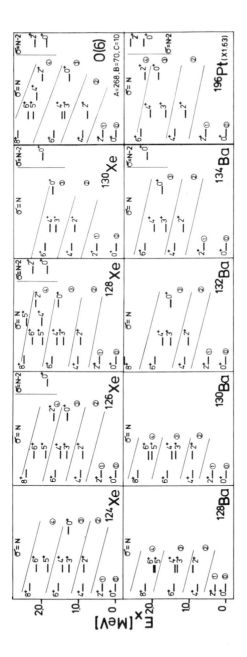

Fig. 5: Positive parity level schemes of different Xe and Ba nuclei and ^{196}Pt compared to the O(6) symmetry of the IBM. The data are from ref. [1] and refs. therein.

Fig. 5 shows the level schemes of different Xe and Ba nuclei in comparison to the predicted level scheme of the O(6)–symmetry of the IBM. One observes, that the general pattern of the level schemes of nuclei in the A = 130 region is similar to the predictions of the O(6) symmetry of the IBM. Moreover, the experimental staggering in the quasi γ bands has the same sequence in all nuclei and agrees with the results of γ soft models excluding the rigid γ deformed model (ARM).

In summary, the decay of the first excited 0^+ states and the the staggering in the quasi γ bands imply γ softness as a common feature of the Xe and Ba nuclei in the A = 130 region.

III. LOW SPIN STATES IN THE NUCLEUS ^{129}Xe

In order to study the γ degree of freedom in an odd nucleus, we investigated the nucleus ^{129}Xe by means of the (α,n) reaction. The γ–γ-coincidence experiment was performed with the OSIRIS–12 spectrometer. Spins and parities were determined with standard γ-ray experiments [20]. All measurements were carried out at the FN–Tandem accelerator of the University of Cologne. The constructed level scheme contains 110 levels.

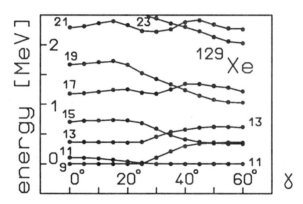

Fig. 6: Calculated γ–dependence of the yrast levels ($\pi = -$) in ^{129}Xe ($\epsilon = 0.23$) [14,20]. Spin: multiplied by 2

Calculations were performed with the rigid triaxial rotor plus particle (RTRP) model [12,13,14], of which a detailed description is given in ref. [12]. In the RTRP model an odd particle is coupled to a rigid triaxial core moving in a harmonic potential with deformation parameters ϵ and γ. A new version of the program developed in Lund and Oxford was used, which includes a variable moment of inertia (VMI).

The calculated γ–dependence of negative parity yrast states for ^{129}Xe is shown in fig. 6. At a value of $\gamma = 30°$ the level ordering is reproduced well, whereas at axial symmetry the small signature splitting can not be reproduced. We found a good agreement of predicted and experimental data at $\gamma = 30°$ and $\varepsilon = 0.23$. In fig. 7 the experimental levels are compared to the predicted ones.

Table 3: Branching-ratios of negative parity states in ^{129}Xe [14,20]. Energies in keV.

level	spin	transition	branching-ratios exp.	RTRP
908.8	$(9/2_2^-)$	634.5→ $9/2_1^-$	25±11	20
		672.6→$11/2_1^-$	100	100
1032.0	$13/2_2^-$	757.8→ $9/2_1^-$	100	100
		795.9→$11/2_1^-$	35±10	23
1144.6	$(11/2_2^-)$	235.8→$(9/2_2^-)$	11±5	≤1
		373.5→$13/2_1^-$	57±20	36
		870.3→ $9/2_1^-$	132±37	909
		908.5→$11/2_1^-$	100	100
1395.5	$15/2_2^-$	572.2→$15/2_1^-$	≤ 10	9
		624.4→$13/2_1^-$	100	100
		1159.4→$11/2_1^-$	not obs.	45
1507.2	$17/2_1^-$	683.9→$15/2_1^-$	100	100
		736.1→$13/2_1^-$	43±19	69
1972.2	$17/2_2^-$	392.2→$19/2_1^-$	not obs.	17
		576.7→$15/2_2^-$	100	100
		940.0→$13/2_2^-$	37±15	81
		1148.9→$15/2_1^-$	not obs.	22
		1201.1→$13/2_1^-$	not obs.	225

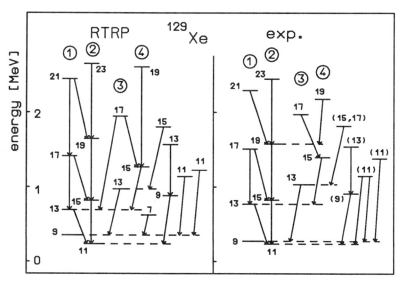

Fig. 7: Experimental and calculated excited negative parity states of ^{129}Xe [14,20], spin: multiplied by 2

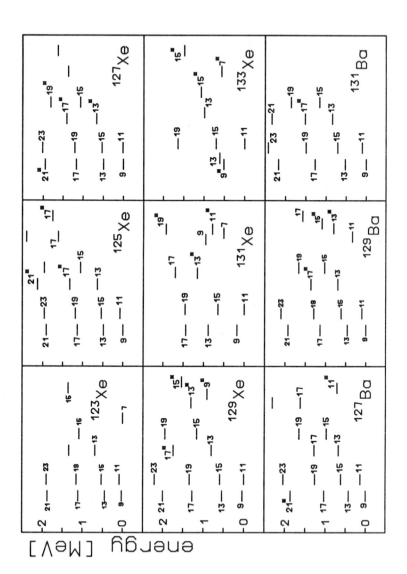

Fig. 8: Partial negative parity level schemes of different Xe and Ba nuclei [14]. Data are from refs. [20–27]. Spin: multiplied by 2, *: no unique spin assignment

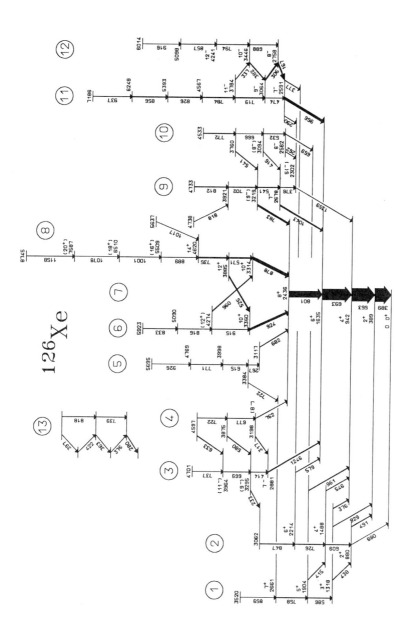

Fig. 9: Level scheme of ^{126}Xe observed with the ^{116}Cd(^{13}C,3n) reaction [28].

The experimental and theoretical branching ratios are listed in comparison in table 3. The agreement is rather good for the yrast band. The experimentally strong branchings of the yrast band are also predicted to be strong and the weak as weak. In contrast there are some discrepancies in the yrare band.

A survey of negative parity states of various odd Xe and Ba nuclei is presented in fig. 8. One notes that the favoured and unfavoured levels of the yrast bands are nearly degenerate in most cases. The first $9/2^-$ states lie below the corresponding $11/2^-$ states for the light Xe isotopes with $N \leq 73$ and for $^{127-131}$Ba. In addition the 13/2 state of the yrare band are close to the corresponding $15/2_1^-$ state. These systematic trends can be reproduced by RTRP calculations with a triaxial deformed potential [20].

IV. HIGH SPIN STATES IN THE NUCLEUS ^{126}Xe

In addition to the low spin measurements the nucleus ^{126}Xe was investigated using the ^{116}Cd(^{13}C,3n) reaction [28]. The coincidence experiment was performed with the OSIRIS-12 spectrometer at the FN–Tandem accelerator of Cologne. Spins and parities were determined by γ–angular distributions measurements, γ–excitation functions and γ–linear polarization measurements. The deduced level scheme is shown in fig. 9. In the figure the bands assigned to ^{126}Xe are labelled with numbers from 1 to 13.

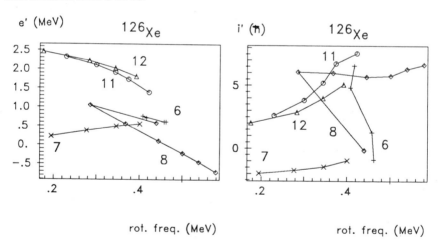

Fig. 10: Experimental routhians e' (left) and alignment i' (right) versus rotational frequency of bands 6, 7, 8, 11 and 12 of ^{126}Xe relative to a reference given by the Harris formula [29] with $J_0 = 17.0\hbar^2/\text{MeV}$ and $J_1 = 25.8\hbar^4/\text{MeV}^3$.

The experimental routhians e' and alignments i' were extracted using a reference rotating frame described by the Harris formula [29] and the parameters $J_0 = 17.0\hbar^2/\text{MeV}$ and $J_1 = 25.8\hbar^4/\text{MeV}^3$. These parameters were successfully applied in similar analyses of high spin bands in the A = 130 region [15]. The routhians and alignments are shown in fig. 10. In fig. 11 the values of the signature splitting S (defined by eq. 2 and 3) are plotted.

The positive parity band 8 crosses the ground state band (7) at a frequency 0.37 MeV. The alignment of band 8 is around $7.5\hbar$. The crossing of the other positive parity band (6) seems to take place at higher rotational frequency. Following ref. [15] the neutron crossing sets in earlier than the proton crossing for $^{124-130}$Xe. We assign $(\nu h_{11/2})^2$ and $(\pi h_{11/2})^2$ configuration for band 8 and 6, respectively. We found 3 pairs of bands with negative parity. One of these pairs (bands 11 and 12) shows a small signature splitting indicating a high value of K. In ref. [30] a life time of 1.5 ns is measured for the 8^- state of band 12. In the neighbouring nucleus ^{128}Xe a similar band with small signature splitting and an isomeric 8^- state has been reported [31,32]. Lifetime and g factor measurements [32] in ^{128}Xe gave a measured half life of $T_{1/2} = 83$ ns and a g-factor (g = -0.039) [32] leading to a $\nu h_{11/2} g_{7/2}$ configuration assignment for the 8^-. In analogy, the state in ^{126}Xe should have the same configuration.

The bands 3, 4 and 9, 10 exhibit a strong signature splitting, displaying contributions from orbitals with low K.

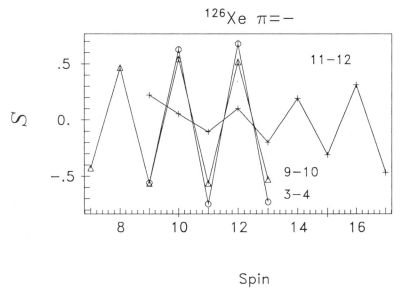

Fig. 11: Experimental signature splitting S (see text) versus spin of bands 3,4 9,10 and 11,12 in ^{126}Xe.

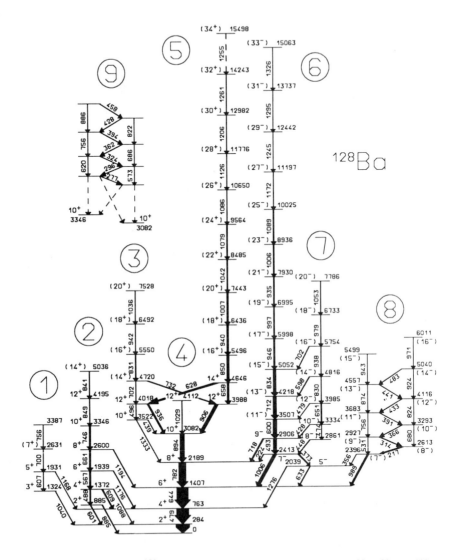

Fig. 12: Level scheme of ^{128}Ba obtained by the Köln group with the ^{96}Zr(^{36}S,4n)^{128}Ba with the OSIRIS-12 spectrometer at the Vicksi accelerator of the Hahn Meitner Institute [33].

V. HIGH SPIN STATES IN THE NUCLEUS ^{128}Ba

In order to investigate the high spin states of ^{128}Ba we have measured a γ–γ coincidence experiment with the OSIRIS-12 spectrometer by means of a 150 MeV ^{36}S beam at the VICKSI facility of the Hahn Meitner institute in Berlin [33].

The constructed level scheme is shown in fig. 12. We were able to extend the positive parity band (no. 5) up to a spin of $34\hbar$, and the negative parity band no. 6 up to a spin of $33\hbar$. In fig. 13 the experimental routhians and alignments are plotted versus rotational frequency. In band 4/5 a crossing occurs at 0.4 and approximately 0.5 MeV with an alignment $7.5\hbar$ and $8-9\hbar$, respectively.

In the negative parity band (no. 6) a crossing is observed at $\hbar\omega = 0.5$ MeV. Since this band is interpreted as being based on a $\pi h_{11/2} d_{5/2}$ configuration with prolate deformation [34,35], one expects the crossing to be due to an alignment of $(\nu h_{11/2})^2$ quasi particles. Also the second crossing in band 5 is then expected to be based on $(\nu h_{11/2})^2$ quasi particles. Consequently a $(\pi h_{11/2})^2$ alignment is responsible for the strong backbend at lower rotational frequency in band 5. Nevertheless the gain in alignment for the second crossing in band 5 seems to be too high for a single crossing and contributions from the alignment of other particles and shape changes may be involved in addition [36].

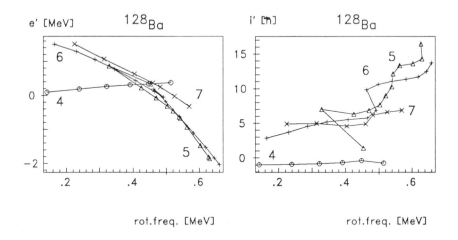

Fig. 13: Experimental routhians e' (left) and alignment i' (right) versus rotational frequency of bands 4-7 of ^{128}Ba. Harris parameters are the same as given in fig. 10.

In fig. 14 the signature splitting S (see eq. 2) of band 6–8 is plotted versus spin. The bands labelled with 8 are strongly coupled bands. Thus these bands involve single particle orbitals with high K values and they exhibit a relatively constant signature splitting indicating that no change of deformation occurs with increasing rotational frequency.

The systematic forking of the ground band in two S–bands in the even even Xe, Ba and Ce nuclei is shown in fig. 15. In ref. [15] the displayed S–bands are analysed in the framework of the cranking model. It was found that the aligned $(\nu h_{11/2})^2$ configuration polarizes the nuclear shape towards oblate deformation and the $(\pi h_{11/2})^2$ configuration drives the nucleus towards prolate deformation.

The knowledge of a few g factors and the analysis of crossing frequencies, alignments and systematics allows configuration assignments. The S–bands which are displayed to the left are assigned as $(\pi h_{11/2})^2$ and those which are displayed to the right of the ground state band have a $(\nu h_{11/2})^2$ assignment [15]. For bands marked with "?" no definite assignment can be made at present.

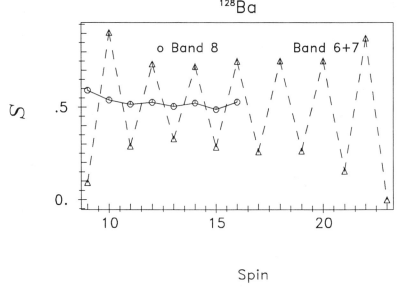

Fig. 14: Experimental signature splitting S (see text) versus spin of bands 6,7 and 8 of ^{128}Ba.

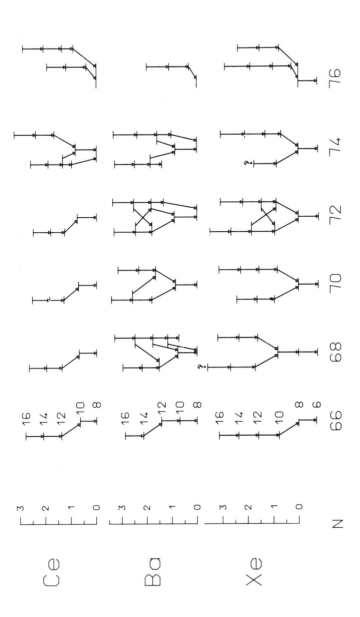

Fig. 15: Observed forking of the ground band in two S bands in Xe, Ba and Ce nuclei (see [15] and refs. therein)

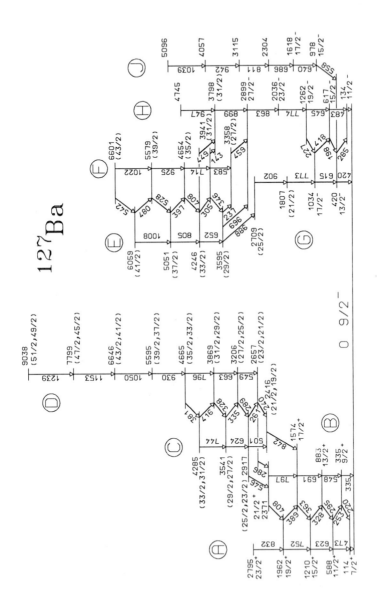

Fig. 16: Level scheme of ^{127}Ba obtained by the Köln group with the $^{96}Zr(^{36}S,5n)^{127}$Ba with the OSIRIS-12 spectrometer at the Vicksi accelerator of the Hahn Meitner Institute [37].

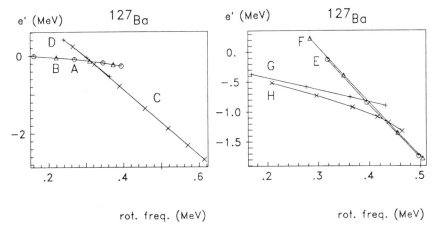

Fig. 17: Experimental routhians e' of positive parity bands (left) and negative parity bands (right) versus rotational frequency of ^{127}Ba. Harris parameters are the same as given in fig. caption 10.

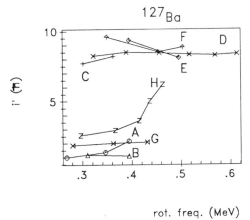

Fig. 18: Experimental alignment i' versus rotational frequency of ^{127}Ba. See also fig. caption 17.

VI. HIGH SPIN STATES IN THE NUCLEUS ^{127}Ba

The high spin states of the nucleus ^{127}Ba were studied with the ^{96}Zr(^{36}S,5n) reaction. The level scheme (fig. 16) was deduced from the γ–γ–coincidence experiment, which was measured with the OSIRIS-12 detectorsystem at the VICKSI facility of the Hahn Meitner Institute in Berlin [37]. We could extend the negative parity yrast band up to a tentative $43/2^-$ state at 6601 keV. The positive parity band is observed up to an excitation energy of 9038 keV with possible spins of 51/2 or 49/2 $\hbar$. Fig. 17 and 18 show the experimental routhians and the corresponding alignments. A backbending of the "$g_{7/2}$" band is observed at $\hbar\omega \approx 0.35$ MeV with an alignment of about $7\hbar$. Thus one assigns for Band C and D $(\nu g_{7/2})(\pi h_{11/2})^2$ or $(\nu g_{7/2})(\nu h_{11/2})^2$ configurations.

In the $h_{11/2}$ quasi particle band a crossing occurs at $\hbar\omega \approx 0.45$ MeV. Again an alignment of around $7\hbar$ indicates a crossing of neutron or proton $h_{11/2}^2$ quasi particles. In this case the signature splitting decreases drastically (fig. 19) after the crossing. This can be understood with a change of γ deformation due to the alignment of $h_{11/2}^2$ quasi particles. Below the crossing frequency (i.e. for states with $I \leq 27/2$) the signature splitting increases with increasing spin.

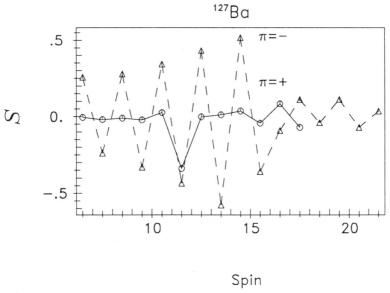

Fig. 19: Experimental signature splitting S (defined in the text) versus spin in ^{127}Ba.

VII. CONCLUSION

We have presented new low spin data of ^{126}Xe and ^{129}Xe, and new high spin data of ^{126}Xe, ^{128}Ba and ^{127}Ba.

The low spin states of ^{126}Xe were described satisfactorily with γ soft models as e.g. the O(6) symmetry of the IBM-1. The data rule out the rigid asymmetric rotor model, which has a rigid γ deformation. A one to one correspondence of the experimental and the predicted levels in ^{126}Xe is found in a certain spin and energy window. This and the properties of the continuum feeding excitation functions indicate the possibility of a "completeness" of the (α,n) reaction.

We have studied the influence of the γ-deformation parameter in ^{129}Xe in the rigid triaxial rotor plus particle model. The level ordering is reproduced nicely, especially the near degeneracy of favoured and unfavoured yrast states. We found that the γ degree of freedom is an essential ingredient for a satisfactory description of the odd Xe and Ba nuclei and in particular for ^{129}Xe.

The high spin states of ^{126}Xe, ^{127}Ba and ^{128}Ba were interpreted as rotational bands built on neutron or proton $h_{11/2}$ quasi particles alignments. This orbital has a strong γ deformation driving force, i.e. changes in γ deformation are expected.

ACKNOWLEDGEMENTS

We would like to thank in particular Drs. I. Ragnarsson, and P.B. Semmes for their contribution to this talk. We further thank Dr. K. Schiffer, Dr. Rikovska, I. Wiedenhöver, R. Wirowski and Dr. K.O. Zell for fruitful discussions. This work was supported by the German Federal Minister for Research and Technology (BMFT) under the contract number 06OK143.

REFERENCES

[1] R.F. Casten and P. von Brentano, Phys. Lett. 152, 22(1985)
[2] A. Arima and F. Iachello, Ann. Phys. 123, 468(1979)
[3] L. Wilets and M. Jean, Phys. Rev. 102, 788(1956)
[4] S.G. Rohoziński, J. Srebrny and K. Horbaczewska, Z. Phys. 268, 401(1974)
[5] A.S. Davydov and G.F. Filippov, Nucl. Phys. 8, 237(1958)
[6] A.S. Davydov and A.A. Chaban, Nucl. Phys. 20, 499(1960)
[7] G. Leander, Program ARM and VRM, unpublished
[8] W. Zipper, A. Dewald, W. Lieberz, R. Reinhardt, F. Seiffert and P. von Brentano, Nucl. Phys. A504, 36(1989)
[9] P. von Brentano, A. Dewald, W. Lieberz, R. Reinhardt, K.O. Zell and W. Zipper, in: Nuclear Structure of the Zr Region, eds. J.Eberth, R.A. Meyer and K. Sistemich, Springer, Berlin 1988, p. 157
[10] A. Faessler, W. Greiner and R.K. Sheline, Nucl. Phys. 80, 417(1965)
[11] J.M. Eisenberg and W. Greiner, Nuclear Models, Vol. 1 (North-Holland, 1987).
[12] S.E. Larsson, G. Leander and I. Ragnarsson, Nucl. Phys. A307, 189(1978)

[13] H. Toki and A. Faessler, Nucl. Phys. A253, 231(1975)
[14] D. Lieberz, A. Gelberg, A. Granderath, P. von Brentano, I. Ragnarsson, and P.B. Semmes, to be published; D. Lieberz, Ph.D Thesis Köln, (1990,1991)
[15] R. Wyss, A. Granderath, R. Bengtsson, P. von Brentano, A. Dewald, A. Gelberg, A. Gizon, S. Harissopulos, A. Johnson, W. Lieberz, W. Nazarewicz, J. Nyberg and K. Schiffer, Nucl. Phys. A505, 337(1989)
[16] W. Lieberz et al., to be published
[17] R.M. Lieder, H. Jäger, A. Neskakis, T. Venkova and C. Michel, Nucl. Instr. Meth. 220, 363(1984)
[18] W. Lieberz, A. Dewald, W. Frank, A. Gelberg, W. Krips, D. Lieberz, R. Wirowski and P. von Brentano, Phys. Lett. B240, 38(1990)
[19] W.P. Alford, R.E. Anderson, P.A. Batay-Csorba, R.A. Emigh, D.A. Lind, P.A. Smith and C.D. Zafiratos, Nucl. Phys. A323, 339(1979)
[20] Z. Zhao, J. Yan, A. Gelberg, R. Reinhardt, W. Lieberz, A. Dewald, R. Wirowski, K.O. Zell and P. von Brentano, Z. Phys. A331 113(1988); J. Yan et al., to be published
[21] A. Luuko, J. Hattula, H. Helppi, O. Knuuttila and F. Dönau, Nucl. Phys. A357, 319(1981)
[22] A. Granderath, D. Lieberz, A. Gelberg, S. Freund, W. Lieberz and P. von Brentano, submitted to Nucl. Phys. A
[23] W. Urban, T. Morek, Ch. Droste, B. Kotliński, J. Srebrny, J. Wrzesiński and J. Styczeń, Z. Phys. A320, 327(1985)
[24] T. Lönroth, J. Kumpulainen and C. Tuokko, Phys. Scr. Vol. 27, 228(1983)
[25] J. Gizon and A. Gizon, Z. Phys. A281, 99(1977)
[26] J. Gizon, A. Gizon and J. Meyer-Ter-Vehn, Nucl. Phys. A277, 464(1977)
[27] R. Ma, Y. Liang, E.S. Paul, N. Xu, D.B. Fossan, L. Hildingsson and R.A. Wyss, Phys. Rev. C41, 717(1990)
[28] W. Lieberz, S. Freund, A. Granderath, A. Gelberg, A. Dewald, R. Reinhardt, R. Wirowski, K.O. Zell and P. von Brentano, Z. Phys. A330, 221(1988)
[29] S.M. Harris, Phys. Rev. C138, 509(1965)
[30] W. Gast, thesis, University of Cologne (1982)
[31] R. Reinhardt, A. Dewald, A. Gelberg, W. Lieberz, K. Schiffer, K.P. Schmittgen, K.O. Zell and P. von Brentano, Z. Phys. A329, 507(1988)
[32] T. Lönroth, S. Vajda, O.C. Kistner, M.H. Rafailovich, Z. Phys. A317, 215(1984)
[33] U. Neuneyer, H. Wolters, A. Dewald, W. Lieberz, A. Gelberg, E. Ott, J. Theuerkauf, R. Wirowski, P. von Brentano, K. Schiffer, A. Alber and K.H. Maier, Z. Phys. A336, 245(1990)
[34] K. Schiffer, A. Dewald, A. Gelberg, R. Reinhardt, K.O. Zell, Sun Xiangfu and P. von Brentano, Nucl. Phys. A458, 337(1986)
[35] H. Wolters, K. Schiffer, A. Gelberg, A. Dewald, J.Eberth, R. Reinhardt, K.O. Zell, P. von Brentano, D. Alber and H. Grawe, Z. Phys. A328, 15(1987)
[36] K. Schiffer et al., to be published
[37] F. Seiffert, A. Granderath, A. Dewald, W. Lieberz, U. Neuneyer, E. Ott, J. Theuerkauf, R. Wirowski, H. Wolters, K.O. Zell, P. von Brentano, K. Schiffer, D. Alber and K.H. Maier, Z. Phys. A336, 237(1990)

153

EFFECTS OF THE RESIDUAL PROTON-NEUTRON INTERACTION IN THE DEVELOPMENT OF COLLECTIVITY IN NUCLEI

R. F. Casten

Brookhaven National Laboratory
Upton, New York, 11973, USA

ABSTRACT

The widespread effects of the residual T=0 proton-neutron (p-n) interaction in the evolution of nuclear structure are discussed. Although these effects in inducing single nucleon configuration mixing, and hence in the development of non-spherical nuclear shapes, collectivity, and the associated shape and phase transitions have been known for four decades, it is only in recent years that their deep ramifications have become more fully appreciated. This has led to a unified phenomenological understanding of the role of the p-n interaction in nuclear collectivity and to, for example, the proposal of the N_pN_n scheme and the associated concept of the P factor, which is a normalized value of N_pN_n reflecting the average number of p-n interactions per valence nucleon. Simultaneously, experimentally-determined p-n matrix elements for many nuclei have been extracted: they disclose striking anomalies for N=Z nuclei, and intriguing microstructure. These developments and empirical results will be discussed along with microscopic calculations that can be used to interpret them.

1. INTRODUCTION

Nearly four decades ago de Shalit and Goldhaber[1] suggested that the p-n interaction could play an important role in the onset of deformation in nuclei. In the 1960's and since then Talmi[2] has repeatedly emphasized and delineated the critical role of the p-n interaction and explored the microscopic basis for single nucleon configuration mixing in the Shell

Model induced by this interaction. In the early 1970's a rapid, enigmatic, new region of deformation was discovered near A=100. In the late 1970's Federman and Pittel[3] published a series of articles explaining the onset of deformation as due, primarily, to enhanced p-n interactions in spin-orbit partner single particle states, in particular $p1g_{9/2}$-$n1g_{7/2}$. This work was extended to the rare earth region by Casten et al.[4]

Traditionally, nuclear systematics are plotted in terms of the nucleon numbers N, Z and A. Typically, they are extremely complex, especially in nuclear transition regions. The importance of the p-n interaction, however, suggests that it might be useful to reconsider the evolution of nuclear structure in terms of some parameter reflecting the strength of this interaction. Such a quantity is the valence product N_pN_n of the number of valence protons times the number of valence neutrons, counted to the nearest closed shell with appropriate regard given to important subshell closures. Nuclear systematics is substantially simplified in the N_pN_n scheme[5]. Moreover, a normalized value of N_pN_n, the so-called P factor[6], allows different mass regions to be compared on an equal footing and shows the similarity of phase transitional regions throughout medium and heavy nuclei.

Recently, empirical techniques have been used[7,8] to extract actual individual p-n matrix elements of the last proton with the last neutron in all even-even nuclei. These interaction matrix elements exhibit fascinating structure including near singularities for light N=Z nuclei, and interesting microstructure in medium and heavy nuclei. Systematic trends in their values also suggest the possibility of distinguishing the roles of the monopole and quadrupole components of the p-n interaction. This is an intriguing development since these two multipoles play different and complementary roles in the evolution of structure: specifically, the monopole p-n interaction acts to shift single particle energy levels[9] while the quadrupole interaction, acting on the space of levels provided in part by the monopole interaction, is critical in the development of collectivity[10].

These various aspects of the p-n interaction, along with some very basic nuclear data that highlights its effects in a very simple way, will be described in the sections to follow. Much of this paper should be viewed as embodying the results from recent collaborative papers[7-8] to which the reader is referred for further details and discussion. I am grateful to my co-authors D. S. Brenner, D. D. Warner, W.-T. Chou, J.-Y. Zhang, and C. Wesselborg for the collaborations that led to these results and for permitting me to discuss them here.

2. MANIFESTATIONS OF THE P-N INTERACTION IN NUCLEAR STRUCTURE

It is in a way remarkable that the p-n interaction has been discussed relatively little in the history of nuclear structure physics while, at the same time, there are some extremely simple, basic sets of nuclear data which expose its role quite directly and dramatically. In this section we will mention a few of these as examples and some others will become evident later on.

Perhaps the most straightforward evidence of the importance of the p-n interaction is provided by the energies of low lying levels in even-even nuclei. The energies of the 2_1^+ levels of nuclei in the Sn region are shown in Fig. 1. Note the constancy of $E(2_1^+)$ in Sn and the rapid decrease for non-singly magic nuclei. Since low 2_1^+ energies are characteristic of the onset of collectivity and deformation, the key role of the p-n interaction is obvious: the valence p-n interaction clearly vanishes for Sn, but not for the other nuclei. It is trivial to show[2,11] that excitation energies of j^n configurations are independent of n in the seniority scheme. And, seniority is a good quantum number for many realistic 2-body residual interactions, such as a δ-function interaction, acting in such configurations. The constancy of $E(2_1^+)$ in Sn is therefore not surprising. The interesting feature of Fig. 1 is the very different behavior for nuclei near Sn with valence protons as well as neutrons. Residual interactions are much more effective in inducing single nucleon configuration mixing[2] in p-n systems than in T=1 states, breaking the seniority scheme, and removing the prescription of constant excitation energies. Thus, the empirical behavior in Fig. 1 is de facto evidence of the role of the p-n interaction and this role is consistent with simple properties of residual 2-body interactions.

One of the simplest aspects of nuclear data is the separation energy for the last proton or neutron. These data are extremely well known and yet

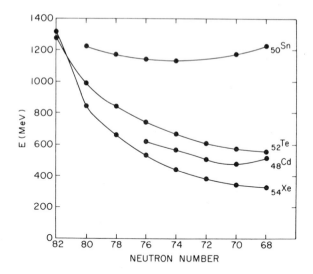

Fig. 1. $E(2_1^+)$ as a function of neutron number for the singly magic nucleus Sn and for nearby elements with a few valence protons.

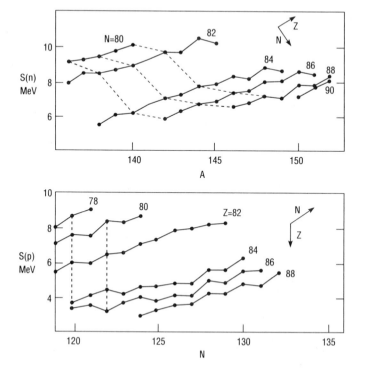

Fig. 2. Single nucleon separation energies.

their deep implications for the p-n interaction, though mentioned rather early by Talmi[2], are surprising little recalled. Figure 2 shows a selection of these for two mass regions. These plots, though simple and familiar, contain an interesting clue to the p-n interaction and an interesting contrast between its effects and those of the like-nucleon interaction. Inspection of Fig. 2 will show that the separation energy of a given type of nucleon *increases* with the addition of nucleons of the other type and *decreases* with the addition of nucleons of the same type. Thus, for example, S(n) increases with increasing proton number but decreases with increasing neutron number: that is, adding neutrons to a nucleus makes the last neutron less bound while adding protons makes it more bound. This has a simple and obvious implication, namely that, on average, the like-nucleon interaction must be repulsive whereas the unlike-nucleon (p-n) interaction is attractive. (This latter, of course, refers to its T=0 component since the T=1 component must be identical to the p-p and n-n interactions.) Correlations and collectivity therefore must stem in large part from the p-n interaction.

The p-n interaction can be viewed in many ways, one of which being its multipole expansion. In a multipole expansion, the monopole and quadrupole components are often dominant. These two components have complementary effects. In has been shown[9] that the effect of the monopole p-n interaction is to shift single particle energy levels of one kind of particle as a function of the number of nucleons of the other type. It is primarily for this reason that the ordering of Shell Model single particle levels is not uniform throughout the nuclear chart. A particularly dramatic example of this is seen in Fig. 3 which compares the single particle energies of ^{91}Zr, ^{131}Sn, and ^{207}Tl. The most dramatic change is the rapid descent of the $1g_{7/2}$ orbit between ^{91}Zr and ^{131}Sn. It is between these two elements, with Z=40 and 50, respectively, that the proton $1g_{9/2}$ orbit is filling. The monopole interaction is constant over all space and therefore the p-n matrix elements depend only on the radial overlaps of the proton and neutron wave functions, being largest for similar orbits. In fact, though rather crude, the approximation that the monopole p-n matrix elements scale as $1/(\Delta n + \Delta l + 1)$ is not terrible. Being attractive, the $p1g_{9/2}$--$n1g_{7/2}$ interaction between Zr and Sn leads to the lowering of these single particle energies, as clearly seen in Fig. 3. As Tl is approached, the next higher neutron and proton shells are

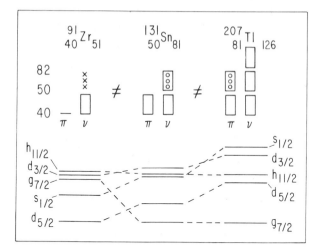

Fig. 3. Single particle energy levels in nuclei with one valence nucleon. The cartoons above each scheme indicate the filled levels in each nucleus. From Ref. 11.

filling and they have, on average, higher j values than the lower shells. Therefore there is a tendency toward more attractive interactions for higher j filled orbits than for lower j ones. This effect is also reflected in Fig. 3 where the higher j orbits are lowered relative to the lower j single particle levels.

The effect of the monopole p-n interaction is also particularly evident[4] in the Z=64 region but, before discussing that, a few comments on the quadrupole component are helpful. This component depends primarily on the angular *separation* of the proton and neutron orbits. Its strength varies as $P_2(\cos\theta)$ or $\cos^2\theta$ where θ is the angle between the proton and neutron orbital planes. Viewed in a deformed, or Nilsson, picture, this means that the quadrupole interaction will be strongest for Nilsson orbits with similar Ω or K values. In terms of the Nilsson diagram, this means that two downsloping or two upsloping orbits will have a larger attractive quadrupole p-n interaction than an upsloping and a downsloping orbit.

The quadrupole p-n interaction also depends however on the *absolute* angle of each orbit relative to the nuclear symmetry axis or nuclear equator. The reason for this is founded in the basic properties of even tensor interactions in the seniority scheme. In this scheme matrix elements of an even tensor operator are negatives of each other about mid-shell and, consequently, vanish at the mid-shell point. Although this applies specifically to a single j shell configuration, the basic effect persists[10] in the

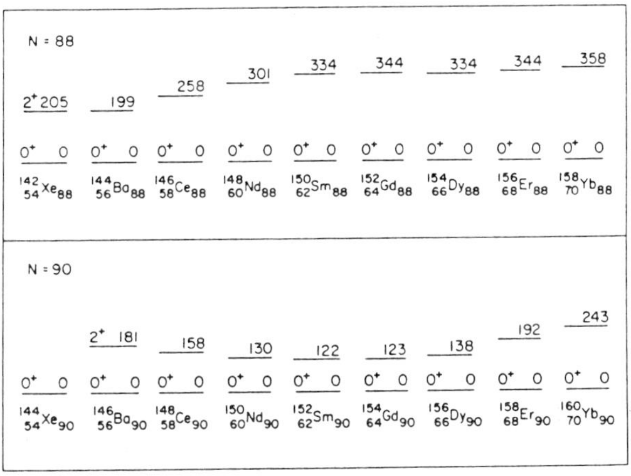

Fig. 4. $E(2^+_1)$ values for N = 88 and 90.

multi-j Nilsson scheme where flat Nilsson orbits have very small quadrupole moments. This variation of the quadrupole p-n interaction with orbit angle will have important consequences later. The quadrupole p-n interaction helps induce collectivity since it is rather long range and therefore grows with increasing number of valence protons and neutrons provided, of course, that the relative orientations of the orbits are suitable.

Finally, with this background, we can now look at the phase transitional region near A=150 and see another striking manifestation of the monopole p-n interaction. Figure 4 shows the systematics of $E(2^+_1)$ for the N=88 and 90 isotones. The development of collectivity toward mid-shell would suggest that $E(2^+_1)$ should decrease from Ba to Gd and this is exactly what happens for N=90. In contrast, however, the N=88 isotones display exactly the opposite behavior with a maximum near Gd. It is hard to reconcile this with the expected behavior across a normal major shell. The explanation, of course, is quite simple and, by now, well known[4]. At the beginning of the Z=50-82 shell there is a substantial subshell gap[12] at Z=64. On account of this, a nucleus such as Sm with Z=62 actually has, in effect, *fewer* valence protons than Ba (Z=56), and, therefore, quite naturally, a higher $E(2^+_1)$. As neutrons fill the $1h_{9/2}$ orbit near N=90, however, the strong attractive monopole interaction with the $p1h_{11/2}$ orbit lowers the energy of the latter and rather suddenly obliterates the Z=64 shell gap. This leads, just at N=90, to a recovery of the normal proton shell 50-82 and therefore to the "normal" behavior of $E(2_1^+)$ seen for N=90 in Fig. 4.

3. THE N_pN_n SCHEME

An attempt to reflect the importance of the valence p-n interaction in nuclear structure is the N_pN_n scheme[5] in which nuclear observables are plotted against this valence product. Examples of N_pN_n plots in comparison with traditional ones are shown in Figs. 5 and 6. The simplication wrought by this scheme is evident. The systematics of each observable fall on a smooth curve throughout a half shell. The rationale behind the N_pN_n scheme is that, if one assumes that the p-n interaction is orbit independent, then its integrated strength should scale as N_pN_n. We will see below an empirical test of this assumption as well as deviations from it near mid-shell. The N_pN_n scheme has been extensively discussed in the literature[5,13,14] and there is no need for a repetition here. Perhaps, however, it would be interesting to present an informal list of some areas in which the N_pN_n scheme has been applied. The list contains only those items known currently to the author:

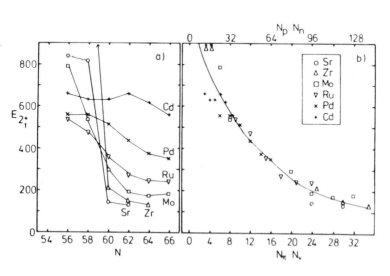

Fig. 5. Normal and N_pN_n plots. From Ref. 5.

- evolution of collectivity and the structure of phase/shape transitions
- subshell effects and effective valence nucleon numbers
- intruder states and shape coexistence
- simplification and parameterization of collective model calculations
- interpretation of Hartree-Fock calculations of phase transitions
- studies of monopole radiation and nuclear radii
- study of isovector M1 excitations, F-spin, the Majorana interaction
- tests of the N_pN_n scheme far from stability
- high spin states
- N_pN_n multiplets and comparisons of exotic pairs of nuclei
- odd-mass nuclei and odd-particle blocking effects in p-n interactions
- r-process nuclei
- heavy ion fusion reactions
- energy weighted sum rules for quadrupole and octupole excitations
- saturation of collectivity in deformed nuclei
- relation of monopole and quadrupole components of p-n interaction

and so on.

Given the recent emphasis and interest in nuclei far off stability both as a test of nuclear structure theories and as input to calculations of the nuclear r-process, the structure of unknown nuclei, currently inaccessible, becomes of some importance.

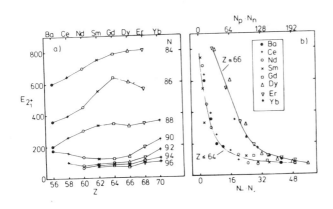

Fig. 6. Normal and N_pN_n plots. From Ref. 5.

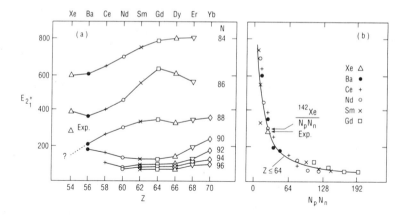

Fig. 7. Normal and N_pN_n predictions for $E(2_1^+)$ in the A = 150 region. A crude extrapolated guess for $E(2_1^+)$ in ^{142}Xe is indicated by the question mark on the left. On the right, the N_pN_n prediction is obtained simply by reading off the curve at N_pN_n = 24. The experimental value for ^{142}Xe is indicated on both sides. From Ref. 15.

Perhaps, therefore, it is useful to illustrate one application of the N_pN_n scheme, namely its use in predicting the properties of such nuclei.

Normally, the prediction of properties of nuclei far from stability is a process of *extrapolation* in N, Z or A with all its attendant risks. The N_pN_n scheme is useful here because it frequently converts the normal process of extrapolation to one of *interpolation* since N_pN_n values for many nuclei far from stability are in the same range as those for known nuclei in the same region. Therefore, the prediction of their properties in the N_pN_n scheme simply involves reading off the appropriate value for some observable on an already existing curve[13]. The idea is illustrated by the recently studied example of ^{142}Xe, whose level scheme has been deduced from fission product studies[15]. Figure 7 shows a normal and a N_pN_n plot for $E(2_1^+)$ in the rare earth region. If one tries to predict the 2_1^+ energy for ^{142}Xe using normal systematics, there arises an immediate ambiguity. The first, most tempting, approach is to follow the N=88 curve, giving the approximate prediction indicated by the dashed line and question mark on the left in Fig. 7. However, this simple extrapolation ignores the fact that, with decreasing Z in this region, one is approaching the Z=50 shell closure and that, at some point, the rigidity associated with the magic numbers should lead to a rise in $E(2_1^+)$, rather than a continued drop. Since, a priori, there is no simple way to anticipate when this will happen in a normal plot, there is considerable uncertainty in the actual value for $E(2_1^+)$: one can only guess that it should fall somewhere between 150-300 keV. The N_pN_n scheme, in contrast, *automatically* balances the competition between increased collectivity and the approach to magicity since this is, in fact, controlled by the p-n interaction. The N_pN_n prediction (see Fig. 7, right) is almost exactly correct.

Clearly, the N_pN_n scheme is very useful in considering the systematic behavior of nuclear structure and observables in a given region: that is, *changes in* N_pN_n are correlated with *changes* in structure. However, N_pN_n is a relative quantity, and this complicates the comparison of different mass regions. To facilitate such comparisons, it is useful to introduce[6] a normalized form of N_pN_n, the so-called P factor defined as

$$P = \frac{N_pN_n}{N_p + N_n} \quad (1)$$

The P factor has an obvious physical interpretation, namely as the number of valence p-n interactions per like-nucleon interaction or, equivalently, the average number of p-n interactions per valence nucleon. Normally, the behavior of different mass regions appears rather different when observables are plotted against N, Z or A. Given the importance of the p-n interaction, though, one might expect an underlying similarity in different transition regions. This similarity emerges when the same data are plotted against the P factor. This is shown for the energy ratio $E(4_1^+)/E(2_1^+)$ in Fig. 8. The figure shows that, not only does each region behave very similarly to the others (with the slight exception of the actinides), but all regions also fall within a relatively narrow envelope and each passes through a phase transition between P=4-5. The similarity of the P-factor curves is interesting in itself but this critical value of P is even more revealing. This is evident if P is expressed, as above, as the ratio of the number of p-n interactions to the number of like-nucleon interactions and if it is recalled that typical p-n interaction strengths are 200-300 keV while the pairing interaction is around 1 MeV. Thus, P_{crit}=4-5 corresponds to exactly that region in a shell where the p-n interaction begins to dominate the pairing interaction. This result[6]

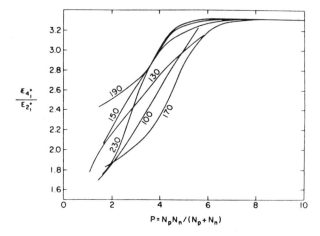

Fig. 8. Plot of $E(4_1^+)/E(2_1^+)$ against the P-factor for six mass regions of heavy nuclei. From Ref. 6.

highlights once again the intimate connection between the p-n interaction and collectivity on the one hand, and the utility of N_pN_n and the P factor as measures of the integrated p-n strength on the other.

4. EMPIRICAL VALIDATION OF THE N_pN_n SCHEME

[Note: The bulk of this section and the next is taken from Refs. 7, 8 and 16. I am grateful to my collaborators D. S. Brenner, J.-Y. Zhang, W.-T. Chou, D. D. Warner, and C. Wesselborg for this material.]

The N_pN_n scheme was originally motivated by the recognition of the importance of the valence p-n interaction and is justified by its practical utility. It is based on the assumption that the integrated p-n interaction strength is a simple function of N_pN_n. Though not rigorously required, it is a common presumption that this dependence is more or less linear. Given the success of the N_pN_n scheme, it would be useful to test this assumption empirically. To do this, one needs to extract empirical values for specific p-n matrix elements and demonstrate their approximate orbit independence or, at least, a scaling of their integral with N_pN_n. This has recently been done. Zhang et al.[8] have discussed the use of a specific double difference of binding energies that isolates the p-n interaction of the last valence proton with the last valence neutron. This p-n interaction, denoted δV_{pn}, is defined by

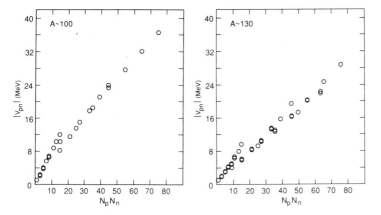

Fig. 9. Plot of integrated empirical p-n interaction V_{pn}, obtained by summing Eq. 2, against N_pN_n. From Ref. 8.

$$\delta V_{pn} = \frac{1}{4}\left[\,(\,B(Z+2,N+2) - B(Z+2,N)\,) - (\,B(Z,N+2) - B(Z,N)\,)\,\right] \quad (2)$$

where $B(Z,N)$ is the (negative) binding energy of the nucleus Z,N. This double difference is designed to subtract off mean field and like-nucleon interactions. It is discussed more extensively in Ref. 8 where its antecedents in existing mass equations are also discussed. Below, we shall present plots of individual δV_{pn} values over the entire Periodic Table. Here, however, we are more interested in sums of δV_{pn} values over all valence protons and neutrons and in plotting these sums as a function of N_pN_n. The data to do this exists in two mass regions. We show these in Fig. 9 where the integral of δV_{pn}, denoted V_{pn}, is plotted against N_pN_n. The result is striking: there is an almost exact linearity in the relationship. [The one exception to this trend are the single points at $N_pN_n = 15$ but these are truly exceptions which prove the rule since these nuclei corresponds to N_p and N_n values of 1 and 15 in comparison to nuclei with N_p and $N_n = 3$ and 5.] Of particular note in Fig. 9 are the nearly overlapping circles occurring for the same N_pN_n values. These correspond to nuclei with different separate values of N_p and N_n but the same valence product. The virtually identical values of V_{pn} strongly support the idea of the approximate linearity of integrated p-n strength with N_pN_n at least up to about the one-third filled part in the shell. These results provide an empirical microscopic, underpinning to the rationale behind the N_pN_n scheme and give it a conceptual credence beyond its mere practical utility.

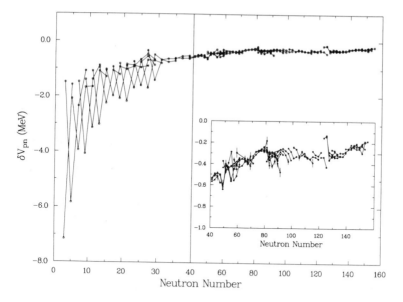

Fig. 10. δV_{pn} data for all even-even nuclei. From Ref. 7.

5. EMPIRICAL P-N INTERACTION MATRIX ELEMENTS

As stated just above, empirical p-n matrix elements of the last proton with the last neutron have been extracted for all even-even nuclei where the mass data are available. The results are plotted in Fig. 10 and show a number of fascinating features[7]. To discuss these, note first that larger (attractive) p-n interaction strengths correspond to *lower* values on the plot. Ignoring, for a moment, the obvious singularities that occur for certain light nuclei, we see that, globally, there is a rather smooth trend toward smaller and smaller values with increasing mass. This has an obvious, and well known, explanation: the p-n matrix elements are sensitive, in one way or another, to the overlaps of the respective orbits. For a residual two-body interaction of constant radius, these overlaps will decrease as the radii of the respective Shell Model orbits increase and the wave functions become more spread out in space. A second reason reinforces this, namely, that the neutron excess in heavy nuclei means that neutrons are filling the next higher shell beyond that for the protons, thereby increasing further the difference in radii and decreasing the overlaps.

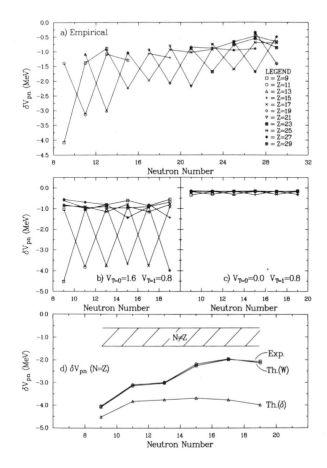

Fig. 11. δV_{pn} values in the region of light N = Z nuclei. From Ref. 7. Top: Expanded view of the empirical δV_{pn} values of Fig. 10. Middle: Calculations with a surface δ interaction. Bottom: Comparison of the locus of N = Z δV_{pn} values with calculations using a surface δ interaction [Th(δ)] and the Wildenthal-USD interaction [Th(W)]. From Ref. 7.

Besides this secular trend there are two obvious structural features of the plot. The most dramatic is the set of enormous spikes in δV_{pn} for certain light nuclei. Inspection of the data shows that, in each case, these are specifically those nuclei with N=Z. An explanation for this apparently anomalous behavior is readily at hand. It is implicitly contained[2] in typical parameterizations of nuclear masses which contain a T(T+1) term, which gives an extremum for T=0 characterizing N=Z nuclei. A more microscopic interpretation can be obtained[7] with simple Shell Model calculations. The region of N=Z nuclei from Fig. 10 is expanded in Fig. 11 on the top. In the middle panels we present the results of two Shell Model calculations, carried out in the 2s-1d shell (N=8-20) with a surface δ interaction. The single particle energies were chosen to approximately fit the spectra of ^{17}O and ^{17}N while the strength of the T=1 part of the two-body interaction approximately reproduces the spectra of ^{18}O and ^{18}N. Calculations are shown for two values of the T=0 strength. On the right the results are shown for $V_{T=0} = 0$: clearly these calculations cannot reproduce the empirical results. This is not surprising since it is well known that the T=0 strength should be substantially greater than the T=1 strength. The calculation on the left uses a more realistic T=0 strength, namely, $V_{T=0} = 2V_{T=1}$. These calculations, even though the space is highly restricted and the force is a schematic surface δ interaction, reproduce the data remarkably well. They yield spikes exactly at N=Z and of roughly in the right order of magnitude. The only discrepancy with the data is that the spikes tend to be of rather constant magnitude whereas the empirical spikes decrease in magnitude with increasing A across the shell.

The bottom panel shows a calculation using a more realistic interaction, namely the Wildenthal universal 2s-1d interaction[17] (USD) which fits experimental data in the sd shell remarkably well. In this panel, the δV_{pn} values for the band of nuclei with N≠Z are schematically indicated by the diagonally marked band while the locus of N=Z spikes is indicated by the points lying below this band. The calculations (just described) corresponding to a surface δ function interaction and those from the Wildenthal interaction are indicated. It is seen that the latter exactly reproduces the empirical results. Partly this is due to an A-dependence of the USD-Wildenthal interaction but partly it is due to a more complex set of two-body matrix elements. Though the USD interaction is already well established[17], these results provide another test of it which is perhaps more sensitive than the binding energies or masses alone since δV_{pn} represents the rather small resultant of a double difference of binding energies from four adjacent even-even nuclei. The main point of this discussion, however, is that it is the T=0 interaction which is primarily responsible for the N=Z spikes. The reproduction of the N=Z singularities both in the highly schematic calculations and in those with a realistic interaction reflects its insensitivity to small details of the wave functions and its origin in very basic features of two-nucleon wave functions, in particular the strong p-n

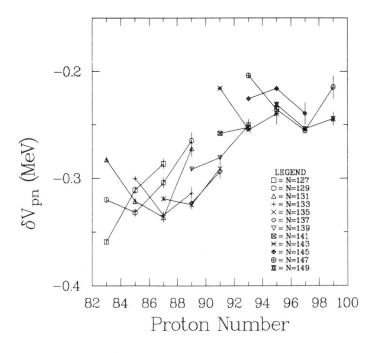

Fig. 12. Experimental δV_{pn} values for the actinides. From Ref. 7.

interactions in N=Z nuclei with protons and neutrons in equivalent orbits characterized by enhanced spatial symmetry of their wave functions.

The microstructure in heavier nuclei, that is, the detailed fluctuations in δV_{pn} is no less interesting and, from it, we can learn much about the residual p-n interaction. We saw earlier that the integrated p-n interaction[8] is almost exactly linear in $N_p N_n$ early in a shell. This reflects the approximate orbit independence of the p-n interaction. Except for the N=Z spikes, this is evident in Fig. 10. However, there is evident microstructure in the figure as well and, moreover, one has an a priori expectation that the p-n interaction will decrease somewhat toward mid-shell. The reason is very simple and related to a point we made earlier about the role of monopole and quadrupole components in this interaction and the behavior of the latter near mid-shell as a reflection of the properties of even tensor operators in a single j seniority scheme. Specifically, one expects the quadrupole component to become extremely small near mid-shell in heavy nuclei. The data in Fig. 10 seem to reflect this phenomena. This is shown in Fig. 12 for the actinide nuclei[7] where one sees a gradual and relatively

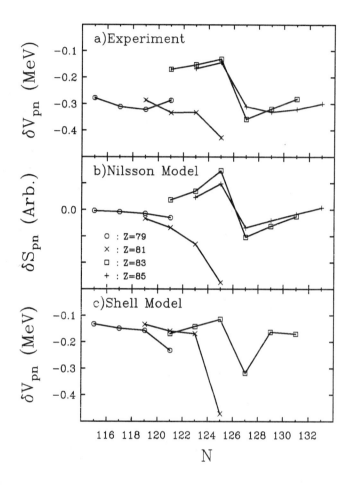

Fig. 13. p-n interaction strengths near ^{208}Pb. Top: Expanded view of the experimental δV_{pn} values from Fig. 10. Middle: Nilsson Model calculations of the quadrupole interaction, δS_{pn}, of the last proton with the last neutron. Bottom: Shell Model calculations of δV_{pn}. From Ref. 16.

systematic decrease in the magnitude of δV_{pn} with increasing proton number and, at the extreme right of the plot, an apparent levelling off to an asymptotic value near-250 keV. Although the data itself does not prove the following interpretation, it seems likely that this behavior reflects the presence of both monopole and quadrupole components near the beginning of a shell and the decrease of the quadrupole component toward mid-shell, leaving the dominant monopole component in the latter region. If this is true, the figure suggests an average value of ~ -250 keV for this component, which is roughly compatible with the known magnitudes of the single particle energy shifts which it induces. The behavior in Fig. 12 for the actinide nuclei is also reflected in the rare earth region[8] where it gives a similar estimate for the monopole interaction.

This is but one aspect of the microstructure in p-n interactions and the kind of information it can provide. Other fluctuations are both more dramatic and more easily interpretable in a quantitative way. Moreover, such an interpretation strongly supports the notion that the p-n interaction matrix elements can be understood in terms of very simple overlap arguments. The fluctuations referred to are near doubly magic nuclei. The basic idea is repeated in several such regions. The most dramatic is the Pb region where the sudden changes in δV_{pn} are even evident in the unexpanded portion of Fig. 10. Figure 13 shows an expanded view of the lead region along with two sets of calculations[16] to be discussed momentarily. The empirical results show dramatic changes in δV_{pn} on different sides of ^{208}Pb. Specifically, when protons and neutrons are either both above or both below the N=82 and Z=126 magic numbers, the p-n interactions are large. When the proton and neutron numbers are on opposite sides of these two magic numbers, the p-n interactions drop markedly. As noted, this has a simple "overlap" explanation. Single particle Shell Model orbits, labelled by the quantum numbers (n,l,j) at the end of a major shell are typified by relatively high n and low l and j whereas, at the beginning of a new shell, the single particle states are of lower n and high l and j. Thus, even using the simple estimate given earlier concerning the magnitudes of p-n matrix elements, it is clear that the overlap of a pair of orbits on one side or on the other of a major shell will be high whereas there will be a small overlap of the low j and high j orbits appearing on opposite sides of a pair of magic numbers.

This qualitative explanation of the fluctuation patterns in p-n matrix elements near ^{208}Pb, and in other doubly magic regions, can be quantified and substantiated by both Nilsson and Shell Model calculations. We illustrate this in Fig. 13. The idea behind the Nilsson Model calculations is that the proton and neutron quadrupole moments, and hence the quadrupole p-n interaction, is largest just below and above closed shells and therefore that the fluctuations in p-n matrix elements near closed shells should be largely due to this component. These nuclei are nearly spherical and the calculations were done with a (nearly arbitrary) deformation parameter of $\varepsilon = 0.01$. For small deformations, the calculated *fluctuations* in p-n matrix elements are nearly independent of ε. Since, however, these matrix elements scale approximately as the square of the deformation for nearly spherical nuclei, the arbitrariness in ε merely means that the p-n matrix elements are calculated on an arbitrary scale. [In any case a negative monopole contribution of approximately -250 keV needs to be added to the Nilsson calculations as well.] The point of these last remarks is simply to emphasize that it is the calculated *fluctuations* in δV_{pn} with the Nilsson Model that are the essential quantities to consider. The fact that they oscillate about 0, however, does mean that, when the monopole contribution is added, average δV_{pn} values of approximately the right magnitude will be obtained. The calculations were done[16] with standard pairing and Nilsson parameters. It is clear that they reproduce empirical values extraordinarily well. Not only are the sudden jumps around ^{208}Pb reproduced, but also many of the fine details of the microstructure are represented in the calculations. It is interesting to compare these calculations with those in the Shell Model since the two approaches are rather complementary: the Nilsson Model embodies a rather large space of single particle levels (at least two major shells) but a highly schematic force, whereas the Shell Model calculations must, of necessity, be truncated to an extremely small space but they incorporate a more realistic, surface δ interaction discussed in Ref. 18. Such Shell Model calculations are presented in the bottom panel of Fig. 13. Below the Pb magic numbers, the basis states include the three low j Shell Model orbits while, above Z=82 and N=126, only a single orbit is incorporated[16]. However, test calculations[16] for different choices of that orbit show that similar results are obtained.

There are two interesting results from these calculations. First, as with Nilsson Model, the Shell Model nicely reproduces most of the effects seen experimentally. The exception is that the gap in δV_{pn} across Z=82 for N<126 is much smaller in the calculations than in the data. Aside from this, though, the empirical features are well reproduced. The second aspect of these calculations, as opposed to those in the Nilsson scheme, is that they are on an absolute scale and automatically include all relevant multipoles in the space considered As a consequence, the calculated absolute magnitudes of δV_{pn} can be compared with the data. One sees that the calculated values between -200 and -400 keV are in excellent agreement with the data. It is interesting to note, in recalling the Shell Model calculations for δV_{pn} in N=Z nuclei discussed above, that we now have calculations in two widely diverse mass regions[7,16] that can explain empirical δV_{pn} values on widely different magnitude scales.

The fact that the Nilsson Model, with extended space and schematic force, and the Shell Model, with restricted space and much more realistic interaction, both reproduce the data and give very similar results one to the other, strongly suggests that the origin of the fluctuations in δV_{pn} in the Pb region resides not in some small details of the wave functions but rather in the substantial changes occurring in the gross proton and neutron wave function overlaps as the doubly magic region is traversed. This supports the qualitative arguments made earlier and confirms the essentially simple nature of the underlying physics behind the fluctuations and microstructure in p-n matrix elements in heavy nuclei. Similar calculations for the A=90 and 140 regions also produce excellent agreement with the data in the Nilsson scheme.

6. CONCLUSIONS

A number of issues relating to the p-n interaction in nuclear structure have been discussed, including its manifestations in very simple nuclear data, the different roles of its separate multipoles, the $N_p N_n$ scheme and the P factor, their validation through empirical integrated valence p-n interaction strengths, the global and regional behavior of individual p-n matrix elements, including singularities for N=Z nuclei, the saturation of quadrupole p-n interactions near mid-shell, and strong fluctuations in closed shell regions, along with their interpretation in terms of Nilsson and Shell Model calculations.

ACKNOWLEDGEMENTS

Research has been performed under contract No. DE-AC02-76CH00016 with the United States Department of Energy. I am grateful to my collaborators in much of this work, D. S. Brenner, W.-T. Chou, J.-Y. Zhang, D. D. Warner, and C. Wesselborg. I would also like to thank K. Heyde, I. Talmi, and E. K. Warburton for many useful discussions.

REFERENCES

1. A. de Shalit and M. Goldhaber, Phys. Rev. **92**, 1211 (1953).
2. I. Talmi, Rev. Mod. Phys. **34**, 704 (1962); *Progress in Particle and Nuclear Physics, Collective Bands in Nuclei*, Vol. 9, ed. D. Wilkinson (Pergamon, Oxford, 1983), p. 27; *Interacting Bosons in Nuclei*, ed. F. Iachello (Plenum Press, New York, 1979), p. 79.
3. P. Federman and S. Pittel, Phys. Lett. **69B**, 385 (1977) and Phys. Lett. **77B**, 29 (1978).
4. R. F. Casten et al., Phys. Rev. Lett. **47**, 1433 (1981).
5. R. F. Casten, Phys. Lett. **152B**, 145 (1985); Phys. Rev. Lett. **54**, 1991 (1985); Nucl. Phys. **A443**, 1 (1985).
6. R. F. Casten, D. S. Brenner, and P. E. Haustein, Phys. Rev. Lett **58**, 658 (1987).
7. D. S. Brenner et al., Phys. Lett. **B243**, 1 (1990).
8. J.-Y. Zhang, R. F. Casten, and D. S. Brenner, Phys. Lett. **B227**, 1 (1989).
9. K. Heyde et al., Nucl. Phys. **A466**, 189 (1987).
10. R. F. Casten, K. Heyde, and A. Wolf, Phys. Lett. **B208**, 33 (1988).
11. R. F. Casten, *Nuclear Structure from a Simple Perspective* (Oxford, New York, 1990), Chapters 4 and 5.
12. M. Ogawa et al., Phys. Rev. Lett. **41**, 1480 (1978); Y. Nagai et al., Phys. Rev. Lett. **47**, 1259 (1981).
13. R. F. Casten, Phys. Rev. **C33**, 1819 (1986).
14. K. Heyde and J. Sau, Phys. Rev. **C33**, 1050 (1986).
15. A. S. Mowbray et al., Phys. Rev. **C42**, 1126 (1990).
16. W.-T. Chou et al., to be published.
17. B. A. Brown and B. H. Wildenthal, Ann. Rev. Nucl. Sci. **38**, 29 (1988); B. H. Wildenthal, private communication to E. K. Warburton.
18. N.A.F.M. Poppelier and P.W.M. Glaudemans, Z. Phys. **A329**, 275 (1988).

Physics of superdeformed bands

WITOLD NAZAREWICZ

Institute of Theoretical Physics, University of Warsaw
ul. Hoża 69, PL-00681 Warsaw, Poland

Institute of Physics, Warsaw University of Technology
ul. Koszykowa 75, PL-00662 Warsaw, Poland

Abstract

In the year 1986 new types of exotic states, the so-called superdeformed (SD) states, were discovered in fast rotating heavy atomic nuclei. They are characterized by an unusually large distortion (comparable to that appearing in fissioning isomers) and occur at high angular momenta. A sizeable amount of high-spin data on SD states makes it possible, for the first time, to test many fine details of the shell structure at very large elongations. Moreover, it offers an opportunity to probe exotic orbitals that appear at normal deformations at completely different regions of particle number and rotational frequency.

1 Introduction

The existence of nuclear rotational excitations is directly related to a nonisotropic distribution of nuclear matter in a nucleus, i.e. the nuclear deformation. Rotational couplings arising from the Coriolis and centrifugal forces acting on the nucleons in the intrinsic (rotating) frame greatly modify the single-particle motion. Consequently the original shape of a nucleus has to change in order to minimize the total nuclear energy.

According to the simplest model of a rotating nucleus, the rotating liquid drop model [1], nuclei at high angular momentum are expected to undergo a sharp shape transition (superbackbending) from moderately deformed oblate shapes to very well deformed (superdeformed) prolate or triaxial shapes before fissioning at still higher spins. This *macroscopic* rotational behaviour which is fully determined by a competition between the surface tension on the one hand and the Coulomb and centrifugal forces on the other hand, is modified by the *microscopic* shell effects. The quantum-mechanical shell correction [2], strongly fluctuating with particle number and angular momentum, plays a very important role in determining the excitation energy of superdeformed (SD) structures. Due to the pronounced shell effects seen in the single-particle spectrum at very elongated shapes [3] certain particle numbers can stabilize SD shapes at high spin. It was pointed out more than decade ago [4-7] that the best prospects where to find SD rotational bands are the light rare-earth nuclei around ^{152}Dy, i.e. the nuclei with the particle numbers $62 \leq Z \leq 68$ and $80 \leq N \leq 88$. In the subsequent papers [8-13] detailed predictions were made about the position and the structure of SD bands expected in the nuclei from this region.

The first reliable hint pointing to the possible existence of high-spin nuclear states at extremely large deformation was obtained by means of experiments using the γ-γ energy correlation method in the so-called quasicontinuum spectroscopy. The γ-γ correlation experiments performed in 1982 [12] and 1984 [14] for the nucleus ^{152}Dy (Z=66, N=86) have indeed shown the existence of two ridges with a small separation in the two-dimensional plot of the γ-γ correlations, suggesting a very large moment of inertia for certain nuclear states. The estimated values of the quadrupole deformation β_2 was of the order of 0.6, i.e. very close to the early-predicted values for superdeformation. These observations were corroborated afterwards by lifetime measurements in the quasicontinuum [15].

At this point one has to mention the advent of a new generation of escape-supressed spectrometers in the beginning of the eighties. The multi-detector systems consisting of germanium detectors (having very high energy resolution for γ-rays) sorrounded by scintillator shields (made of NaI, BGO or BaF$_2$ scintillators) operating in anti-coincidence with the germanium detectors (and thus reducing the Compton background) have revolutionized the field of γ-ray spectroscopy [16]. In the above-mentioned quasicontinuum study [14] one of the first such spectrometers, the Daresbury detector TESSA2, was used. A real turning-point in the field of nuclear superdeformation was the direct observation of a discrete SD rotational band in ^{152}Dy by the Daresbury-Liverpool-Copenhagen group in 1986 [17] using the spectrometer TESSA3. The distance between two adjacent γ-rays coming from a SD nuclear band in ^{152}Dy is almost constant and equal to $\Delta E_\gamma \approx 47$ keV ($\beta_2 \approx 0.6$). This result has turned out to be in a perfect agreement with the result of a previous estimate derived from the γ-γ correlations.

The recent discovery of superdeformed (SD) bands in ^{191}Hg [18] was a nice confirmation of various theoretical predictions (see e.g. [19-23]) suggesting the presence of strongly elongated configurations in the mass A$\sim$190 region. Contrary to the A$\sim$150 mass region (where the SD bands are built upon aligned many-quasiparticle configurations) the calculations reveal the appearance of SD minima already at $I=0$. Consequently, in the Hg-Pb mass region the lowest observed spins in SD bands are as low as about 10 $\hbar$. The observation of SD states at low spins inspired many experimental groups to study the properties of well-deformed shape isomers in this region.

To date the SD rotational bands have been observed in a number of nuclei from the A$\sim$150 and A$\sim$190 mass regions. The systematics of experimental dynamical moments of inertia, $\Im^{(2)} = dI/d\omega \approx 4/\Delta E_\gamma$ versus rotational frequency, $\hbar\omega = \frac{1}{2}E_\gamma$, is

177

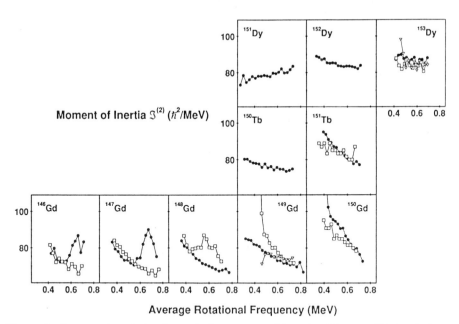

Figure 1: Systematics of the dynamical moments of inertia of known SD bands in the A~150 mass region. (From [24].)

shown in fig. 1 (Gd-Dy region) and fig. 2 (Hg-Pb region). Although for such strongly deformed systems one would expect a rather similar, rigid rotor like, rotational behaviour, it is seen that the observed rotational patterns show pronounced variations with particle number and angular momentum.

In the state of angular momentum $60\hbar$ the nucleus makes around 10^{20} rotations per second. In order to slow down the rotational motion it emits the very fast electric quadrupole (E2) γ-radiation. In 1987 it became possible to determine experimentally the reduced transition probability B(E2) within a SD band in ^{152}Dy [25]. The quadrupole transitions connecting SD states turned out to be super-collective, of the order of 2500 W.u. and the resulting value of the quadrupole deformation β_2 deduced from the transition quadrupole moment $Q_t \approx 19$ eb has coincided very well with the previous estimates. The transition quadrupole moments of SD bands have also been found in ^{149}Gd [26], ^{150}Gd [27] and ^{190}Hg [28], ^{191}Hg [18,29] and ^{192}Hg [30]. The quadrupole deformations extracted from measured transition quadrupole moments are consistent with theoretical predictions for equilibrium shapes in SD configurations [30,31].

New possibilities in the field of superdeformation have been opened up by the discovery of multiple SD bands within one nucleus [32]. The excited SD configu-

Superdeformation : A=190 Region.

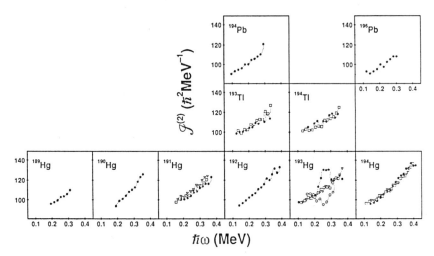

Figure 2: Systematics of the dynamical moments of inertia of known SD bands in the A~190 mass region (courtesy of D.M. Cullen).

rations turned out to have markedly different structures as compared with the SD yrast bands, suggesting a change in the intrinsic configurations. However, the most remarkable and puzzling issue is the recent observation of identical γ-ray sequences in the pairs of nuclei ^{151}Tb-^{152}Dy, ^{150}Gd-^{151}Tb [33] and ^{150}Tb-^{149}Gd [34]. Such *twinned* SD bands have also been found in the Hg-Pb region [35,36].

The main objective of this paper is to discuss recent developments in the physics of nuclear superdeformation. In particular, some questions related to the structure of twinned superdeformed bands are addressed. It is demonstrated that the spectroscopy of superdeformed nuclei is a very powerfull tool for probing certain Nilsson levels. A comment is made on the possible quenching of pairing correlations at large elongations. Finally, evidences for the presence of low-lying octupole correlations at 2:1 shapes are briefly discussed.

2 Shell structure of deformed harmonic oscillator at very deformed configurations

The most important source of trends leading to the formation of nuclear SD states is essentially derived from the quantal nature of the nucleus which manifests itself through the shell structure in the nucleus. This phenomenon is best understood in terms of the irregular distribution of the quantal energy levels in the nucleus forming "shells", i.e. bunches of close-lying states.

As is well known, the nuclear shell structure succesfully explains most nuclear features such as the occurence of especially stable systems (magic nuclei), the abundance of isotopes in nature, etc. What is, however, crucial in the search for the origin of superdeformation is the observation that pronounced shell structure may also appear in the extreme conditions of very large distortions. It has turned out that some of the idealized quantal many-body systems may also favour the very distorted shapes in addition to the well-known shell structure at the spherical shape.

In this section I shall discuss some properties of the single-particle shell structure of an axially-deformed harmonic oscillator. In particular, I will show how the "supershell" structure, which appears at the $k:l$ shape, may be understood and in particular how it can be related to the multi–"cluster" picture. The following discussion is based on the results of refs. [3,11], but some new aspects will also be pointed out.

The single-particle Hamiltonian of the axially-symmetric oscillator potential is given by

$$\hat{H} = \hbar\omega_\perp(\hat{a}_x^+\hat{a}_x + \hat{a}_y^+\hat{a}_y + 1) + \hbar\omega_z(\hat{a}_z^+\hat{a}_z + \frac{1}{2}). \tag{1}$$

The single-particle energies of (1) are

$$e_{n_\perp,n_z} = \hbar\omega_\perp n_\perp + \hbar\omega_z n_z + \frac{3}{2}\hbar\omega_o, \tag{2}$$

where $\omega_o = \frac{1}{3}(2\omega_\perp + \omega_z)$. Due to the axial symmetry each level is $(2s+1)(n_\perp+1)$ times degenerate. By introducing the deformation parameter $\varepsilon=(\omega_\perp - \omega_z)/\omega_o$ eq. (2) can be written as

$$e_{n_\perp,n_z} = \hbar\omega_o\Big[N + \frac{3}{2} - \frac{1}{3}\varepsilon(2n_z - n_\perp)\Big]. \tag{3}$$

Here $N = n_\perp + n_z$ is the total number of oscillator quanta (the principal oscillator quantum number). The single-particle diagram of the axially-deformed harmonic oscillator is displayed in fig. 3. According to eq. (3) the slope of the single–particle energy level versus deformation, the single-particle quadrupole moment, is proportional to $2n_z - n_\perp = 3n_z - N$. Consequently, the state with $n_z=N$ carries the greatest quadrupole moments within a given oscillator shell.

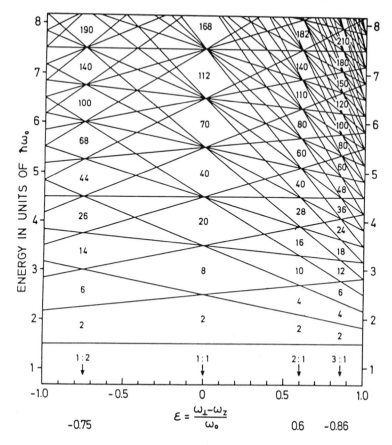

Figure 3: Single-particle level spectrum of the axially symmetric harmonic oscillator shown as a function of quadrupole deformation ε. The orbital degeneracy is $2(n_\perp +1)$. The arrows indicate the characteristic deformations corresponding to the ratio of $\omega_\perp : \omega_z$ = 1:2, 1:1, 2:1 and 3:1. (From [11].)

In the following, the oscillator deformation ε will be expressed in terms of the ratio ζ:

$$\zeta = \frac{\omega_\perp}{\omega_z}, \qquad \varepsilon = \frac{3(\zeta - 1)}{2\zeta + 1}. \tag{4}$$

It is immediately seen from eq. (2) that orbitals with the same value of $\zeta n_\perp + n_z$ are degenerate. The strongest degeneracy occurs when ζ is a rational number [3,37,38]. Since our discussion is focused on superdeformed ($\zeta=2$) or hyperdeformed ($\zeta=3$) prolate states it will be assumed that $\zeta=k$ is an integer number (the generalisation

to other cases is straightforward).
Let us classify the eigenstates (2) using the new quantum numbers Λ and Λ_o:

$$\Lambda = \frac{1}{2}(n_\perp + n_z), \quad \Lambda_o = \frac{1}{2}(n_z - n_\perp). \qquad (5)$$

For each value of k the single-particle levels (2) are bunched forming equidistant shells characterized by deformed principal quantum numbers $N_{shell}^{(k)}$:

$$N_{shell}^{(k)} = (k+1)\Lambda - (k-1)\Lambda_o, \qquad (6)$$

where, according to eq.(5),

$$-\Lambda \le \Lambda_o \le \Lambda. \qquad (7)$$

If $k > 1$ each shell $N_{shell}^{(k)}$ contains states $|\Lambda, \Lambda_o\rangle$ originating from different spherical shells N. It follows immediately from eq. 6 that for each $N_{shell}^{(k)}$ there always exists a state with the maximum value of $\Lambda = \Lambda_{max}$ and $\Lambda_o = \Lambda_{max}$, where

$$2\Lambda_{max} = N_{shell}^{(k)} = N. \qquad (8)$$

It is seen, therefore, that the deformed shell $N_{shell}^{(k)}$ always contains the orbital $|\Lambda_{max}, \Lambda_{max}\rangle$ originating from the spherical shell $N = N_{shell}^{(k)}$. This orbital, labelled alternatively as $|n_\perp = 0, n_z = N\rangle$ has the greatest quadrupole moment within $N_{shell}^{(k)}$.

The deformed shell (6) obviously contains states with various values of Λ and Λ_o. Since the variation of Λ and Λ_o within the deformed shell is equal to $(k-1)/2$ and $(k+1)/2$, respectively, all states belonging to $N_{shell}^{(k)}$ are characterized by

$$\Lambda = \Lambda_{max} - \frac{k-1}{2}r, \quad \Lambda_o = \Lambda_{max} - \frac{k+1}{2}r, \qquad (9)$$

where $r = 0, 1, ..., M$. The maximum value of r, denoted as M, is given by the condition (7), which leads to

$$M = \left[\frac{N_{shell}^{(k)}}{k}\right] \qquad (10)$$

($[a]$ indicates, as usual, an integer part of a). Consequently, $N_{shell}^{(k)}$ can be expressed as

$$N_{shell}^{(k)} = kM + \kappa, \quad \kappa = 0, 1, ..., k-1. \qquad (11)$$

In view of the above it is clear that for the $k:1$ shape characteristic modulations of the shell structure would occur. The number of $N_{shell}^{(k)}$-shells having the same value of M is equal to k, see eq. (10). Each shell contains $M+1$ levels labelled with the index r. The total number of states within $N_{shell}^{(k)}(M)$ is thus equal:

$$n(k, M) = \sum_{r=0}^{M} 2\left(\Lambda_{max} - r\frac{k-1}{2} - \Lambda_{max} + r\frac{k+1}{2} + 1\right) = (M+1)(M+2). \qquad (12)$$

This *supershell* structure can be nicely illustrated in terms of classical closed trajectories [3]. By assuming the volume conservation, $\omega_\perp^2 \omega_z = \bar{\omega}_o^3$, the difference between neighbouring oscillator shells

$$\hbar\omega_{shell} = \hbar\omega_z = \hbar\omega_o\left(1 - \frac{1}{3}\varepsilon\right) = \bar{\omega}_o k^{-2/3}, \tag{13}$$

decreases smoothly with deformation. This indicates that the overall magnitude of the shell effects is expected to be strongest at the spherical shape (k=1).

We are now in a position to calculate the total number of particles in the levels from M=0 up to and including the κ sets with quantum number M:

$$\begin{aligned}\mathcal{N}^{(k)}(M,\kappa) &= k\sum_{m=0}^{M}(m+1)(m+2) - (k-\kappa-1)(M+1)(M+2) \\ &= (k-\kappa-1)N_{M-1} + (\kappa+1)N_M,\end{aligned} \tag{14}$$

where

$$N_M = \frac{1}{3}(M+1)(M+2)(M+3) \tag{15}$$

is the number of particles in a spherical magic nucleus with levels up to and including the N=M oscillator shell filled (N_M=2, 8, 20, 40, 70, 112, 168, ...). Consequently, the total number of particles $\mathcal{N}^{(k)}(M,\kappa)$ is formally equal to sum of the total particle number of (k-κ-1) individual spherical nuclei N_{M-1} and (κ+1) spherical nuclei N_M. Of course, the total number of spherical "clusters" is k. This result offers a nice interpretation to the deformed gaps occuring at $k : 1$ shapes. Let us consider some examples.

Superdeformed shapes, k=2, fig. 4. In this case κ=0 or 1. Systems with κ=1 correspond to two aligned spherical "clusters" of *equal size** (symmetric necking: 4=2+2, 16=8+8, 40=20+20, 80=40+40, etc.).

Systems with $\kappa = 0$ correspond to two sperical "clusters" of *unequal size* in a row (10=8+2, 28=20+8, 60=40+20, 110=70+40, etc.). This indicates that the every second SD magic number should be unstable with respect to mass-assymetric distortions [11], see sect. 9.

Hyperdeformed shapes, k=3, fig. 5. Here we have to consider three situations corresponding to κ=0, 1, or 2. For κ=2 the hyperdeformed particle number can be associated with three spherical "clusters" of equal size in a row (6=2+2+2, 24=8+8+8, 60=20+20+20, 120=40+40+40, etc.). The κ=1 can be formed from two identical large (L) spherical "clusters" and one smaller (S) (18=8+8+2, 48=20-+20+8, 100=40+40+20, etc.). In this case there are two possibilities: in the symmetric variant, L-S-L, there is a strong tendency for neck formation, whilst the second variant, L-L-S, corresponds to mass-asymmetric shapes. Finally, for κ=0

---

*Here and in the following the term "cluster" should not be understood in the most direct sense of a *real* spherical cluster, since in medium mass and heavy nuclei the probability of clustering into large fragments is strongly inhibited by the Pauli principle. It rather means a *tendency* to develop an overall equilibrium shape described by a set of spherical clusters. This average shape is illustrated schematically by the dashed line in figs. 4 and 5.

$\kappa = 0$: $\mathcal{N}^{(2)} = 2, 10, 28, 60, 110, ...$

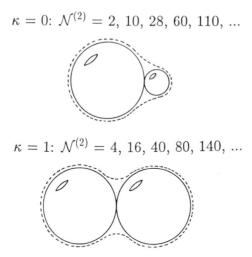

$\kappa = 1$: $\mathcal{N}^{(2)} = 4, 16, 40, 80, 140, ...$

Figure 4: Simplistic geometrical interpretation of superdeformed shells occuring at the oscillator frequency ratio $k=2$. The dashed line indicates an average equilibrium shape.

two identical small spherical "clusters" and one larger (12=2+2+8, 36=8+8+20, 80=20+20+40, etc.) can be aligned either symetrically, S+L+S, or asymetrically, S+S+L. In the first case no necking is expected but rather a "diamond"-like shape, characteristic of a large positive hexadecapole deformation. In the latter case, on the other hand, there is a strong tendency towards reflection asymmetry, see sect. 9.

Let us also comment on the structure of very deformed oblate shapes ($\omega_z/\omega_\perp = l$). In this case one can introduce the quantum number λ [11] which takes the values $0 \leq \lambda < l$ (λ is an oblate analogon of κ). Contrary to the case of prolate shapes the shell degeneracy (12) depends explicitly on λ:

$$n(l, M, \lambda) = (M + 1)[lM + 2(\lambda + 1)]. \tag{16}$$

The total number of states is then equal

$$\mathcal{N}^{(l)}(M, \lambda) = \frac{(\lambda+1)(\lambda+2)}{2} N_M + [\frac{l(l-1)}{2} + (\lambda + 2)(l - \lambda - 1)]N_{M-1}$$
$$+ \frac{(l-\lambda-1)(l-\lambda-2)}{2} N_{M-2}. \tag{17}$$

As we can see on the oblate side there are three kinds of spherical "clusters" with particle numbers N_M, N_{M-1} and N_{M-2} that form the deformed intrinsic state. The total numbers of "clusters" is independent on λ and is equal to l^2. For $l=2$, superdeformed oblate states the third term in eq. (17) vanishes since $\lambda=0$ or 1. For $\lambda=1$ (the whole M-supershell is filled) this would be attributed to three larger spherical

$\kappa = 0$: $\mathcal{N}^{(3)} = 2, 12, 36, 80, 150, ...$

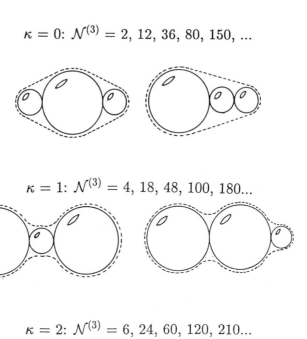

$\kappa = 1$: $\mathcal{N}^{(3)} = 4, 18, 48, 100, 180...$

$\kappa = 2$: $\mathcal{N}^{(3)} = 6, 24, 60, 120, 210...$

Figure 5: Similar to fig. 4 but for hyperdeformed shapes, $k=3$.

"clusters" in a triangle and one smaller in the middle (e.g. 6=2+2+2, 26=8+8+8+2, 68=20+20+20+8, etc.). For $\lambda=0$ (the upper M-shell is empty) the situation is reversed, i.e. there are three smaller "clusters" and one larger (e.g. 14=2+2+2+8, 44=8+8+8+20, 100=20+20+20+40, etc.). It is clear that in both situations the system is unstable with respect to mass asymmetric distortions - there will always be a tendency to develop a stable octupole deformation and/or hexadecapole-type deformations.

The classification scheme discussed above works surprisingly well for realistic nuclei where the oscillator scheme is violated, see sect. 3, 9 and ref. [11].

3 Superdeformed light nuclei

Although for light nuclei the whole concept of the intrinsic system may be questionable, it has proven to be quite useful in identifying minima in potential energy surfaces and associating these with experimental states. In light nuclei the spin–orbit interaction is relatively weak and, in addition, the diffuseness of the nuclear surface is comparable with the nuclear radius. Consequently, the simple harmonic oscillator model gives a fairly good approximation to the nuclear average potential.

A particular feature of light nuclei is that the rearrangement of a few particles may drastically change the shape. Among many well-deformed configurations in light nuclei there are several good examples nicely illustrating the simple oscillator scheme discussed in sect. 2, namely:

- The superdeformed ground state of ^{8}Be, which corresponds to two alpha particles side by side (fig. 6).

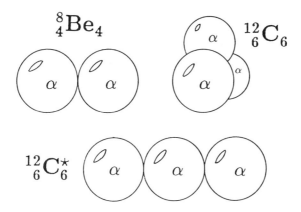

Figure 6: Schematic representation of multi α-cluster configurations in ^{8}Be and ^{12}C.

- The ground state of ^{12}C, which resembles three α-particles in a triangle, see fig. 6. This SD oblate shape ($k=1$, $l=2$) can be associated with the SD oblate gap at $\mathcal{N}^{(l=2)}(0,1)=6$. The calculated ground–state quadrupole moment, -22 efm^2 [39,40], agrees well with the experimental value, -20.1 efm^2 [41]. At $I^\pi=4^+$ the hyperdeformed state (three aligned alpha particles) becomes lowest in energy. The band-head of this very elongated structure is, most likely, the 0^+ resonant state at 10.3 MeV (cf. discussion in refs. [8,42]).

- The ground-state SD reflection asymmetric configuration of ^{20}Ne, which can be well described as arising from an ^{16}O–^{4}He di-nucleus configuration [43-45]. As noticed above the $\mathcal{N}^{(2)}(1,1)=10$ system can be viewed as a combination of two spherical "clusters" with particle numbers 2 (alpha particle) and 8 (spherical ^{16}O).

- A 4:1 state in ^{16}O, which can be described in terms of the four aligned alpha particles [46-48].

- The calculated low–lying reflection–asymmetric hyperdeformed minimum ($\varepsilon = 1$, $\varepsilon_3=0.3$) in ^{24}Mg [39] can be associated with the $\mathcal{N}^{(k=3)}(1,0)=12$ gap. According to the simple oscillator scheme this configuration can be associated with the symmetric $\alpha+^{16}O+\alpha$ (see [49]) or asymmetric $^{16}O+\alpha+\alpha$ or $^{16}O+^{8}Be$ structures. Experimentally, resonances in the asymmetric fission ^{24}Mg$\rightarrow^{16}$O + ^8Be (or ^{20}Ne+α) have been observed in the energy region of $14\,MeV \lesssim E_x \lesssim 28\,MeV$, cf. ref. [49] and refs. quoted therein.

- The ground state of ^{28}Si, which can be associated with the SD oblate gap at $\mathcal{N}^{(l=2)}(1,0) = 14$. The calculated ground–state quadrupole moment, -66 efm^2 [39,40], agrees well with the experimental value, -64 efm^2 [50].

- The weakly-deformed triaxial ground state of ^{32}S coexists with the low-lying SD minimum ($\varepsilon=0.7$) [39]. As demonstrared in [39] the SD configuration, corresponding to the $\mathcal{N}^{(k=2)}(1,1)=16$ gap, is very stable with respect to reflection-asymmetric distortions. This state can be considered in terms of clustering into two ^{16}O substructures [51,52,49].

4 Pseudo-SU(3) scheme and the Nilsson model

As shown by Elliott [53], the spectra of light nuclei from the sd shell can be successfully classified using the leading SU(3) representations. As already mentioned in sect. 3. the reason for this is twofold. Firstly, the spin-orbit interaction in light nuclei is very weak. Secondly, their mean field potential can be well approximated by a simple harmonic oscillator. However, for heavier systems the SU(3) coupling scheme breaks down since the spin-orbit splitting increases and the average nuclear field has a flat bottom. Consequently, the nucleon single-particle spectrum calculated using realistic average potentials does not have the simple structure of the harmonic oscillator.

Fig. 7 illustrates the independent-proton energy levels of the Woods-Saxon average potential plotted as functions of the nuclear quadrupole deformation β_2 (roughly proportional to ε). At first sight the single-particle spectrum looks rather "messy" with many levels crossing each other at different deformations. However, one can attempt to simplify this picture by introducing specific auxiliary quantum numbers. Firstly, one can observe that the single-particle orbitals can be divided into two groups, namely:

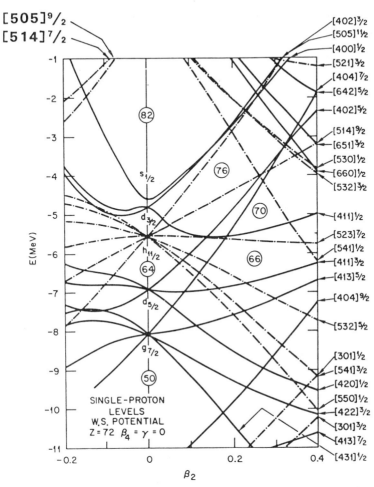

Figure 7: Proton single-particle levels (representative for rare-earth nuclei) versus quadrupole deformation β_2. (From [54].)

- The "normal-parity" states strongly mixed by the deformed mean field (in fig. 7 these are the $N=4$ states, except the [404]9/2 and [413]7/2 orbitals).

- The "abnormal-parity" (unique-parity) intruder orbitals originating from the high-j, high-N subshell (in fig. 7 these are the $h_{11/2}$ $N=5$ states). This subshell is moved down in energy partly due to the spin-orbit interaction and partly due to the flat bottom of the average field. Because of their opposite parity the intruder states do not mix with the normal-parity orbitals and therefore can be approximately interpreted as pure-j shell model states. Because of its large single-particle angular momentum the intruder states play an essential role in the high-spin behaviour of a nucleus and the onset of nuclear deformation [55].

Interestingly enough, the combined effect of the spin-orbit interaction and flat bottom (or the ℓ^2-term in the Nilsson model) leads to a very small energy splitting between certain spherical normal-parity subshells. For instance, it is seen in fig. 7 that certain pais of states, i.e. [411]3/2 and [413]5/3 or [530]1/2 and [532]3/2 are very close to each other and have very similar deformation dependence. These orbitals originating from the normal parity subshells with $j = \ell - \frac{1}{2}$ and $j' = (\ell - 2) + \frac{1}{2}$ can thus be viewed as pseudo spin-orbit doublets with $j = \tilde{\ell} \pm \frac{1}{2}$ and identical pseudo-orbital angular momentum $\tilde{\ell}$ [56,57]. For example, the $2d_{5/2}$ and $1g_{7/2}$ subshells form a $\tilde{f}_{5/2,7/2}$ doublet. Unfortunately, a solid microscopic foundation for this apparent symmetry does not exist. It has been demonstrated, however, that certain residual interactions, such as the surface delta interaction or surface multipole moments, are invariant with respect to the pseudo-spin transformation [56,57].

Formally, the transformation to the pseudo-spin representation is given by [58]

$$U = e^{i\pi \hat{h}} = 2i\hat{h}, \qquad (18)$$

where $\hat{h} = \frac{1}{r}\boldsymbol{r}\boldsymbol{s}$ is the helicity operator. Since this transformation commutes with r the possible inclusion of deformation degrees of freedom associated with multipole-multipole interactions does not destroy the pseudo spin-orbit degeneracy.

In an attempt to find a powerful coupling scheme for shell-model-like calculations one can go even further and ignore the spherical splitting between natural-parity states. (Because of the large energy splitting between natural-parity states and the unique-parity high-j intruder subshell with $j=N$ the latter orbitals do not belong to this coupling scheme.) This symmetry, referred to as pseudo-SU(3), proved to be very powerful in microscopic calculations. Indeed it has been demonstrated, see e.g. [59], that natural-parity orbitals can indeed be reasonably classified using the leading pseudo SU(3) representations [60].

In the pseudo SU(3) limit the natural parity Nilsson orbitals form a pseudo-oscillator spectrum which can be labelled using the pseudo-asymptotic quantum numbers $\tilde{N}=N$-1, $\tilde{n}_z=n_z$. Moreover, the pairs of Nilsson levels $[Nn_z\Lambda]\Omega = \Lambda + \frac{1}{2}$, $[Nn_z\Lambda + 2]\Omega = \Lambda + \frac{3}{2}$ can be viewed as pseudo spin-orbit doublets $[\tilde{N}\tilde{n}_z\tilde{\Lambda}]\Omega = \tilde{\Lambda} \pm \frac{1}{2}$ with $\tilde{\Lambda} = \Lambda + 1$. Such pairs of orbitals are usually very close to each other due to the very small pseudo spin-orbit splitting, and they have very similar particle number dependence and deformation behaviour. Experimentally, pseudo spin-orbit doublets show up clearly in the experimental systematics of single-particle excitations [54].

At the large deformations characterizing SD configurations the pseudo SU(3) scheme is still expected to be valid since the high-j intruder orbitals seem to be well separated from the natural parity states, see also [61,62]. Since the normal-parity orbitals form a pseudo-oscillator pattern they form regions of high level density at 2:1 shapes. The unique-parity intruder orbitals cross the pseudo-oscillator shell closures forming the actual superdeformed gaps seen in realistic single-particle spectra. This

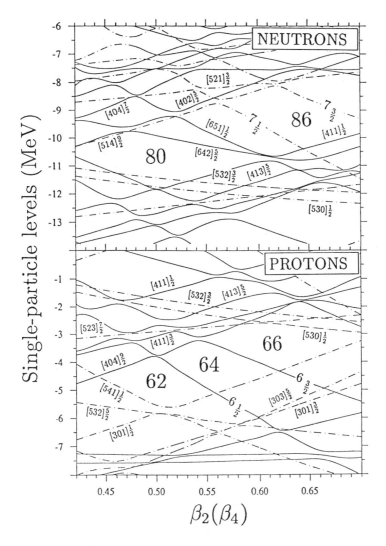

Figure 8: Single-particle Woods-Saxon levels for neutrons (top) and protons (bottom) plotted versus the $\beta_2(\beta_4)$ trajectory corresponding to the path of equilibrium deformations of SD bands in the A~150 mass region. (From ref. [31].)

simple mechanism explains in a natural way the abundance and systematics of large single-particle gaps at SD shapes [63].

In the analysis of rotational SD bands built upon many-quasiparticle configurations the correct order of the single-particle Nilsson states is of crucial importance. Fig. 8 displays the single-particle Woods-Saxon levels for the A$\sim$150 mass region plotted versus β_2 [31]. It is easily seen that the regions of the low level density occur at the specific particle numbers, for example at neutron number N=86 and proton numbers Z=64,66 at $\beta_2\approx$0.6. Thus gadolinium and dysprosium nuclei with N$\approx$86 have been suggested theoretically since the mid-seventies as very good candidates to observe the low-lying SD states.

The single-particle Woods-Saxon spectrum appropriate for the superdeformed Hg and Pb nuclei is shown in fig. 9. The largest single-particle gaps seen in fig. 8 correspond to particle numbers N=112, Z=80 ($\beta_2\approx$0.47), N=104 ($\beta_2\approx$0.40), and N=110, Z=88 ($\beta_2\approx$0.74). The nucleus ^{192}Hg can thus be considered as a "doubly-magic" SD system in the A$\sim$190 region.

5 High-N classification scheme

The SD bands observed so far are by no means identical. They show significant variations both with particle number and angular momentum, see figs. 1 and 2. The calculations based on the deformed shell model theory [31,65,66] explain many of the observed properties of SD bands in terms of intruder orbitals originating from the high-N oscillator shells and approaching the Fermi surface at large deformations. These states, i.e. N=7 neutrons and N=6 protons, carry large intrinsic quadrupole moments and therefore they lead to significant shape-polarisation effects causing deformation variations between different SD bands. Moreover, because of their large intrinsic angular momentum they are strongly influenced by the Coriolis force. The Nilsson model cranking calculations by Bengtsson et al. [66] were able to attribute the observed variations in the dynamical moments of inertia in the SD bands to the number of high-N intruder states occupied. The different high-N occupations have also been shown to explain deformation variations with the particle number.

Fig. 10 shows the comparision between experimental dynamical moments of inertia for the selected SD bands from the A$\sim$150 mass region with the Woods-Saxon cranking calculations [31]. A change in the high-N occupation clearly results in a change of the $\Im^{(2)}$ behaviour.

The experimental dynamical moments of inertia in SD bands in the A$\sim$190 region reveal a steady increase with rotational frequency, see fig. 2. As previously mentioned [29,67,68] this effect can be attributed to the successive alignment of $\nu j_{15/2}$ (N=7) and $\pi i_{13/2}$ (N=6) intruder orbitals in the presence of dynamical pairing correlations. The observed small deviations between the moments of inertia are consistent with the changes in the high-N occupation.

In short, the high-N scheme explains the rotational dependence of the moments of inertia in SD bands. Therefore, high-spin spectroscopy of SD bands provides us with a valuable information about the intruder content of SD intrinsic states.

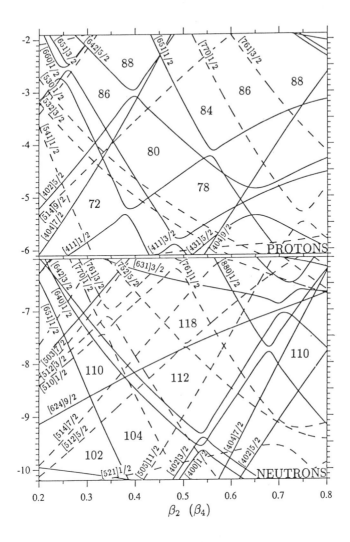

Figure 9: Single–particle Woods–Saxon levels (in MeV) for protons (top) and neutrons (bottom) plotted versus the average $\beta_2(\beta_4)$ equilibrium trajectory corresponding to the path of equilibrium deformations of SD bands in the $A \sim 190$ mass region. (From ref. [64].)

6 Strong coupling approach to excited SD bands

The physics of twinned SD bands is certainly one of the most exciting recent issues in low–energy nuclear structure. Actually, very similar rotational bands in even–

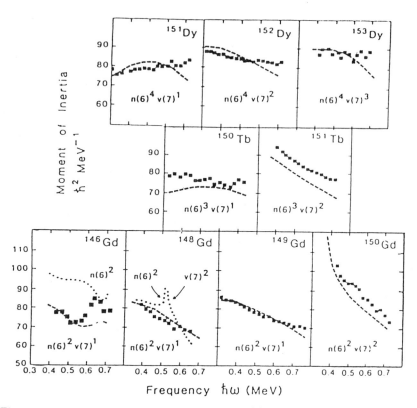

Figure 10: Experimental moments of inertia, $\Im^{(2)}$, versus rotational frequency for superdeformed bands around ^{152}Dy. The results of Woods-Saxon cranking model [31] are also shown for comparison. This diagram nicely illustrates the correlation between $\Im^{(2)}$ and the number of high-N particles occupied.

even and odd-A nuclei are not unknown in nuclear structure. For example, in the limit of total rotational alignment [69] the lowest-lying rotational states in an odd-A system can be associated with a high-j valence particle whose spin is oriented perpendicular to the symmetry axis while the core is completely decoupled and rotates with collective angular momentum $R=I-j$. This gives rise to a spectrum which is equivalent to that of the neighbouring even-even nucleus [70]. Such a situation takes place at relatively small deformations where the deformation splitting of the high-j subshell is not too large. However, in the case of SD bands we have to employ a different limit of the particle-rotor model, namely the strong coupling approach, which describes the coupling of the independent particle motion to the deformed core. The valence particles moving in the Nilsson orbitals with energies ϵ_n are characterised by the single-particle angular momentum j. The remaining nucleons form a core with angular momentum $\boldsymbol{R}$ and a moment of inertia $\Im$. The total angular momentum $\boldsymbol{I}$ is then equal to $\boldsymbol{I}=\boldsymbol{R}+\boldsymbol{j}$ and the rotor part of the nuclear Hamiltonian is given by

$$H_{ROT} = \frac{1}{2\Im}\boldsymbol{R}^2 = \frac{1}{2\Im}\bigl[\boldsymbol{I}^2 + \boldsymbol{j}^2 - I_z j_z - (I_+ j_- + I_- j_+)\bigr]. \qquad (19)$$

If the coupling of the odd particle to the deformation is much stronger than the perturbation of the single-particle motion by the Coriolis interaction [last term in eq. (19)] the odd particle adiabatically follows the deformation of the even-even core. This strong coupling limit is expected to work particularly well at large deformations (e.g. superdeformations) where the deformation splitting of Nilsson levels (proportional to β_2) is large and the Coriolis term (proportional to $\hbar^2/2\Im$) is small.

In the case of axial symmetry the strong-coupling energy spectrum in first order perturbation theory is given by [3]

$$E_K(I) = \epsilon_K + \frac{1}{2\Im}\Bigl[I(I+1) - K^2 + a(I+\frac{1}{2})(-1)^{I+1/2}\delta_{K,1/2}\Bigr], \qquad (20)$$

where $K = \Omega$ is the third component of $\boldsymbol{j}$ and a is the decoupling parameter. It is instructive to consider three limits of eq. (20):

(i) No initial angular momentum alignment, $a=0$ (e.g. $K > 1/2$). In the perfect case of strong coupling the signature splitting between rotational bands disappears, i.e. both signature partners with I, $I+2,...$ and $I+1$, $I+3,...$ lie on the same "effective" $I(I+1)$ parabola (the adjective "effective" means that the moment of inertia dependence on angular momentum is very smooth). In other words, the energy difference between gamma rays in the favoured and unfavoured band, ΔE_γ^{uf}, disappears:

$$\Delta E_\gamma^{uf} = \frac{1}{2}\Bigl[E_\gamma^u(I+1) + E_\gamma^u(I-1)\Bigr] - E_\gamma^f(I) \approx 0, \qquad (21)$$

where $E_\gamma^{band}(I) = E^{band}(I) - E^{band}(I-2)$.

Secondly, the energy transitions within deformation–aligned excited many-quasiparticle bands fulfil the simple rule

$$\Delta E_\gamma^\star = \frac{1}{2}\Bigl[E_\gamma^f(R \pm \frac{1}{2}) + E_\gamma^u(R \mp \frac{1}{2})\Bigr] - E_\gamma^{core}(R) \approx 0 \qquad (22)$$

where R is the angular momentum of a core.

(ii) $a=1$ (e.g. coupling of a $j=1/2$ particle to the rotor). In this case the two sequences with favoured signature r=−i ($I=\frac{1}{2},\frac{5}{2},...$) and unfavoured signature r=+i ($I=\frac{3}{2},\frac{7}{2},...$) are degenerate, i.e. the two states $\mathrm{E}(I = |R \pm \frac{1}{2}|)$ with $R=2,4,6,...$ have exactly the same energies and, in addition, $E_\gamma^{core}(R) \approx E_\gamma(I = |R \pm \frac{1}{2}|)$.

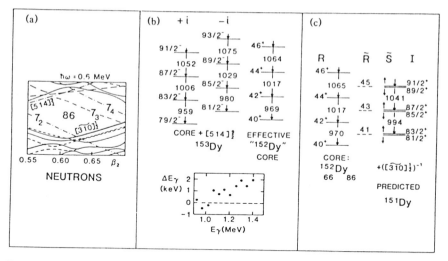

Figure 11: (a) Calculated single-neutron routhians at $\hbar\omega=0.6$ MeV as functions of β_2. (b) The effective "^{152}Dy" core transitions have been extracted from the [514]9/2 bands in ^{153}Dy and compared to the experimental transition energies in the SD band in ^{152}Dy. (c) The predicted [411]1/2 ($[\widetilde{310}]1/2$) SD band in ^{151}Dy shown together with the experimental transition energies in SD band of ^{152}Dy. (From ref. [71].)

(iii) $a=-1$. Here the situation is similar to the case (ii), but the favoured band has signature r=+i. The rotational band consists of doublets $(\frac{1}{2},\frac{3}{2})$, $(\frac{5}{2},\frac{7}{2})$,..., i.e. the states $E(I - |\tilde{R} \pm \frac{1}{2}|)$ with $\tilde{R}=R+1=1,3,5,\ldots$ have exactly the same energies and, in addition, $E_\gamma^{core}(\tilde{R}) \approx E_\gamma(I = |\tilde{R} \pm \frac{1}{2}|)$.

Let us come back to superdeformed configurations. Because of the large single-particle SD gaps at Z=66 and N=86 the nucleus ^{152}Dy is expected to be a very good "doubly-magic" SD core. Moreover, the pairing correlations in the SD band in ^{152}Dy are very weak (see discussion in sect. 8), which leads to a "rigid-like" rotational pattern which reasonably obeys the $I(I+1)$ rule.

In the nucleus ^{153}Dy there are three known SD bands [32]. The lowest one can be associated with the 7_3 intruder state originating from the $j_{15/2}$ subshell (fig. 8). The remaining two bands form a strongly coupled structure with no signature splitting. The calculations suggest that they can be built upon the high-Ω [514]9/2 orbital [31]. The effective core for the $|^{152}Dy \otimes [514]9/2\rangle$ bands can be calculated from eq. (22) assuming the scheme (i) above. The resulting gamma transitions in the "^{152}Dy" core for $40 \leq I \leq 46$ are shown in fig. 11 (left) and the deviations, ΔE_γ, from the ^{152}Dy spectrum are shown in the insert. It is seen that ΔE_γ is around 1 keV. This demonstrates that the change in the moment of inertia when adding the [514]9/2 neutron to the ^{152}Dy core is around 0.1 $\hbar^2/MeV$, i.e. $\Delta\Im/\Im \approx 10^{-3}$!

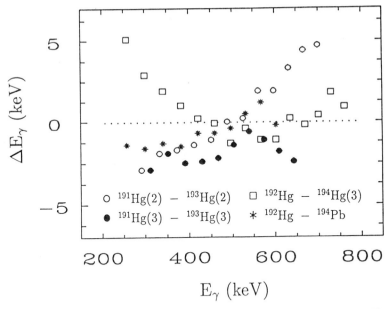

Figure 12: Difference in gamma-ray energy, ΔE_γ, between: ○ ^{191}Hg (2) - ^{193}Hg (2) bands; ● ^{191}Hg (3) - ^{193}Hg (3) bands; □ ^{192}Hg - ^{194}Hg (3) bands; ∗ ^{192}Hg - ^{194}Pb bands. (From [64].)

As noticed by Stephens et al. [35,36] many SD bands in the A∼190 region exhibit remarkable degeneracies expressed either in terms of absolute gamma-ray energies, or in terms of the so-called incremental alignment [35].

Most of the single-particle orbitals which appear near the Fermi surface at N≈112 and Z≈80, like the [512]5/2, [624]9/2, [752]5/2 neutron states or the [642]5/2, [514]9/2 proton states, have relatively large values of the Ω quantum number, see fig. 9. This suggests that the strong coupling limit should work particularly well in this case [64]. Indeed, many examples analogous to the case of "twinned" bands in ^{153}Dy–^{152}Dy have been found in the A∼190 mass region.

For Hg isotopes with N=109 and N=111 the ground-state configuration can be associated with the N=7 [761]3/2 level and the corresponding $\nu 7^3 \pi 6^4$ bands have indeed been found in ^{189}Hg (ref. [28]) and ^{191}Hg (refs. [18,29]). Almost degenerated with this orbital is the [642]3/2 Nilsson state. The [512]5/2 and [624]9/2 Nilsson orbitals are very close in energy in all nuclei discussed. They are supposed to form the lowest 1-q.p. excitations in ^{193,195}Hg$_{113,115}$ isotopes. In fact, strongly coupled bands with almost no signature splitting recently found in ^{193}Hg (ref. [72]) have been associated with these high-K states. Fig. 12 shows the difference in E_γ between the strongly-coupled bands labelled 2 and 3 in ^{191}Hg and ^{193}Hg, respectively. As

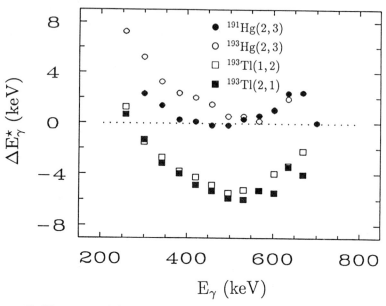

Figure 13: The energy difference ΔE_γ^*, eq. (22), versus E_γ for strongly coupled bands in ^{191}Hg (•), ^{193}Hg (○), and ^{193}Tl (□) plotted with respect to the ^{192}Hg reference. (From [64].)

mentioned in [72] they are very similar (the average deviation in energy is around $2\,keV$). Fig. 13 displays ΔE_γ^* as a function of E_γ for coupled bands in ^{191,193}Hg. For these bands the relation (22) holds up to 2-4 keV, i.e. they can be understood in terms of a strong coupling of the odd neutron (most likely the [642]3/2 orbital in ^{191}Hg [29] and the [624]9/2 Nilsson orbital in ^{193}Hg, see ref. [72]) to the core of ^{192}Hg.

For ^{194}Hg (N=114) the lowest neutron excitations are predicted to involve the [624]9/2, [512]5/2 and [752]5/2 orbitals. Experimentally, two excited SD bands in ^{194}Hg (ref. [68]) are, most likely, signature partners with $\Delta E_\gamma^{uf} \approx 0$, fig. 14. One of these bands (labelled as band 3 in [68]) is a twinned band to the SD yrast band in ^{192}Hg, see fig. 12. This is again consistent with the strong coupling of two valence neutrons to the ^{192}Hg core. In ref. [68] the excited SD bands in ^{194}Hg have been assigned to the [624]9/2⊗[512]5/2 (K=7 or K=2) neutron configuration (there is some recent evidence, however, that the [512]5/2⊗[752]5/2 configuration might also be a good candidate [73]).

The [642]5/2 level is expected to form the SD ground state of Tl-isotopes (Z=81), see fig. 9. Consequently, these bands are expected to contain five $N=6$ protons. Experimentally, two SD rotational bands have been found in ^{193}Tl (ref. [74]). As

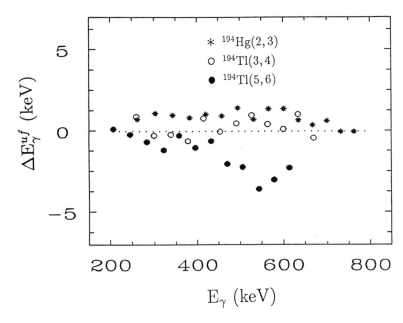

Figure 14: Signature splitting energy, eq. (21), between coupled bands in ^{194}Hg (bands 2 and 3, $\star$) and ^{194}Tl (bands 3 and 4, $\circ$, and bands 5 and 6, $\bullet$). (From [64].)

seen in fig. 13 they are not simply related to the ^{192}Hg core. At lowest spins the signature splitting between these bands is very small, but it increases with rotational frequency. This is consistent with properties of the [642]5/2 high-N orbital [74].

The recent observation of six SD bands in ^{194}Tl (ref. [75]) brings new information about the nature of the lowest proton and neutron excitations. These six bands can be grouped into three strongly coupled sequences with $\Delta E_\gamma^{uf} \approx 0$. Contrary to the SD bands in Hg isotopes, which can be nicely related to the very stable core of ^{192}Hg, the SD bands in ^{194}Tl do not have such a property. However, it has been demonstrated [74,75] that the yrast SD band in ^{193}Tl can be employed as a good reference for Z=81. Indeed, bands 3 and 4 of ref. [75] (in ref. [75] they are labelled as 2a and 2b) are identical in the sense of eq. (22) with the unfavoured [642]5/2 band in ^{193}Tl while bands 5 and 6 (labelled as 3a and 3b in ref. [75]) form a twinned structure with the favoured SD band in ^{193}Tl. This suggests that the eighty first proton occupies an orbital, which exerts a change in $\Im^{(2)}$ with respect to the ^{192}Hg core. Such a feature is again consistent with the $N=6$ [642]5/2 intruder orbital. Therefore, the most likely candidates for bands (3,4) and (5,6) are the $\pi[642]5/2 \otimes \nu[624]9/2$ or $\pi[642]5/2 \otimes \nu[512]5/2$ excitations.

The nature of bands 1 and 2 in ^{194}Tl is more difficult to explain. A plausible scenario involves the proton excitation from the [642]5/2 state to the [514]9/2 state coupled to either the [752]5/2 neutron orbital or the [624]9/2 (or [512]5/2) neutron. In this case the corresponding high-K band is expected to contain four $N=6$ protons, i.e. it should be similar to the high-K 2-q.p. bands in even-even Hg isotopes. In fact, the bands 1 and 2 in ^{194}Tl have almost identical transition energies to the bands 2 and 3 in ^{194}Hg.

In the SD band of ^{194}Pb [76,77] both [642]5/2 proton levels are filled. Nevertheless, this band is degenerate with the yrast band of ^{192}Hg, see fig. 12. This indicates that either the *two* occupied [642]5/2 protons have a very weak influence on the $\Im^{(2)}$ moment of inertia or that the [514]9/2 proton orbital lies slightly lower than predicted in the calculations (see discussion above about bands 1 and 2 in ^{194}Tl), which would lead to the reduced occupation of the [642]5/2 level.

The almost perfect degeneracy between the strongly coupled bands in ^{191,193,194}Hg and ^{194}Tl and the SD cores of ^{192}Hg and ^{193}Tl, respectively, suggests that the latter nuclei play the role of very good SD references in the A$\sim$190 mass region (cf. [35]). What can be the reason for this unususal stability ? Within the standard mean field theory the shape polarisation effects induced by the odd particle, together with the variations in the pairing field due to blocking, lead to changes in the moments of inertia which are usually much larger. Moreover, the nuclear radius effect itself is expected to give a typical deviation of the order of 10^{-2}. It seems that the possible explanation should involve almost exact cancellations between several contributions responsible for changes in the moment of inertia due to the polarisation effects exerted by valence particles, see sect. 10.

7 Pseudo-SU(3) scheme and SD bands

In the pseudo-spin regime the single-particle angular momentum can be expressed as a sum of the pseudo-orbital angular momentum $\tilde{\ell}$ and pseudo-spin $\tilde{s}$, $\boldsymbol{j}=\tilde{\boldsymbol{\ell}}+\tilde{\boldsymbol{s}}$. The pseudo-orbital angular momenta of the valence particles are strongly coupled to the angular momentum of the core to form the total pseudo-orbital angular momentum $\tilde{\boldsymbol{R}}=\boldsymbol{R}+\tilde{\boldsymbol{\ell}}$, and the pseudo-spins are then added to form the total angular momentum of the core, $\boldsymbol{I}=\tilde{\boldsymbol{R}}+\tilde{\boldsymbol{s}}$ (for more discussion we refer the reader to [60]).

At high rotational frequencies the rotational coupling, $-\omega \tilde{s}_x$, is expected to align the pseudo-spin along the axis of total angular momentum [58]. Consequently, again in this symmetry limit, the angular momentum alignment of a given normal-parity state is expected to be $\pm\frac{1}{2}$. As demonstrated in ref. [58] based on cranking model the projection of pseudo-spin onto the axis of rotation, $\langle \tilde{s}_x \rangle$, can be treated as a fairly good quantum number.

A similar conclusion can be drawn on the basis of the particle-plus-rotor model. In the limit of pseudo-SU(3) symmetry the rotor Hamiltonian can be expressed in terms of corresponding Casimir operators, C_2, and it reads [78]

$$H_{ROT} = A_{\tilde{R}} C_{2\ SO_{\tilde{R}}(3)} + A_I C_{2\ SU_I(2)} = A_{\tilde{R}} \tilde{\boldsymbol{R}}^2 + A_I \boldsymbol{I}^2 =$$

$$(A_{\tilde{R}} + A_I)\boldsymbol{I}^2 + A_{\tilde{R}}\tilde{\boldsymbol{s}}^2 - 2A_{\tilde{R}}I_z\tilde{s}_z - A_{\tilde{R}}\left[I_+\tilde{s}_- + I_-\tilde{s}_+\right]. \qquad (23)$$

The last term in square brackets in eq. (23) represents the so-called pseudo Coriolis interaction. In contrast to the usual Coriolis interaction, see eq. (19), the pseudo Coriolis interaction involves the pseudo-spin operator rather than the total angular momentum of the odd-particle [58,59,78].

Let us now apply the concept of pseudo spin to twinned SD bands. The single-proton Woods-Saxon routhians at rotational frequency $\hbar\omega=600$ keV are shown in fig. 15. The lowest SD band (+,-i) in ^{151}Tb can be associated with the 6_4 hole in the ^{152}Dy core [31]. The excited $\pi=-$ bands can be labelled as $|^{152}Dy \otimes ([\widetilde{200}]1/2)^{-1}\rangle$ where $[\widetilde{200}]1/2$ is the pseudo-asymptotic label for the [301]1/2 Nilsson state. The

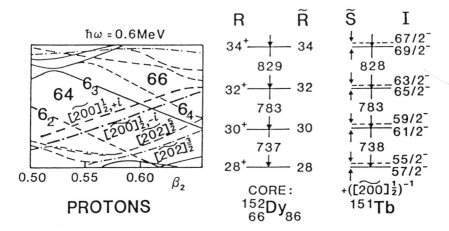

Figure 15: Calculated single-proton routhians at $\hbar\omega=0.6$ MeV and the expected deformations of SD states. The experimental transitions in the $[\widetilde{200}]1/2$ SD band in ^{151}Tb are identical with the transitions in the SD core of ^{152}Dy. (From [71].)

decoupling parameter for a $K = \frac{1}{2}$ rotational band has the value $a = (-1)^N \delta_{\Lambda 0}$ in the normal asymptotic limit. Assuming that the real Nilsson quantum numbers are good for the [301]1/2 orbital the corresponding decoupling parameter should be zero. On the other hand, in the pseudo-asymptotic limit the decoupling parameter would have the value $a = (-1)^{\tilde{N}} \delta_{\tilde{\Lambda}0}=+1$ (see also [59,58] for more discussion concerning this point).

Assuming (a) that the $[\widetilde{200}]1/2$ hole is coupled to a very stable core of ^{152}Dy and (b), that the pseudo-asymptotic labelling is valid ($a \approx 1$, or alignment $i \approx \frac{1}{2}$), one approaches the coupling scheme (ii) discussed in sect. 6, i.e. the transition energies in the excited SD band of ^{151}Tb *are expected to be equal* to γ-ray energies in the lowest SD band in ^{152}Dy [71]. The experimental data [33] show that this is indeed the case. According to fig. 15 the mean deviation ΔE_γ between the two sequences is around $2\,keV$. To what extent should the conditions (a) and (b) be fulfilled to obtain the degeneracy seen experimentally ? As discussed above in connection with the ^{153}Dy data, the "effective" core moment of inertia of ^{151}Tb is practically equal to that of ^{152}Dy (at least up to $\Delta\mathfrak{F} \approx 0.1\hbar^2/MeV$). For the decoupling parameter this condition is not that rigorous. The estimate based on eq. (20) gives $\Delta E_\gamma(I)/E_\gamma(I) \approx \Delta a/2I$ which yields $\Delta a \approx 0.1$ at $I \approx 50$. The deduced value of the decoupling parameter for the $[\widetilde{200}]1/2$ band in ^{151}Tb is thus $a = 1.0 \pm 0.1$.

In the limit of pseudo SU(3) symmetry the $[\widetilde{200}]1/2$ orbital should be pure even at very high spins. In fact, the Coriolis coupling to the closest $\pi = -$ levels, i.e. $[\widetilde{202}]3/2$ and $[\widetilde{202}]5/2$, is zero. In the realistic calculations [31] some Coriolis coupling is present, but the extracted decoupling parameter is around 0.85. Similar conclusions have also been drawn within the analysis based on the modified-oscillator potential [79,80].

The recently observed excited SD band in ^{150}Gd [33] can be given a very similar interpretation. The lowest particle-hole excitation corresponds to the promotion of the $[\widetilde{200}]1/2$ proton to the third $N=6$ routhian, 6_3. The resulting $(-,-1)$ band can be labelled as $|^{151}Tb_{SD,yrast} \otimes ([\widetilde{200}]1/2)^{-1}\rangle$ and, consequently, it is expected to be identical with the lowest SD band in ^{151}Tb, which involves three $N=6$ protons. This scenario is indeed confirmed experimentally - the two bands are identical within an accuracy of about 2 keV [33]. As we see, the excited band in ^{150}Gd can be interpreted in terms of a decoupling (or: a full alignment) of the pseudo-spin from the rotational motion of the effective odd-A core of ^{151}Tb, which has total pseudo-orbital angular momentum $\tilde{R} = \frac{1}{2}, \frac{5}{2}, \frac{9}{2}, \ldots$ This twinned pair of bands can also be given an alternative strong-coupling explanation: the excited SD band in ^{150}Gd is nothing else but the perfect example of a semidecoupled band discussed in refs. [81,82].

Another example of identical bands has recently been discovered in ref. [34]. The second excited band in ^{149}Gd is identical to the SD yrast band in ^{150}Tb [83] labelled as $\nu 7^1 \pi 6^3$. Here, in a full analogy to the previous cases, the $[\widetilde{200}]1/2$ proton hole is coupled to the odd-odd core of ^{150}Tb, which gives rise to the excited $|^{150}Tb \otimes (\pi[\widetilde{200}]1/2)^{-1}\rangle$ band in ^{149}Gd.

Experimentally, the SD yrast bands in some even-even (^{152}Dy), odd-even (^{151}Tb), and odd-odd (^{150}Tb) nuclei, having different high-N occupations and thus different $\mathfrak{F}^{(2)}$-behaviour, can be considered as ideal cores for a $[\widetilde{200}]1/2$ proton state. To close the loop one should look for an example of an ideal even-odd core. Such a core is the yrast SD band $\nu 7^1 \pi 6^4$ in ^{151}Dy [84]. By making a hole in the $[\widetilde{200}]1/2$ proton level one obtains an excited twinned SD band in ^{150}Tb, labelled as $|^{151}Dy \otimes (\pi[\widetilde{200}]1/2)^{-1}\rangle$. This configuration, yet undiscovered, would be the second example of the perfect SD semidecoupled band.

As in the case of ^{151}Tb the nucleus ^{151}Dy can be viewed as a one hole system with respect to the ^{152}Dy core. The observed SD band [84] can be associated with the $(-,+i)$ $|^{152}Dy \otimes (7_2)^{-1}\rangle$ configuration, see fig. 11. The lowest positive parity SD band in ^{151}Dy is formed by promoting one neutron from the $[\widetilde{310}]$ ($[411]1/2$) level to the 7_2 routhian. In the resulting $|^{152}Dy \otimes (\pi[\widetilde{310}]1/2)^{-1}\rangle$ configuration the favoured signature is $+i$, as shown in the single-routhian diagram of fig. 11. In fact, the $[\widetilde{310}]1/2$ neutron orbital in ^{151}Dy plays a role very similar to that of the $[\widetilde{200}]1/2$ proton level in ^{151}Tb. Both appear just below the SD subshell closure, both have a particle character, and both carry negative quadrupole moments. However, the decoupling parameter of the $[\widetilde{310}]1/2$ state is $a=-1$ in the pseudo-asymptotic limit. In this symmetry limit the spectrum of the SD $[\widetilde{310}]1/2$ band should follow the coupling scheme indicated schematically in the right portion of fig. 11; the transition energies in this band are then expected to lie just half way between the gamma-ray energies of the SD band in ^{152}Dy. However, this band is calculated [31] to be slightly disturbed by the Coriolis interaction with the $[\widetilde{312}]3/2$ orbital lying about 1 MeV below. This coupling is expected to push the r=+i member slightly up in energy and increase the transition energies by a few keV.

As discussed in sect. 6 the degenerated bands in the Hg-Pb region can be consistently explained in terms of the strong coupling approach. Such a scenario does not involve any quantized angular momentum alignment associated with an aligned pseudo–spin as suggested by Stephens et al. [35], but is a natural consequence of a simple strong coupling scheme. (For more discussion concerning this point we refer the reader to refs. [64,85,86]). Let us also note that most of the orbitals expected to occur close to the Fermi level in the nuclei discussed *do not belong* to the pseudo-SU_3 coupling scheme, which, according to Stephens et al., is responsible for the quantized alignment.

8 Pairing correlations at superdeformed shapes

It is well known that the nuclear moments of inertia are, at low spins, much lower than the corresponding rigid-body values. The reason for this phenomenon is the existence of a short-range two-body residual pairing interaction that couples the pairs of nucleons in the time-reversed orbits and leads to the presence of nuclear superfluidity characterized by the order parameter Δ, the pairing gap. At high spins the Coriolis force tries to align the single-particle angular momenta along the axis of nuclear rotation. Therefore, it is expected that at very high rotational frequencies the nuclear superfluidity should break down. On the other hand, at large deformations pairing correlations are expected to be very weakly affected by the Coriolis force because of a simple deformation-rotation scaling [87,88]. Consequently they should persist to higher frequencies than in normal-deformation bands. Because of the very large values of rotational frequency and the near-rigid rotational pattern of the observed SD bands one can expect that these states can give us some information about the nuclear superfluidity at extreme conditions.

Surprisingly enough, the influence of large shape-elongations on pairing field is still not well understood. For example, in the "doubly-magic" nucleus ^{152}Dy pairing is expected to play a minor role [31,89-91]. Indeed, due to the very low level density of single-particle states the superfluid-type correlations in this band are seriously quenched and they are mainly of a dynamical character, i.e. the pairing interaction cannot form a boson condensate. In ^{150}Gd the large increase of $\Im^{(2)}$ in the lower part of SD band [92,27], see fig. 1, has been interpreted as a *paired* band crossing associated with an alignment of the $N=7$ neutron pair [93,31]. A similar crossing has been found in the first excited SD band in ^{149}Gd [34]. Another piece of experimental evidence suggesting the presence of pairing at SD shapes is a steady increase of $\Im^{(2)}$ in the SD bands in the A$\sim$190 region, which can be attributed [67,68] to the alignment of $N=7$ neutrons and $N=6$ protons. Calculations without pairing yield fairly constant moments of inertia.

However, there is evidence that pairing correlations are extremely weak at large elongations. For instance, the moments of inertia in SD bands are very close to their rigid-body values and even a strongly reduced pairing field yields too much quasiparticle alignment [68,94]. Moreover, no strong evidence for blocking effects in SD configurations has been found in the A$\sim$190 region. Let us finally mention, that the crossing between the [651]1/2 and [642]5/2 routhians resulting in the observed "hump" in the $\Im^{(2)}$ moment of inertia in the SD band of ^{146}Gd [95,42] and in the excited SD band of ^{147}Gd [96] is considerably reduced if some pairing is present [31].

One possible explanation for these anomalies can be given in terms of the T=0 proton-neutron interaction between high-j intruder orbitals [97,94,55]. On the other hand, the T=1 pairing model itself is still rich enough to provide an alternative explanation. Let us consider, for instance, the standard pairing interaction with a state-dependent pairing force:

$$H_{pair} = -\sum_{i,j>0} G_{i,j} c_i^+ c_{\bar{i}}^+ c_{\bar{j}} c_j, \qquad (24)$$

where $G_{i,j} = <i\bar{i}|v|j\bar{j}>_{AS}$ is the antisymmetrized matrix element of the two-body pairing interaction. The equation for a state-dependent gap parameter is now given by

$$\Delta_i = \frac{1}{2} \sum_j \frac{G_{i,j}}{\sqrt{(\epsilon_j - \lambda)^2 + \Delta_j^2}} \Delta_j. \qquad (25)$$

The matrix elements $G_{i,j}$ have been calculated by many authors using various residual interactions like the delta force, the surface delta interaction, the Skyrme force, or a density dependent delta interaction [98,99,100] (see also [101,102]). It has been found that the pairing matrix elements are relatively enhanced for orbitals with similar values of $\langle n_z \rangle / \langle N \rangle$, i.e. orbitals with good angular overlap. In particular it has been observed that the pairing matrix elements between the high-j intruder orbitals (like [660]1/2 and [651]3/2) are rather large, as are those between high-j intruder states and the natural-parity orbitals with $j=N-1$ and similar Ω-values (like [660]1/2 and [541]1/2). At normal deformations the single-particle unique-parity orbitals are relatively close to each other and to normal-parity states with similar spatial overlap. However, at large deformations states originating from completely different shells approach the Fermi level. These states are very weakly coupled through pairing interaction. Moreover, the "favoured" coupling between unique-parity levels is diminished because of their large deformation splitting; see the energy denominator in eq. (25). In view of the above it is clear that the pairing correlation energy should decrease with deformation - an effect that is analogous to the fragmentation of pairing matrix elements caused by the Coriolis force. This effect should have many important consequences, i.e. it should influence the calculated half-lives of fission isomers, decays from SD bands, etc.

Recent calculations utilising the zero-range delta interaction or the finite-range D1S interaction [103] indeed suggest a sizeable reduction in pairing correlations for fission configurations. However, at SD shapes the predicted quenching of pairing is found to be rather weak. Unfortunately, the only two examples discussed in ref. [103], i.e. ^{194}Hg and ^{232}Th, do not allow us to draw more general conclusions concerning the deformation dependence of the effective pairing force strength.

9 Low-energy octupole correlations at superdeformed shapes

In sect. 2 it has been demonstrated that for the axially-deformed harmonic oscillator model the shell energy should decrease with increasing reflection-asymmetry for SD gaps that correspond to the particle number of two unequal spherical shapes ($k=2$, $\kappa=0$), i.e. for N or Z equal to 28, 60, 110, etc. On the other hand, for the particle numbers 40, 80, 140 the nuclear shape is expected to be stable with respect to reflection asymmetric distortions. Calculations based on the realistic mean-field potentials do confirm this general tendency, i.e. regions of particle numbers, which favour reflection-symmetric or reflection-asymmetric shapes alternate [11,42,104].

The shell energy of the Nilsson model is shown in fig. 16 (top) as a function of neutron number and octupole deformation, ε_3. Other deformation parameters are kept fixed at the values representative for superdeformed shapes, $\varepsilon_2=0.6$, $\varepsilon_4=0.09$. It is seen that the mass-asymmetry is strongly favoured at particle numbers around 28, 64 and 114 whilst for particle numbers around 38, 84 and 144 the minimum shell correction energy is found at $\varepsilon_3=0$, i.e. the microscopic calculations follow quite closely the harmonic oscillator pattern.

The microscopic mechanism behind reflection asymmetry at certain SD shapes is twofold. The octupole interaction Y_{30} couples the pseudo-SU$_3$ orbitals with asymptotic quantum numbers $[N\ n_z\ \Lambda]\Omega$ and $[N+1\ n_z\pm 1\ \Lambda]\Omega$. The largest number of such matrix elements corresponds to states with the highest possible value of $n_\perp$, i.e. for $n_z=0$. In particular, SD shells with $\kappa=0$ ($N^{(2)}_{shell}$-even) contain the $|n_\perp = N^{(2)}_{shell}, n_z = 0\rangle$ orbital, which is $N^{(2)}_{shell}+2$ folded degenerate. This orbital interacts via the octupole potential with the corresponding $n_z=1$ level belonging to the next SD shell ($\kappa=1$), but it does not have a partner to interact with in the lower shell $N^{(2)}_{shell}-1$. Consequently, all these $n_z=0$ states are pushed down by octupole interaction and the system becomes unstable with respect to ε_3. This tendency has been discussed long ago in the context of the fission barrier asymmetry [105,106]. The second mechanism behind the octupole instability in SD states is the octupole

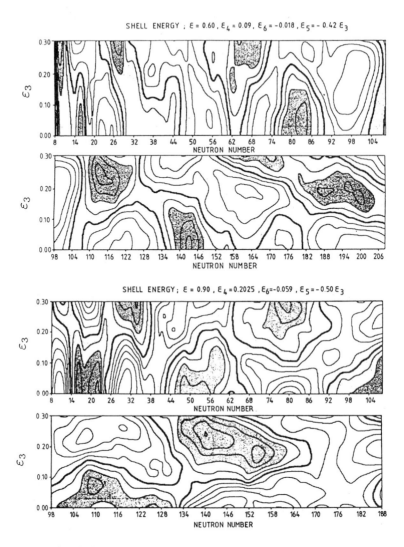

Figure 16: The shell correction of the modified oscillator model as a function of neutron number and octupole deformation ε_3 for superdeformed (top) and hyperdeformed (bottom) shapes. (From [11].)

interaction between the high-N intruder orbitals and specific pseudo-oscillator levels. For example, the same pairs of orbitals, such as ([660]1/2–[530]1/2) or ([770]1/2–[640]1/2), which are responsible for octupole deformations in the light actinides appear close to the Fermi level in SD configurations around ^{148}Gd and ^{192}Hg.

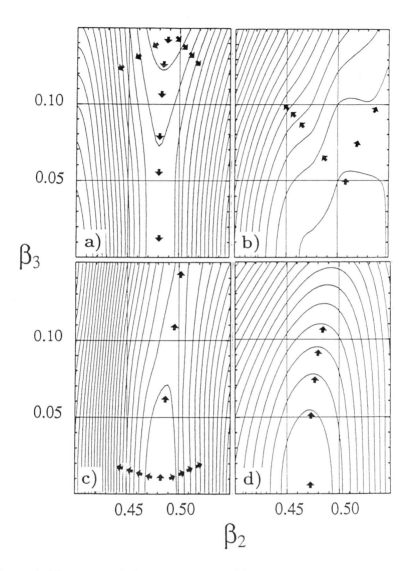

Figure 17: The neutron shell correction energy (a), the proton shell correction energy (b), the total shell energy (c) and the total potential energy (d) for ^{192}Hg as a function of β_2 and β_3. Arrows indicate the direction of growing energy. The distance between the contour lines is 200 keV. (From [64].)

For SD bands around ^{152}Dy the low-energy octupole collectivity can be attributed to the "octupole-driving" proton number Z=64. On the other hand there is no such a tendency for the neutron system. Indeed, the particle number N=84 has been found to strongly favour reflection-symmetric shapes. The opposite is true for SD configurations in the Hg-Pb region, i.e. octupole correlations have neutron origin (because of the "optimal" neutron number N=114). As an illustrative example the results of Woods-Saxon-Strutinsky calculations for ^{192}Hg are shown in fig. 17. Whilst the neutron shell correction strongly favours octupole distortions the proton shell correction drives the system towards β_3=0. The resulting shell correction is almost insensitive to β_3 and, thanks to the very shallow macroscopic energy, the total potential energy reveals a pronounced octupole softness.

Recently, the octupole susceptibility in SD configurations has been investigated within the parity-projected Skyrme-Hartree-Fock model [107]. The selfconsistent calculations do confirm the predictions of models based on the shell-correction approach, i.e. they indicate quite a sizeable lowering of the octupole excitations built on the SD intrinsic state. Octupole softness (but not *octupole instability*) in superdeformed nuclei in the A$\sim$190 region is expected to persist at high angular momenta [42,108,104]. This means that collective octupole vibrational excitations can be mixed with low-lying one- and two-particle states thus modifying the excitation pattern near the yrast line.

According to the calculations the first excited state in the doubly-magic SD configurations in ^{152}Dy and ^{192}Hg should have a collective octupole character, in a nice analogy to the well-known collective 3$^-$ state in the doubly-magic spherical nucleus ^{208}Pb. Moreover, the B(E1) rates for depopulating the SD octupole band should be markedly enhanced because of the reduced excitation energy of the giant dipole resonance built on the SD state [109]. Octupole correlations are expected to reduce single-particle alignments, increase band interactions, and lead to a more collective pattern for the lowest rotational bands [110-112]. All these effects are particularly strong for nuclei that are predicted to be octupole-deformed. It should be emphasized, however, that even in the case of dynamic octupole correlations the zero-point octupole energy, E_{zp}^{oct}, contributes to the total angular momentum of a nucleus, and the corresponding alignment reads:

$$I_x^{oct} = -\frac{dE_{zp}^{oct}}{d\omega}, \qquad (26)$$

where

$$E_{zp}^{oct} = \frac{1}{2}\left[\sum_\alpha W_\alpha^{RPA} - \sum_{p,h}(\epsilon_p - \epsilon_h)\right] \qquad (27)$$

is calculated within the RPA formalism and W_α^{RPA} are the corresponding one-phonon excitations. It remains to be investigated, however, whether the rotational dependence of the zero-point octupole energy can account for present discrepancies between experimental and calculated angular momentum alignment in SD bands in Hg, Tl and Pb isotopes.

Recent experimental data on ^{193}Hg [72] show a low-frequency pseudo crossing in one of the observed SD bands as well as dipole transitions, in one direction only, between one SD band and another. An admixture of an octupole phonon built on the [624]9/2 ground-state into the [512]5/2 band can explain these effects [72]. The experimental data for ^{193}Hg, together with the observed reduction in alignments and the unusual similarity of SD bands in the A~190 region, are the first pieces of experimental evidence supporting the presence of strong octupole correlations in SD configurations.

For hyperdeformed shapes, $k=3$, the harmonic oscillator model discussed in sect. 2 suggests that the strongest tendency for reflection asymmetry should be expected for $\kappa=0$ hyperdeformed shells, i.e. for the particle numbers $12, 36, 80, 150$. Particle numbers that are most "rigid" with respect to octupole distortion correspond to $\kappa=2$ shells, i.e. they are equal to 24, 60, 120. The shell correction landscape computed within the Nilsson model is shown in fig. 16 (bottom) at deformation parameters, $\varepsilon_2=0.9$, $\varepsilon_4=0.203$, representative for hyperdeformed shapes. Again, like for the 2:1 shapes, simple predictions of the harmonic oscillator model are corroborated by the results of microscopic calculations, see also [108].

In the heavier actinide nuclei a third hyperdeformed minimum around the fission barrier has been calculated [114-116]. It corresponds to large elongations, $\beta_2 \approx 0.9$, and pronounced reflection-asymmetry, $\beta_3 \approx 0.2$. The third minimum is very localized in particle number and it is expected to be very shallow (the deepest minimum, about 1.5 MeV, has been predicted [115] for very neutron-rich nuclei around Z=86, N=148, i.e. exactly around the octupole-driving particle numbers 80 and 150). Experimentally, the third minimum shows up in the (n, f) and (d, pf) reactions as an alternating-parity microstructure of resonances [116,117].

10 Summary

The recent observations of extremely elongated, rapidly rotating configurations of the nucleus, the SD states, open a new area in the physics of atomic nuclei – high-spin spectroscopy at 2:1 shapes. Indeed, a measurement of the discrete SD states in the 20-60 $\hbar$ spin range is slowly becoming a "standard" experiment for the nuclear γ-ray spectroscopist, thanks to both the high skills of the experimentalists and the advent of very powerful γ-ray Compton-supressed multi-detector arrays. However, without undetermining the unquestionable experimental success, one has to stress that theoretical predictions for the presence and properties of high-spin SD states preceded the discovery of these states by almost a decade. A sizeable amount of the high-spin discrete–line spectroscopic data on SD states taken up to now makes it possible, for the first time, to test fine details of the underlying shell structure. The recent experimental facts have greatly stimulated many theoretical works on the nature of very elongated shapes.

We have learnt, for example, that the spectroscopy of SD bands can give us some insight into the structure of unique-parity high-j, high-N states that govern the rotational properties at 2:1 shapes. In particular, a correlation has been found between the number of these intruder orbitals and the behaviour of the moments of inertia.

The intruder orbitals which become occupied in the SD bands lie very high in energy (around 8-10 MeV) above the Fermi surface at typical ground-state deformations. This means that the spectroscopy of SD bands offers a very unique possibility for probing the relatively pure shell-model configurations at extreme conditions. By extracting the positions of high-N levels at large deformations and extrapolating them back to the normally-deformed or spherical shapes we may in the future gain some information about the energy position of the high-j subshells that otherwise cannot be accessed.

We can also see that the existence of superdeformed shapes also enables us to probe the structure of natural-parity states that carry very little angular momentum. In addition, because of the very weak Coriolis force these states have extremely pure configurations; thus they can be used to test fundamental symmetries of the mean field, such as the pseudo-spin or pseudo $SU(3)$.

Although some of the questions concerning the structure of SD states have been already answered, many problems still remain open. One of the most puzzling questions is the population mechanism of the SD bands. Why are just these special states so favoured among many other possible configurations that appear at excitation energies around 8 MeV (a typical neutron separation energy) ? On the other hand, we have already learnt that as soon as the SD states are formed the relative in-band intensity is almost constant. This suggests that these states hardly mix with (very many) close-lying levels because of their very distinct intrinsic structure. Another interesting problem is the very rapid feeding out of the SD bands (the decay path out of the SD band has not been determined experimentally for any of the observed cases !).

However, the most puzzling issue is the observed unusual stability of certain SD cores, like ^{152}Dy or ^{192}Hg, with respect to polarisation effects exerted by valence particles. This strinking stability manifests itself by the presence of several pairs of identical, "twinned", rotational bands in different nuclei. The striking similarity between the bands, $\Delta\Im\approx 0.1$ $\hbar^2/MeV$ or $\Delta E_\gamma\approx 2$ keV, imply that the SD cores are insensitive to removal (or addition) of the low-j particles such as a $[301]1/2$ proton or $[514]9/2$ ($[624]9/2$) neutron. This phenomenon can hardly be explained quantitatively by the standard models, which always yield some changes in the mean field caused by polarisation effects; they certainly are not able to reproduce the energy within an accuracy of 1 keV. Qualitatively, the change in the moment of inertia within a n-q.p. configuration caused by occupation of orbitals ν $(=\nu_1\,\nu_2,...)$ is given by:

$$\delta\Im_\nu = \frac{\partial\Im}{\partial A}dA + \left(\frac{\partial\Im}{\partial\bar{\beta}}\right)_\nu d\bar{\beta} + \left(\frac{\partial\Im}{\partial\omega}\right)_\nu d\omega + \left(\frac{\partial\Im}{\partial\bar{\Delta}}\right)_\nu d\bar{\Delta} + ... \qquad (28)$$

where $\bar{\beta}$ and $\bar{\Delta}$ stand for deformation and pairing fields, respectively. Since, as discussed in sect. 8, the pairing correlations in SD bands are probably strongly quenched, $d\bar{\Delta}\approx 0$. Assuming an $A^{5/3}$-dependence of $\Im$ the average change of radius [first term in eq. (28)] yields $\frac{5}{3}\Im/AdA\approx 10\,keV$.

Recently, there have been some attempts to understand the configuration dependence of the $\Im$-polarisation, $d\Im_\nu$, using the harmonic oscillator model. Ragnarsson [79,80] demonstrated that for certain orbitals such as [301]1/2 or [514]9/2 the three first terms on the r.h.s. of eq. (28) cancel each other. Dudek and Szymański pointed out [118] that $d\Im_\nu$ generally decreases with deformation and its magnitude depends crucially on the n_z quantum number of an orbital ν. States with small n_z-values, like [301]1/2 or [411]1/2, cause very small $\Im$-polarisation at SD shapes while intruder states charecterized by large-n_z components lead to significant changes in the moment of inertia. These explanations, which account for some important effects but obviously ignore many others (like e.g. octupole and hexadecapole correlations or pairing) do not give us a complete account the physics of super-stable SD cores. From this point of view the fact of an unusual stability of certain SD cores is still an open challenging question that remains to be answered.

Calculations predict the occurence of low-lying collective octupole vibrations built upon the intrinsic SD state. In the Gd-Dy region they can be attributed to the "octupole-driving" SD oscillator gap at Z=60, whilst in the Hg-Pb region they can be associated with the neutron SD gap at N=110. Experimentally, there are some recent suggestions [72] that such collective excitations are indeed present in the superdeformed mercury isotopes.

Theory shows that very elongated structures should also exist in other regions of the periodic table, for example in the A≈80, A≈110 and A≈140 nuclei. Will all these predictions prove to be as successful as those from the seventies ?

A new generation of spectrometer arrays will soon become operational in Europe (EUROGAM) and the U.S.A. (GAMMASPHERE). These new magnificent instruments will certainly shed new light on many of the basic questions surrounding the physics of superdeformation and give rise to the prospect of observing many new phenomena.

Acknowledgements Part of the presented material has been obtained in collaboration with Paul Fallon, Mark Riley, Wojtek Satuła, Zdzisław Szymański, Ramon Wyss, Stefan Ćwiok and Arne Johnson. This work was partly supported by the Polish Ministry of National Education under contract CPBP 01.09

References

[1] S. Cohen, F. Plasil and W.J. Swiatecki, Ann. Phys. (N.Y.) **82** (1974) 557.

[2] V.M. Strutinsky, Nucl. Phys. **A95** (1967) 420.

[3] A. Bohr and B.R. Mottelson, Nuclear Structure, vol. 2 (W.A. Benjamin, New York, 1975).

[4] R. Bengtsson et al., Phys. Lett. **57B** (1975) 301.

[5] K. Neergård and V.V. Pashkevich, Phys. Lett. **B59** (1975) 218.

[6] K. Neergård, V.V. Pashkevich and S. Frauendorf, Nucl. Phys. **A262** (1976) 61.

[7] C.G. Andersson et al., Nucl. Phys **A268** (1976) 205.

[8] I. Ragnarsson, S.G. Nilsson and R.K. Sheline, Phys. Rep. **45** (1978) 1.

[9] I. Ragnarsson, T. Bengtsson, G. Leander and S. Åberg, Nucl. Phys. **A347** (1980) 287.

[10] C.G. Andersson et al., Phys. Scr. **24** (1981) 266.

[11] T. Bengtsson et al., Phys. Scr. **24** (1981) 200.

[12] Y. Schutz et al., Phys. Rev. Lett. **48** (1982) 1535.

[13] J. Dudek and W. Nazarewicz, Phys. Rev. **C31** (1985) 298.

[14] B.M. Nyakó et al., Phys. Rev. Lett. **52** (1984) 507.

[15] P.J. Twin et al., Phys. Rev. Lett. **55** (1985) 1380.

[16] J. Gizon, *these proceedings*.

[17] P.J. Twin et al., Phys. Rev. Lett. **57** (1986) 811.

[18] E.F. Moore et al., Phys. Rev. Lett. **63** (1989) 360.

[19] E.M. Rastopchin, G.N. Smirenkin and V.V. Pashkevich, *Proc. Workshop on Microscopic Theories of Superdeformation in Heavy Nuclei at Low Spin*, Lyon, May 1990; V.V. Pashkevich, Preprint JINR (Dubna) P4-4383 (1969).

[20] C.F. Tsang and S.G. Nilsson, Nucl. Phys. **A140** (1970) 275.

[21] U. Götz, H.C. Pauli, K. Adler and K. Junker, Nucl. Phys. **A192** (1972) 1.

[22] M. Cailliau, J. Letessier, H. Flocard and P. Quentin, Phys. Lett. **46B** (1973) 11.

[23] R.R. Chasman, Phys. Lett. **219B** (1989) 227.

[24] P. Fallon, in: *Proc. Int. Conf. on High Spin Physics and Gamma Soft Nuclei*, Pittsburgh, September, 1990; Univ. of Liverpool Preprint NSG 90/10.

[25] M.A. Bentley et al., Phys. Rev. Lett. **59** (1987) 2141.

[26] B. Haas et al., Phys. Rev. Lett. **60** (1988) 503.

[27] P. Fallon et al., Phys. Lett. **B** (1991) in press.

[28] M.W. Drigert et al., to be published, 1990.

[29] M.R. Carpenter et al., Phys. Lett. **240B** (1990) 44.

[30] E.F. Moore et al., Phys. Rev. Lett. **64** (1990) 3127.

[31] W. Nazarewicz, R. Wyss and A. Johnson, Nucl. Phys. **A503** (1989) 285.

[32] J.K. Johansson et al., Phys. Rev. Lett. **63** (1989) 2200.

[33] T. Byrski et al., Phys. Rev. Lett. **64** (1990) 1650.

[34] B. Haas et al., Phys. Rev. **C42** (1990) R1817.

[35] F.S. Stephens et al., Phys. Rev. Lett. **64** (1990) 2623; Phys. Rev. Lett. **65** (1990) 301.

[36] F.S. Stephens et al., *Int. Conf. on Nuclear Structure in the Nineties*, Oak Ridge, 1990; to appear in Nucl. Phys. (1990).

[37] B.T. Geilikman, in *Proc. Int. Conf. on Nuclear Structure*, Kingston, Canada, eds. D.A. Bromley and E.W. Vogt (Univ. of Toronto Press, Toronto) p. 874.

[38] C.Y. Wong, Phys. Lett. **32B** (1970) 668.

[39] G.A. Leander and S.E. Larsson, Nucl. Phys. **A239** (1975) 93.

[40] Y. Abgrall, B. Morand and E. Caurier, Nucl. Phys. **A172** (1972) 372.

[41] A, Nakada et al., Phys. Rev. Lett. **27** (1971) 745.

[42] S. Åberg, Proc. Int. Conf. on Nuclear Structure in the Nineties, Oak Ridge, 1990; to appear in Nucl. Phys. **A**.

[43] F. Nemoto and H. Bando, Prog. Theor. Phys. **47** (1972) 1210.

[44] S. Marcos, H. Flocard and P.-H. Heenen, Nucl. Phys. **A410** (1083) 125.

[45] D. Proovost, F. Grümmer, K. Goeke and P.-G. Reinhardt, Nucl. Phys. **A431** (1984) 139.

[46] Y. Fujiwara et al., Prog. Theor. Phys. Suppl. **68** (1980) 29.

[47] S. Åberg, I. Ragnarsson, T. Bengtssson and R.K. Sheline, Nucl. Phys. **A391** (1982) 327.

[48] D. Auverlot, P. Bonche, H. Flocard and P.-H. Heenen, Phys. Lett. **149B** (1984) 6.

[49] H. Flocard, P.H. Heenen, S.J. Krieger and M. Weiss, Prog. Theor. Phys. **72** (1984) 1000.

[50] Y. Horikawa et al., Phys. Lett. **36B** (1971) 9.

[51] D. Baye and G. Reidemeister, Nucl. Phys. **A258** (1976) 157.

[52] D. Baye and P.H. Heenen, Nucl. Phys. **A276** (1977) 354.

[53] J.P. Elliott, Proc. Roy. Soc. (London) **A245** (1958) 128, 562.

[54] W. Nazarewicz, M.A. Riley and J.D. Garrett, Nucl. Phys. **A512** (1990) 61.

[55] W. Nazarewicz, in *Contemporary Topics in Nuclear Structure Physics* eds. R.F. Casten, A. Frank, M. Moshinsky and S. Pittel (World Scientific, Singapore, 1988) 467.

[56] K.T. Hecht and A. Adler, Nucl. Phys. **A137** (1969) 129.

[57] A. Arima, M. Harvey and K. Shimizu, Phys. Lett. **30B** (1969) 517.

[58] A. Bohr, I. Hamamoto and B.R. Mottelson, Phys. Scr. **26** (1982) 267.

[59] R.D. Ratna-Raju, J.P. Draayer and K.T. Hecht, Nucl. Phys. **A202** (1973) 433.

[60] D. Warner, *these proceedings*.

[61] J.P. Draayer, O. Castaños and S.C. Park, in *Contemporary Topics in Nuclear Structure Physics* eds. R.F. Casten, A. Frank, M. Moshinsky and S. Pittel (World Scientific, Singapore, 1988) 345.

[62] J.P. Draayer, *Proc. Int. Conf. on Nuclear Structure in the Nineties*, Oak Ridge, 1990, p. 116.

[63] J. Dudek, W. Nazarewicz, Z. Szymański and G.A. Leander, Phys. Rev. Lett. **59** (1987) 1405.

[64] W. Satuła, S. Ćwiok, W. Nazarewicz, R. Wyss and A. Johnson, submitted to Nucl. Phys. **A** (1990).

[65] I. Ragnarsson and S. Åberg, Phys. Lett. **B180** (1986) 191.

[66] T. Bengtsson, S. Åberg and I. Ragnarsson, Phys. Lett. **208B** (1988) 39.

[67] D. Ye et al., Phys. Rev. **C41** (1990) R13.

[68] M.A. Riley et al., Nucl. Phys. **A512** (1990) 178.

[69] F.S. Stephens, R.M. Diamond and S.G. Nilsson Phys. Lett. **44B** (1973) 429.

[70] F.S. Stephens, R.M. Diamond, J.R. Leigh, T. Kammuri and N. Nakai, Phys. Rev. Lett. **29** (1972) 438.

[71] W. Nazarewicz, P.J. Twin, P. Fallon and J.D. Garrett, Phys. Rev. Lett. **64** (1990) 1654.

[72] D.M. Cullen et al., Phys. Rev. Lett. **65** (1990) 1547.

[73] W. Nazarewicz and M.A. Riley, to be published.

[74] P.B. Fernandez et al., Nucl. Phys. **A517** (1990) 386.

[75] F. Azaiez eyt al., preprint 1990.

[76] K. Theine et al., Z. Phys. **A336** (1990) 113.

[77] M.J. Brinkman et al., Z. Phys. **A336** (1990) 115.

[78] A. Frank, S. Pittel, D. Warner and B. Engel, Phys. Lett. **182B** (1986) 233.

[79] I. Ragnarsson, Copenhagen workshop, October 1989; Contribution to the Oak Ridge conference *Nuclear Structure in the Nineties*, 1990; Nucl. Phys., in press.

[80] I. Ragnarsson, submitted to Phys. Lett. **B** (1990).

[81] A.J. Kreiner, Phys. Rev. **C38** (1988) 2486.

[82] W. Nazarewicz, in: *Selected Topics in Nuclear Structure*, ed. by J. Styczen and Z. Stachura, (World Scientific Publ. 1990) vol. 2, p. 53.

[83] M.A. Deleplanque et al., Phys. Rev. **C39** (1989) 1651.

[84] G.-E. Rathke et al., Phys. Lett. **209B** (1988) 177.

[85] C.-L. Wu, D.H. Feng and M.W. Guidry, submitted to Phys. Rev. Lett. 1990.

[86] R.Wyss and S. Pilotte, submitted to Phys. Rev. C; the JIHIR preprint 90-02.

[87] W. Nazarewicz, Z. Szymański and J. Dudek, Phys. Lett. **196B** (1987) 404.

[88] W. Satuła, Z. Szymański and W. Nazarewicz, Phys. Scripta **42** (1990) 515.

[89] Y.R. Shimizu, E. Vigezzi and R.A. Broglia, Phys. Lett. **198B** (1987) 33.

[90] J. Dudek, B. Herskind, W. Nazarewicz, Z. Szymański and T. Werner, Phys. Rev. **C38** (1988) 940.

[91] Y.R. Shimizu, E. Vigezzi and R.A. Broglia, Nucl. Phys. **A509** (1990) 80.

[92] P. Fallon et al., Phys. Lett. **218B** (1989) 137.

[93] W. Nazarewicz, R. Wyss and A. Johnson, Phys. Lett. **225B** (1989) 208.

[94] S.M. Mullins et al., submitted to Phys. Rev. Lett.; Univ. of Liverpool preprint 90/8.

[95] G. Hebbinghaus et al., Phys. Lett. **240B** (1990) 311.

[96] K. Zuber et al., submitted to Phys. Lett. B (1990).

[97] R.Wyss and A. Johnson, to be published.

[98] R.R. Chasman, Phys. Rev. C14 (1976) 1935.

[99] R.R. Chasman, I. Ahmad, A.M. Friedman and J.R. Erskine, Rev. Mod. Phys. 49 (1977) 833.

[100] J. Dobaczewski, H. Flocard and J. Treiner, Nucl. Phys. A422 (1984) 103.

[101] D. Glas and U. Mosel, Nucl. Phys. A216 (1973) 563.

[102] R.E. Griffin, A.D. Jackson and A.B. Volkov, Phys. Lett. 36B (1971) 281.

[103] S.J. Krieger, P. Bonche, H. Flocard, P. Quentin and M.S. Weiss, Nucl. Phys. A517 (1990) 275.

[104] J. Dudek, T. Werner and Z. Szymański, Phys. Lett. 248 (1990) 235.

[105] S.A.E. Johansson, Nucl. Phys. A22 (1961) 529.

[106] C. Gustafsson, P. Möller and S.G. Nilsson, Phys. Lett. 34B (1971) 349.

[107] P. Bonche, S.J. Krieger, M.S. Weiss, J. Dobaczewski, H. Flocard and P.-H. Heenen, Saclay preprint SPhT/90/145, 1990.

[108] J. Höller and S. Åberg, Z. Phys. A336 (1990) 363.

[109] G.A. Leander, W. Nazarewicz, G.F. Bertsch and J. Dudek, Nucl. Phys. A453 (1986) 58.

[110] W. Nazarewicz, P. Olanders, I. Ragnarsson, J. Dudek and G.A. Leander, Phys. Rev. Lett. 52 (1984) 1272, 53 (1984) 2060.

[111] S. Frauendorf and V.V. Pashkevich, Phys. Lett. 141B (1984) 23.

[112] W. Nazarewicz and P. Olanders, Nucl. Phys. A441 (1985) 420.

[113] V.V. Pashkevich, Nucl. Phys. A169 (1971) 275.

[114] P. Möller, Nucl. Phys. A192 (1972) 529.

[115] R. Bengtsson, I. Ragnarsson, S. Åberg, A. Gyurkovich, A. Sobiczewski and K. Pomorski, Nucl. Phys. A473 (1987) 77.

[116] J. Blons, C. Mazur, D. Paya, M. Ribrag and H. Weigmann, Phys. Rev. Lett. 41 (1978) 1282.

[117] B. Fabbro et al., J. Physique Lett. 45 (1984) L-843.

[118] Z. Szymański, Int. Conf. on Nuclear Structure in the Nineties, Oak Ridge, 1990; Nucl. Phys. A, in print; J. Dudek and Z. Szymański, to be published.

QUASIPARTICLE-PHONON NUCLEAR MODEL AND STRUCTURE OF NONROTATIONAL STATES IN DEFORMED NUCLEI

V.G.Soloviev
Joint Institute for Nuclear Research, Dubna, USSR

ABSTRACT

The Hamiltonian of the quasiparticle-phonon nuclear model is constructed which contains an average field, monopole and quadrupole pairing and isoscalar and isovector multipole, spin-multipole and tensor particle-hole and particle-particle interactions. New phonon operators are introduced for describing deformed nuclei which consist of the electric and magnetic parts. It is shown that calculations of one-phonon states in the RPA encounter some difficulties due to the finite rank $n_{max} > 1$ separable interaction of the electric and magnetic type. The use of complex separable $n_{max} > 1$ interactions does not lead to complication of the QPNM equations for describing fragmentation of vibrational states. It is stated that the QPNM can serve as a basis for calculating many properties of excited states of deformed nuclei.

1. INTRODUCTION

The energies and wave functions of two-quasiparticle and one-phonon states of doubly even deformed nuclei were calculated in 1960-1975. Quadrupole and octupole states were calculated in the random phase approximation (RPA). The results of calculatins are collected in refs.$^{/1-3/}$. Good enough description of the available at that time experimental data was obtained and predictions were made most of which were later confirmed experimentally. In recent years, many new and more complete experimental data on the energies, $B(E\lambda)$-values and spectroscopic factors of one- and two-nucleon transfer reactions were obtained. Many experimental data are expected at the new generation of accelerators and detectors. Therefore, new, more exact and complete microscopic calculations of vibrational states of deformed nuclei are needed.

All subsequent calculations of nonrotational states of deformed nuclei were made in the quasiparticle-phonon nuclear model (QPNM) the basic assumptions of which are formulated in refs. /4-6/. In those calculations the wave functions of excited states were represented as sums of one-phonon and two-phonon terms. It was shown in /7/ that due to the shift of two-phonon poles caused by the allowance of the Pauli principle, the energy centroids of collective two--phonon states exceed 2.5 MeV. At these energies the fragmentation of the two phonon strength should take place. Therefore, it was concluded in /7/ that two-phonon collective states are absent in deformed nuclei. If the contribution of the two-phonon component exceeds 50%, this state is thought to be two-phonon one.

The calculations used not only quadrupole and octupole states. In ref. /8/, the calculations of hexadecapole vibrational states showed that collective states with $K^{\pi}= 3^+$ should exist in the isotopes of Er, Yb, Hf and with $K^{\pi}= 4^+$ in the isotopes of Os. It was shown in ref. /9/ that in some cases multipole interactions with λ =5÷9 led to the mixing of two-quasiproton and two-quasineutron states in doubly even deformed nuclei and they should be taken into account. As is known, the system of coordinates coupled to a deformed nucleus is specified by an axial symmetry, and the angular momentum projection onto the symmetry axis of a nucleus, K, and parity $\tilde{\pi}$ are good quantum numbers. We shall not consider the Coriolis interaction mixing states with different K and fixed parity. Thus, we shall restrict our consideration to the internal wave function with a good quantum number K.

The specific feature of deformed nuclei is that one-phonon states with a fixed $K^{\tilde{\pi}}$ can be determined by different multipole and spin-multipole interactions. Thus, the $K^{\tilde{\pi}} = 0^+$ states are determined by the monopole pairing and quadrupole particle-particle interactions. To them multipole interactions with $\lambda\mu$ =40,60, etc. must be added. One-phonon states of the electric type with fixed $K^{\tilde{\pi}}$ can be described by the multipoles $\lambda\mu = KK, K+2K, K+4K$, etc. and by the spin-multipoles $\lambda\lambda\mu = KKK, K+2 K+2 K$, etc. Introducing one phonon for λK and another for $\lambda\lambda K$ we shall have a double number of states. To avoid this, a common phonon is introduced, and taking account of different λ the corresponding secular equation is derived (see $^{/6/}$). Thus, the influence of hexadecapole interactions with $\lambda\mu$ =42 on the $K^{\tilde{\pi}}=2^+$ states is studied by a simultaneous inclusion of $\lambda\mu$ =22+42 interactions. In $^{/10/}$, interactions with λ =1 and λ =3 were taken into account in studying E1 transitions from octupole to ground states.

One-phonon states of the magnetic type are described by the spin-multipole interactions $\lambda' L K$ with $\lambda' = L-1$ and $L+1$. In spherical nuclei, one-phonon states of the electric type with $I^{\tilde{\pi}} = 2^+, 3^-, \ldots$ and magnetic type with $I^{\tilde{\pi}} = 1^+, 2^-, 3^+ \ldots$ are described independently. In deformed nuclei, for instance, the $K^{\tilde{\pi}} = 2^-$ state can be treated as an electric octupole one with $\lambda\mu$ =32 and a magnetic quadrupole one with $\lambda' L K$ =122 and 322. The states with $K^{\tilde{\pi}}=1^+$, which are described excluding spurious states$^{/11,12/}$, connected with rotations, are treated with the spin-spin and quadrupole interactions. If in deformed nuclei, as in spherical nuclei, one introduces independent phonons of the electric and magnetic type, the number of states will be doubled. Therefore, it is necessary to construct a common phonon for a state with a fixed $K^{\widetilde{\pi}}$.

Another stage of calculations is the inclusion of particle-particle (pp) interactions alongside with particle-hole (ph) ones /13-15/. The calculated quadrupole, octupole and hexadecapole states in ^{168}Er, ^{170,172,174}Yb and ^{178}Hf are in good agreement with experimental data. These calculations are limited by the states with $K^{\widetilde{\pi}} \neq 0^+$. In ref. /15/, equations for describing $K^{\widetilde{\pi}}=0^+$ states with inclusion of ph and pp interactions were derived within the QPNM. From the condition of exclusion of spurious 0^+ states in the RPA there were derived equations for monopole and quadrupole pairing which have been studied in /16/. This allows one to include 0^+ states in the general scheme of calculations in the QPNM of all states of the electric type.

In the lectures I have expounded the basic assumptions of the QPNM for deformed nuclei and gave the description of some characteristics of nonrotational states of doubly even deformed nuclei in the rare-earth and actinide regions. The published text contains only the basic assumptions of the QPNM.

2. THE QPNM HAMILTONIAN

The QPNM Hamiltonian for nonrotational states of deformed nuclei contains an average field of neutron and proton systems in the form of the axial-symmetric Saxon-Woods potential, monopole pairing, isoscalar and isovector particle-hole (ph) and particle-particle (pp) multipole, spin-multipole and tensor interactions between quasiparticles. The wave functions of excited states of deformed nuclei have the form

$$\Psi_{MK}^{I}(\nu) = \sqrt{\frac{2I+1}{16\pi^2}} \left\{ D_{MK}^{I} \Psi_{\nu}(K,\sigma=+) + (-)^{I+6} D_{M-K}^{I} \Psi_{\nu}(K,\sigma=-) \right\}$$ (1)

In these lectures we study the internal wave functions $\Psi_{\nu}(K^{\pi}\sigma)$ of excited nonrotational states of doubly even deformed nuclei.

Interactions between quasiparticles in the separable form, usually of the rank n_{max} =1, are used for calculations in the QPNM. As is known, separable interactions of the rank $n_{max} > 1$ are widely used in describing nucleon-nucleon interactions, three-body nuclear systems and light nuclei, i.e. they are used in the cases where the results of calculations are more sensitive to the form of radial dependence of forces in comparison with the QPNM calculations of the properties of complex nuclei. Therefore, the use of separable interactions of the rank $n_{max} \geqslant 1$ in the QPNM calculations is justified.

Let us introduce a separable interaction of the rank $n_{max} > 1$ for deformed nuclei. Expand over multipoles the central spin independent interaction and write it as

$$\sum_{\substack{q_1 q_2 q_1' q_2' \\ \sigma_1 \sigma_2 \sigma_1' \sigma_2'}} <q_1\sigma_1, q_2\sigma_2| \sum_{\lambda\mu} (x_0^{\lambda\mu} + x_1^{\lambda\mu}(\vec{\tau}^{(1)}\vec{\tau}^{(2)})) R^{\lambda\mu}(\tau_1,\tau_2)$$

$$\sum_{\sigma=\pm1} Y_{\lambda\sigma\mu}(\theta_1,\varphi_1) Y_{\lambda-\sigma\mu}(\theta_2,\varphi_2) |q_2'\sigma_2', q_1'\sigma_1'> a_{q_1\sigma_1}^+ a_{q_2\sigma_2}^+ a_{q_2'\sigma_2'} a_{q_1'\sigma_1'}$$

If a separable interaction of the rank $n_{max} > 1$ is taken in the form

$$R^{\lambda\mu}(\tau_1,\tau_2) = \sum_{n=1}^{n_{max}} R_n^{\lambda\mu}(\tau_1) R_n^{\lambda\mu}(\tau_2)$$

then the expansion over multipoles becomes

$$\sum_{\lambda\mu} \sum_{n=1}^{n_{max}} \left\{ \sum_{\tau\rho=\pm 1} (\mathcal{X}_0^{\lambda\mu} + \rho \mathcal{X}_1^{\lambda\mu}) \sum_{\sigma} M_{n\lambda\mu\sigma}^+ (\tau) M_{n\lambda\mu\sigma} (\rho\tau) + \right.$$

$$\left. + \sum_{\tau\sigma} G^{\lambda\mu} P_{n\lambda\mu\sigma}^+ (\tau) P_{n\lambda n\sigma} (\tau) + \cdots \right\}$$

Introduction of a separable interaction of the finite rank $n_{max} > 1$ in comparison with $n_{max} = 1$ leads to summation over n. Introduction of a separable interaction of the rank n_{max} is meaningful if n_{max} is much smaller than the rank of the determinant of the RPA secular equation for a nonseparable interaction.

The starting Hamiltonian of the QPNM is

$$H = \sum_{\tau} \left\{ \sum_{q\sigma}^{\tau} [E'(q) - \lambda_\tau] a_{q\sigma}^+ a_{q\sigma} - G_\tau \sum_{qq'} a_{q+}^+ a_{q-}^+ a_{q'-} a_{q'+} - \right.$$

$$- \frac{1}{2} \sum_{\lambda\mu\sigma} \sum_{n=1}^{n_{max}} [\sum_{\rho=\pm 1} (\mathcal{X}_0^{\lambda\mu} + \rho \mathcal{X}_1^{\lambda\mu}) M_{n\lambda\mu\sigma}^+ (\tau) M_{n\lambda\mu\sigma} (\rho\tau) +$$

$$+ G^{\lambda\mu} P_{n\lambda\mu\sigma}^+ (\tau) P_{n\lambda\mu\sigma} (\tau)] - \frac{1}{2} \sum_{LK\sigma} \sum_{\lambda=L, L\pm 1} \sum_{n=1}^{n_{max}} [\sum_{\rho=\pm 1} (\mathcal{X}_0^{\lambda'LK} \quad (2)$$

$$+ \rho \mathcal{X}_1^{\lambda'LK})(S_{nLK\sigma}^{\lambda'}(\tau))^+ S_{nLK\sigma}^{\lambda'}(\rho\tau) + G^{\lambda'LK}(P_{nLK\sigma}^{\lambda'}(\tau))^+ P_{nLK\sigma}^{\lambda'}(\tau)] +$$

$$+ \frac{1}{2} \sum_{LK\sigma} \sum_{\rho=\pm 1} \sum_{n=1}^{n_{max}} (\mathcal{X}_{T0}^{LK} + \rho \mathcal{X}_{T1}^{LK}) [(S_{nLK\sigma}^{L-1}(\tau))^+ S_{nLK\sigma}^{L+1}(\rho\tau) +$$

$$\left. + (S_{nLK\sigma}^{L+1}(\tau))^+ S_{nLK\sigma}^{L-1}(\rho\tau)] \right\} .$$

Here $q\sigma$ are quantum numbers of single-particle states, q equals to K^π and asymptotic quantum numbers $Nn_z\Lambda\uparrow$

at $K = \Lambda + 1/2$ and $Nn_z \Lambda \downarrow$ at $K = \Lambda - 1/2$, $\sigma = \pm 1$; $E(q)$ are the single-particle energies, λ_τ is the chemical potential; $\sum_{qq'}^\tau$ means summation over single-particle states of the proton at $\tau = p$ and neutron at $\tau = n$ systems. Then, $G_\tau^{\lambda\mu}$ are the monopole pairing constants, $G_\tau^{\lambda'LK}$ and $G^{\lambda'LK}$ are the constants of pp interactions; $\varkappa_0^{\lambda\mu}$, $\varkappa_0^{\lambda'LK}$, $\varkappa_{T0}^{LK}$ and $\varkappa_1^{\lambda\mu}$, $\varkappa_1^{\lambda'LK}$, $\varkappa_{T1}^{LK}$ are the isoscalar and isovector constants of ph multipole, spin-multipole and tensor interactions.

Let us perform the canonical Bogolubov transformation

$$a_{q\sigma} = u_q \alpha_{q\sigma} + \sigma v_q \alpha_{q-\sigma}^+$$

and get

$$M_{n\lambda\mu\sigma}(\tau) = \frac{1}{2}\sum_{q_1,q_2}^\tau f_n^{\lambda\mu}(q_1,q_2)\left\{u_{q_1,q_2}^{(+)}[A^+(q_1,q_2;\mu\sigma) + A(q_1,q_2;\mu-\sigma)] + 2v_{q_1,q_2}^{(-)} B(q_1,q_2;\mu\sigma)\right\},$$

$$P_{n\lambda\mu\sigma}(\tau) = \frac{1}{2}\sum_{q_1,q_2}^\tau f_n^{\lambda\mu}(q_1,q_2)\left\{v_{q_1,q_2}^{(+)}[A(q_1,q_2;\mu\sigma) - A^+(q_1,q_2;\mu-\sigma)] + v_{q_1,q_2}^{(-)}[A(q_1,q_2;\mu\sigma) + A^+(q_1,q_2;\mu-\sigma)] - 4u_{q_2}v_{q_1}B(q_1,q_2;\mu-\sigma)\right\},$$

$$S_{nLK\sigma}^{L\pm 1}(\tau) = \frac{1}{2}\sum_{q_1,q_2}^\tau f_n^{L\pm 1LK}(q_1,q_2)\left\{v_{q_1,q_2}^{(-)}[\mathcal{O}^+(q_1,q_2;K\sigma) + \mathcal{O}(q_1,q_2;K-\sigma)] + 2v_{q_1,q_2}^{(+)} \mathcal{B}(q_1,q_2;K\sigma)\right\}$$

and other formulae. Here

$$A^+(q_1 q_2; \mu\sigma) = \sum_{\sigma'} \delta_{\sigma'(K_1,-K_2),\sigma\mu} \sigma' \alpha^+_{q_1,\sigma'} \alpha^+_{q_2,-\sigma'} \text{ or } \delta_{K_1+K_2,\mu} \alpha^+_{q_2,\sigma} \alpha^+_{q_1,\sigma},$$

$$\mathcal{O}\ell^+(q_1 q_2; \mu\sigma) = \sum_{\sigma'} \delta_{\sigma'(K_1,-K_2),\sigma\mu} \alpha^+_{q_1,\sigma'} \alpha^+_{q_2,-\sigma'} \text{ or } \delta_{K_1+K_2,\mu} \sigma \alpha^+_{q_2,\sigma} \alpha^+_{q_1,\sigma},$$

$$B(q_1 q_2; \mu\sigma) = \sum_{\sigma'} \delta_{\sigma'(K_1,-K_2),\sigma\mu} \alpha^+_{q_1,\sigma'} \alpha_{q_2,-\sigma'} \text{ or } \delta_{K_1+K_2,\mu} \sigma \alpha^+_{q_1,\sigma} \alpha_{q_2,-\sigma},$$

$$\mathcal{B}(q_1 q_2; \mu\sigma) = \sum_{\sigma'} \delta_{\sigma'(K_1,-K_2),\sigma\mu} \sigma' \alpha^+_{q_1,\sigma'} \alpha_{q_2,\sigma'} \text{ or } \delta_{K_1+K_2,\mu} \alpha^+_{q_1,\sigma} \alpha_{q_2,-\sigma},$$

$$u^{(\pm)}_{q_1 q_2} = u_{q_1} v_{q_2} \pm u_{q_2} v_{q_1}, \quad v^{(\pm)}_{q_1 q_2} = u_{q_1} u_{q_2} \pm v_{q_1} v_{q_2}.$$

The matrix elements of the multipole and spin-multipole operators are expressed through

$$f_n^{\lambda\mu}(q_1 q_2) = \langle q_1 | R_n^{\lambda\mu}(\tau) Y_{\lambda\mu}(\theta,\varphi) | q_2 \rangle$$

$$f_n^{\lambda'LK}(q_1 q_2) = \langle q_1 | R_n^{\lambda'LK}(\tau) \{\sigma Y_{\lambda'}(\theta\varphi)\}_{LK} | q_2 \rangle, \tag{3}$$

their characteristics are given in /1,6/.

Using the operators $A^+(qq'; \mu\sigma)$ and $A(qq'; \mu\sigma)$ to construct phonons of the electric type, as in /1,5,6/, and the operators $\mathcal{O}\ell^+(qq'; \mu\sigma)$ and $\mathcal{O}\ell(qq'; \mu\sigma)$ to construct phonons of the magnetic type, then in contrast with the spherical nuclei we shall have a doubled number of states. Consider, for instance, the $K^\pi = 2^-$ states shown in fig. 1. They can be described as one-phonon octupole states with $\lambda\mu = 32$ and as a rule, with the enhanced E3 transition from the $I^\pi K_i = 3^- 2_1$ to the ground states. Between the first and second poles there is a second $K_i^\pi = 2_2^-$ state whose energy is determined as the second $i = 2$ root of the RPA secular equation. At the same time, these $K_i^\pi = 2_1^-$ and 2_2^- states can be described as one-phonon quadrupole states

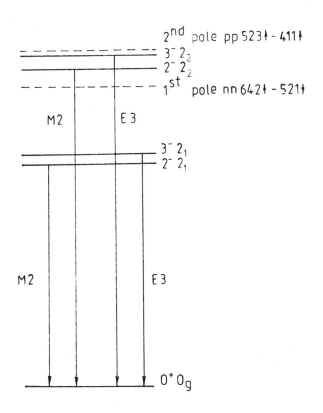

Fig. 1

The first two states with $K^{\pi}=2^-$ described either as quadrupole magnetic states with enhancement of M2 transitions or as octupole electric states with enhancement of E3 transitions and the first two two-quasiparticle states.

of the magnetic type and with the enhanced M2 transition from the $I^\pi K_i = 2^- 2_1$ to the ground states. If the set of one-phonon states is chosen as a basis, as in the QPNM, then the number of one-phonon states with a fixed value of K^π equals the number of two-quasiparticle poles. If phonons of the electric and magnetic type are introduced in the deformed nuclei, then the number of states will be doubled. Therefore, it is necessary to construct a common phonon operator consisting of the electric and magnetic parts. The phonon creation operator consisting of the electric and magnetic parts and with a fixed value of K^π can be written as follows /17/:

$$Q^+_{Ki_2\sigma} = \frac{1}{2} \sum_{qq'} \left\{ \psi^{Ki_2}_{qq'} \left[f^{\lambda K}(qq') u^{(+)}_{qq'} A^+(qq'; K\sigma) + \right. \right.$$

$$\left. + i f^{LK}(qq') u^{(-)}_{qq'} \alpha^+(qq'; K\sigma) \right] - \tag{4}$$

$$- \varphi^{Ki_2}_{qq'} \left[f^{\lambda K}(qq') u^{(+)}_{qq'} A(qq'; K-\sigma) + \right.$$

$$\left. + i f^{LK}(qq') u^{(-)}_{qq'} \alpha(qq'; K-\sigma) \right] \right\} .$$

Here $\psi^{Ki_2}_{qq'} = \psi^{Ki_2}_{q'q}$ and $\varphi^{Ki_2}_{qq'} = \varphi^{Ki_2}_{q'q}$ are common for the electric and magnetic parts, which indicates the existence of a one-phonon state with a fixed value of the state number i_2 where $i_2 = 1, 2, 3, \ldots$ By $f^{LK}(qq')$ we denote the matrix element (3) with $\lambda' LK = L - 1LK$ or $L + 1LK$. When we use a separable interaction with $n_{max} > 1$ by $f^{\lambda K}(qq')$ and $f^{LK}(qq')$ we denote the matrix elements with $n = 1$. It is seen from formula (4) that the electric part of the phonon creation operator $Q^+_{Ki_2\sigma}$ is chosen to be real and the magnetic part to be imaginary.

The one-phonon state in the RPA is described by the wave function

$$Q^+_{Ki_2\sigma} \Psi_0 \, , \qquad (5)$$

where Ψ_0 is the ground state wave function of a doubly even nucleus determined as a phonon vacuum. The normalisation condition of the wave function (5) has the form

$$\frac{1}{2} \sum_{qq'} [(\varphi^{Ki_2}_{qq'})^2 - (\varphi^{Ki_2}_{qq'})^2] \, \gamma^K(qq') = 1 \, ,$$

$$\gamma^K(qq') = | f^{\lambda K}(qq') u^{(+)}_{qq'} + i f^{LK}(qq') u^{(-)}_{qq'} |^2 = \qquad (6)$$

$$= [f^{\lambda K}(qq') u^{(+)}_{qq'}]^2 + [f^{LK}(qq') u^{(-)}_{qq'}]^2 \, ,$$

$$A^+(qq'; K\sigma) = f^{\lambda K}(qq') u^{(+)}_{qq'} \sum_{i_2} [\varphi^{Ki_2}_{qq'} Q^+_{Ki_2\sigma} + \varphi^{Ki_2}_{qq'} Q_{Ki_2-\sigma}] \, ,$$

$$\mathcal{H}^+(qq'; K\sigma) = -i f^{LK}(qq') u^{(-)}_{qq'} \sum_{i_2} [\varphi^{Ki_2}_{qq'} Q^+_{Ki_2\sigma} - \qquad (7)$$

$$- \varphi^{Ki_2}_{qq'} Q_{Ki_2-\sigma}] \, .$$

We do not take into account noncoherent terms. One can easily show that the phonon operators $Q^+_{Ki_2\sigma}$ and $Q_{Ki_2\sigma}$ obey the conditions which are usually imposed on the RPA phonons.

Using formulae (7) we express the operators $M_{n\lambda\mu\sigma}(\tau)$, $P_{n\lambda\mu\sigma}(\tau)$, $S_{nLK\sigma}^{\lambda'}(\tau)$ and others through the phonon operators. After simple transformations the QPNM Hamiltonian becomes

$$H = \sum_{q\sigma} \mathcal{E}_q \alpha_{q\sigma}^+ \alpha_{q\sigma} + H_\vartheta + H_{\vartheta q} , \qquad (8)$$

where the first two terms describe quasiparticles and phonons, and $H_{\vartheta q}$ describes the quasiparticle-phonon interaction. They have the following form:

$$H_\vartheta = H_\vartheta^{00} + \sum_\lambda H_\vartheta^{\lambda 0} + \sum_K H_\vartheta^K , \qquad (9)$$

$$H_\vartheta^{00} = -\frac{1}{2} \sum_\tau \sum_{ii'} G_\tau [d_{g\tau}^i d_{g\tau}^{i'} + d_{w\tau}^i d_{w\tau}^{i'}] Q_{20i}^+ Q_{20i'} \qquad (10)$$

$$H_\vartheta^{\lambda 0} = -\sum_{ii'} W_{ii'}^{\lambda 0} Q_{\lambda 0 i}^+ Q_{\lambda 0 i'}$$

$$H_\vartheta^K = -\sum_{i_1 i_2 \sigma} W_{i_1 i_2}^K Q_{K i_1 \sigma}^+ Q_{K i_2 \sigma} , \qquad (10')$$

$$W_{i_1 i_2}^K = \sum_\lambda [W_{i_1 i_2}^{\lambda K} + W_{i_1 i_2}^{\lambda\lambda K}] + \sum_L [\sum_{\lambda'=L\pm 1} W_{i_1 i_2}^{\lambda' L K} + W_{i_1 i_2}^{TLK}] , \qquad (10'')$$

$$W_{i_1 i_2}^{\lambda K} = \frac{1}{4} \sum_{n=1}^{n_{max}} \sum_\tau \{ \sum_{\rho=\pm 1} (\mathcal{H}_0^{\lambda K} + \rho \mathcal{H}_1^{\lambda K}) D_{n\tau}^{\lambda K i_1} D_{n\rho\tau}^{\lambda K i_2} +$$

$$+ G^{\lambda K} [D_{ng\tau}^{\lambda K i_1} D_{ng\tau}^{\lambda K i_2} + D_{nw\tau}^{\lambda K i_1} D_{nw\tau}^{\lambda K i_2}] \} (1 + \delta_{K0})^2 ,$$

$$W_{i_1 i_2}^{\lambda' L K} = \frac{1}{4} \sum_{n=1}^{n_{max}} \sum_\tau \{ \sum_{\rho=\pm 1} (\mathcal{H}_0^{\lambda' L K} + \rho \mathcal{H}_1^{\lambda' L K}) D_{n\tau}^{\lambda' L K i_1} D_{n\rho\tau}^{\lambda' L K i_2} +$$

$$+ G^{\lambda K} [D_{ng\tau}^{\lambda' L K i_1} D_{ng\tau}^{\lambda' L K i_2} + D_{nw\tau}^{\lambda' L K i_1} D_{nw\tau}^{\lambda' L K i_2}] \} ,$$

$$W_{i_1 i_2}^{TLK} = -\frac{1}{2} \sum_{n=1}^{n_{max}} \sum_{\tau,\rho=\pm 1} (\mathcal{X}_{T0}^{LK} + \rho \mathcal{X}_{T1}^{LK}) D_{n\tau}^{L-1LKi_1} D_{n\rho\tau}^{L+1LKi_2} \ .$$

$$H_{vq} = H_{vq}^{00} + \sum_{\lambda} H_{vq}^{\lambda 0} + \sum_{K} \left\{ \sum_{\lambda} (H_{vq}^{\lambda K} + H_{vq}^{\lambda\lambda K}) + \right.$$
$$\left. + \sum_{L} [\sum_{\lambda'=L\pm 1} H_{vq}^{\lambda' LK} + H_{vq}^{TLK}] \right\} \quad (11)$$

$$H_{vq}^{00} = -\sum_{\tau i} G_\tau \sum_{qq'}^\tau (u_q^2 - v_q^2) u_{q'} v_{q'} \{ [\psi_{qq'}^{20i} Q_{20i}^+ + \varphi_{qq'}^{20i} Q_{20i}] \sum_\sigma \alpha_{q\bar\sigma}^+ \alpha_{q'\bar\sigma} + h.c.\},$$

$$H_{vq}^{\lambda K} = -\frac{1}{4} \sum_{n i_2 \tau \sigma} \sum_{qq'}^\tau f_n^{\lambda K}(qq') V_{n\tau}^{\lambda K i_2}(qq') [(Q_{K i_2 \sigma}^+ +$$
$$+ Q_{K i_2 -\sigma}) B(qq'; K-\sigma) + B(qq'; K\sigma)(Q_{K i_2 -\sigma}^+ + Q_{K i_2 \sigma})], \quad (12)$$

$$H_{vq}^{\lambda 0} = -\sum_{i\tau} \sum_{qq'}^\tau V_\tau^{\lambda 0i}(qq') f^{\lambda 0}(qq') \{(Q_{\lambda 0i}^+ + Q_{\lambda 0i}) B(qq'; \mu=0) + h.c.\},$$

$$H_{vq}^{\lambda' LK} = \frac{i}{4} \sum_{n i_2 \tau \sigma} \sum_{qq'}^\tau f_n^{\lambda' LK}(qq') V_{n\tau}^{\lambda' LK i_2}(qq') [(Q_{K i_2 \sigma}^+ - Q_{K i_2 -\sigma}) \cdot$$
$$\cdot \mathcal{B}(qq'; K-\sigma) + \mathcal{B}(qq'; K\sigma)(Q_{K i_2 -\sigma}^+ - Q_{K i_2 \sigma})], \quad (12')$$

$$H_{vq}^{TLK} = -\frac{i}{4} \sum_{n i_2 \tau \rho} (\mathcal{X}_{T0}^{LK} + \rho \mathcal{X}_{T1}^{LK}) \sum_{qq'}^\tau [D_{\rho\tau}^{L-1LK i_2} f^{L+1LK}(qq') +$$
$$+ D_{\rho\tau}^{L+1LK i_2} f^{L-1LK}(qq')] v_{qq'}^{(+)} [(Q_{K i_2 \sigma}^+ - Q_{K i_2 -\sigma}) \mathcal{B}(qq'; K-\sigma) + h.c.] ,$$

$$V_{n\tau}^{\lambda K i_2}(qq') = \sum_{\rho=\pm 1}(\varkappa_0^{\lambda K}+\rho\varkappa_1^{\lambda K}) v_{qq'}^{(-)} D_{n\rho\tau}^{\lambda K i_2} -$$
$$- G^{\lambda K} u_{qq'}^{(+)} D_{ng\tau}^{\lambda K i_2} , \qquad (13)$$
$$V_{n\tau}^{\lambda' L K i_2}(qq') = \sum_{\rho=\pm 1}(\varkappa_0^{\lambda' L K}+\rho\varkappa_1^{\lambda' L K}) v_{qq'}^{(+)} D_{n\rho\tau}^{\lambda' L K i_2} . \qquad (13')$$

Here $\mathcal{E}_q$ is the quasiparticle energy with the monopole and quadrupole pairing; the operator $Q_{\lambda 0 i}$ is given in /6/.

$$d_{g\tau}^i = \sum_q^\tau \frac{E(q)-\lambda_\tau}{\mathcal{E}_q} g_{qq}^{20i} , \quad d_{n\tau}^i = \sum_q^\tau w_{qq}^{20i} ,$$
$$D_{n\tau}^{\lambda K i_2} = \sum_{qq'}^\tau f^{\lambda K}(qq') f_n^{\lambda K}(qq')(u_{qq'}^{(+)})^2 g_{qq'}^{K i_2} ,$$
$$D_{ng\tau}^{\lambda K i_2} = \sum_{qq'}^\tau f^{\lambda K}(qq') f_n^{\lambda K}(qq') u_{qq'}^{(+)} v_{qq'}^{(-)} g_{qq'}^{K i_2} ,$$
$$D_{nw\tau}^{\lambda K i_2} = \sum_{qq'}^\tau f^{\lambda K}(qq') f_n^{\lambda K}(qq') u_{qq'}^{(+)} v_{qq'}^{(+)} w_{qq'}^{g i_2} ,$$
$$D_{n\tau}^{\lambda' L K i_2} = \sum_{qq'}^\tau f^{\lambda' L K}(qq') f_n^{\lambda' L K}(qq')(u_{qq'}^{(-)})^2 g_{qq'}^{K i_2} , \qquad (14)$$
$$D_{ng\tau}^{\lambda' L K i_2} = \sum_{qq'}^\tau f^{\lambda' L K}(qq') f_n^{\lambda' L K}(qq') u_{qq'}^{(-)} v_{qq'}^{(+)} g_{qq'}^{K i_2} ,$$
$$D_{nw\tau}^{\lambda' L K i_2} = \sum_{qq'}^\tau f^{\lambda' L K}(qq') f_n^{\lambda' L K}(qq') u_{qq'}^{(-)} v_{qq'}^{(-)} w_{qq'}^{K i_2} ,$$
$$g_{qq'}^{K i_2} = \psi_{qq'}^{K i_2} + \varphi_{qq'}^{K i_2} , \quad w_{qq'}^{K i_2} = \psi_{qq'}^{K i_2} - \varphi_{qq'}^{K i_2} ,$$

$E(q)$ are single-particle energies, λ_τ are chemical potentials.

One can easily verify that the Hamiltonian (8) and its parts
(9), (10), (11), (12) and (12^1) are Hermitian.

3. THE RPA EQUATION

Let us derive the RPA equations for the energies ω_{Ki_0}
and wave functions (5) of one-phonon states. The RPA equations for the $K^{\tilde{\pi}}=0^+$ states are given in /15/. To describe
the states with $K^{\tilde{\pi}} \neq 0^+$ we use the following part of the
Hamiltonian (9)

$$\sum_{q\sigma} \varepsilon_q \alpha^+_{q\sigma} \alpha_{q\sigma} + H^K_v \qquad (15)$$

Now, we find an average value (15) over the state (5)
and using the variational principle

$$\delta\left\{ <Q_{Ki_0\sigma} [\sum_{q\sigma}\varepsilon_q \alpha^+_{q\sigma}\alpha_{q\sigma} + H^K_v] Q^+_{Ki_0\sigma}> - \right.$$

$$\left. - \frac{\omega_{Ki_0}}{2}[\sum_{qq'} g^{Ki_0}_{qq'} w^{Ki_0}_{qq'} \gamma^K(qq') - 2]\right\} = 0$$

we get the following equations:

$$\varepsilon_{qq'} \gamma^K(qq') g^{Ki_0}_{qq'} - \omega_{Ki_0} \gamma^K(qq') w^{Ki_0}_{qq'} -$$

$$-\sum_{n=1}^{n_{max}}\left\{\sum_{\rho=\pm 1}(x^{\lambda K}_0 + \rho x^{\lambda K}_1) f^{\lambda K}(qq') f^{\lambda K}_n(qq')(u^{(+)}_{qq'})^2 D^{\lambda Ki_0}_{n\rho\tau} + \right.$$

$$+ G^{\lambda K} f^{\lambda K}(qq') f^{\lambda K}_n(qq') u^{(+)}_{qq'} v^{(-)}_{qq'} D^{\lambda Ki_0}_{ng\tau} +$$

$$+ G^{\lambda\lambda K} f^{\lambda\lambda K}(qq') f^{\lambda\lambda K}_n(qq') v^{(-)}_{qq'} v^{(-)}_{qq'} D^{\lambda\lambda Ki_0}_{ng\tau} +$$

$$+ \sum_{\lambda'=L\pm 1} [\sum_{\rho=\pm 1} (\mathcal{X}_o^{\lambda'LK} + \rho \mathcal{X}_1^{\lambda'LK}) f^{\lambda'LK}(qq') f_n^{\lambda'LK}(qq')(u_{qq'}^{(-)})^2 \cdot$$

$$\cdot D_{n\rho\tau}^{\lambda'LKi_o} + G^{\lambda'LK} f^{\lambda'LK}(qq') f_n^{\lambda'LK}(qq') u_{qq'}^{(-)} v_{qq'}^{(+)} \cdot$$

$$\cdot D_{ng\tau}^{\lambda'LKi_o}]\} = 0 \qquad (16)$$

$$\mathcal{E}_{qq'} \mathcal{Y}^K(qq') w_{qq'}^{Ki_o} - \omega_{Ki_o} \mathcal{Y}^K(qq') g_{qq'}^{Ki_o} -$$

$$- \sum_{n=1}^{n_{max}} \{ G^{\lambda K} f^{\lambda K}(qq') f_n^{\lambda K}(qq') u_{qq'}^{(+)} v_{qq'}^{(+)} D_{nw\tau}^{\lambda Ki_o} -$$

$$- \sum_{\rho=\pm 1} (\mathcal{X}_o^{\lambda\lambda K} + \rho \mathcal{X}_1^{\lambda\lambda K}) f^{\lambda\lambda K}(qq') f_n^{\lambda\lambda K}(qq')(u_{qq'}^{(-)})^2 D_{n\rho\tau}^{\lambda\lambda Ki_o} +$$

$$+ G^{\lambda\lambda K} f^{\lambda\lambda K}(qq') f_n^{\lambda\lambda K}(qq') u_{qq'}^{(-)} v_{qq'}^{(+)} D_{nw\tau}^{\lambda\lambda Ki_o} +$$

$$+ \sum_{\lambda'=L\pm 1} G^{\lambda'LK} f^{\lambda'LK}(qq') f_n^{\lambda'LK}(qq') u_{qq'}^{(-)} v_{qq'}^{(-)} D_{nw\tau}^{\lambda'LKi_o} \} = 0. \quad (17)$$

where $\mathcal{E}_{qq'} = \mathcal{E}_q + \mathcal{E}_{q'}$ and $\langle \cdots \rangle$ means averaging over the phonon vacuum.

From eqs. (16) and (17) we get the functions $g_{qq'}^{Ki_o}$ and $w_{qq'}^{Ki_o}$ and substitute them into formulae for $D_{n\tau}^{\lambda Ki_o}$, $D_{ng\tau}^{\lambda Ki_o}$, $D_{nw\tau}^{\lambda Ki_o}$, $D_{n\tau}^{\lambda\lambda Ki_o}$,

$$D_{n g \tau}^{\lambda \lambda K i_0}, \quad D_{n w \tau}^{\lambda \lambda K i_0}, \quad D_{n \tau}^{L \pm 1 L K i_0}, \quad D_{n g \tau}^{L \pm 1 L K i_0}$$

and $D_{n w \tau}^{L \pm 1 L K i_0}$. With the allowance made for $\tau = \rho, n$ and $n = 1, 2, \ldots n_{max}$ the secular equation for the energies $\omega_{K i_0}$ has the form of the determinant of the rank $24 \cdot n_{max}$, i.e.

$$det \| 24 \cdot n_{max} \| = 0. \tag{18}$$

If the spin-multipole interactions of the electric type $\lambda \lambda K$ are disregarded, the rank of the determinant is $18 \, n_{max}$. If only ph interactions are taken into account, the rank of the determinant is $8 \, n_{max}$. The most interesting case is when ph and pp multipole and ph $L - 1LK$ spin-multipole interactions are taken into consideration; then, the rank of the determinant is $8 \, n_{max}$. It is to be noted that a particular case of eqs. (16) and (17) for $n_{max} = 1$ and ph, pp multipole interactions is given in[14]. The tensor forces being added in the Hamiltonian won't change the rank of the determinant (18).

4. THE QPNM EQUATIONS FOR DOUBLY EVEN DEFORMED NUCLEI

Here we give formulae for describing nonrotational states with $K^\pi \neq 0^+$ in the QPNM with the new phonons $Q_{K i_2 6}^+$. The wave function can be written in the form

$$\Psi_\nu (K_0^{\pi_0} 6_0) = \Big\{ \sum_{i_0} R_{i_0}^\nu Q_{K_0 i_0 6_0}^+ + \sum_{\substack{K_1 i_1 6_1 \\ K_2 i_2 6_2}} \frac{1}{2} (1 + \delta_{K_1 i_1, K_2 i_2})^{1/2} \cdot$$

(19)

$$\cdot \delta_{6_1 K_1 + 6_2 K_2, 6_0 K_0} P_{K_1 i_1, K_2 i_2}^\nu Q_{K_1 i_1 6_1}^+ Q_{K_2 i_2 6_2}^+ \Big\} \Psi_0 \quad ,$$

where $\nu = 1,2,3,\ldots$ is the number of the state with $K_0^{\pi_0}$.
To take the Pauli principle into account in two-phonon terms of the wave function (19) we introduce the function

$$\mathcal{K}^{K_0}(K_2 i_2, K_1 i' | K_1 i_1, K_2 i_2) = (1 + \delta_{K_1 i_1, K_2 i_2})^{-1} \cdot$$
(20)
$$\cdot \sum_{\sigma_1 \sigma_2} \delta_{\sigma_1, K_1 + \sigma_2 K_2, \sigma_0 K_0} \langle Q_{K_2 i_2 \sigma_2} [[Q_{K, i' \sigma_1}, Q^+_{K_1 i_1 \sigma_1}] Q^+_{K_2 i_2 \sigma_2}] \rangle ,$$

$$\mathcal{K}^{K_0}(K_1 i_1, K_2 i_2) \equiv \mathcal{K}^{K_0}(K_2 i_2, K_1 i_1 | K_1 i_1, K_2 i_2) .$$

The normalisation condition of the wave function (19) in the diagonal in $\mathcal{K}^{K_0}$ approximation has the form

$$\sum_{i_0}(R^\nu_{i_0})^2 + \sum_{K_1 i_1 \leq K_2 i_2}(P^\nu_{K_1 i_1, K_2 i_2})^2 [1 + \mathcal{K}^{K_0}(K_1 i_1, K_2 i_2)] = 1 .$$ (21)

Now, let us find an average value of the Hamiltonian (8) over the state (19) and using the variational principle derive the following equations for the energies η_ν and wave function (19)

$$(\omega_{K_0 i_0} - \eta_\nu) R^\nu_{i_0} - \sum_{K_1 i_1 \leq K_2 i_2}(1 + \delta_{K_1 i_1, K_2 i_2})^{-1/2} P^\nu_{K_1 i_1, K_2 i_2} \cdot$$
$$\cdot U^{K_0 i_0}_{K_1 i_1, K_2 i_2}[1 + \mathcal{K}^{K_0}(K_1 i_1, K_2 i_2)] = 0 ,$$ (22)

$$[\omega_{K_1 i_1} + \omega_{K_2 i_2} + \Delta\omega(K_1 i_1, K_2 i_2) - \eta_\nu] P^\nu_{K_1 i_1, K_2 i_2} -$$
$$- \sum_{i_0}(1 + \delta_{K_1 i_1, K_2 i_2})^{-1/2} R^\nu_{i_0} U^{K_0 i_0}_{K_1 i_1, K_2 i_2} = 0 .$$ (22')

Hence, we get the secular equation

$$\det \left\| (\omega_{K_0 i_0} - \eta_\nu) \delta_{i_0 i'} - \sum_{K_1 i_1 \leq K_2 i_2} (1 + \delta_{K_1 i_1, K_2 i_2})^{-1} \right.$$

$$\left. \cdot \frac{U^{K_0 i_0}_{K_1 i_1, K_2 i_2} U^{K_0 i'_0}_{K_1 i_1, K_2 i_2} [1 + \mathcal{K}^{K_0}(K_1 i_1, K_2 i_2)]}{\omega_{K_1 i_1} + \omega_{K_2 i_2} + \Delta \omega (K_1 i_1, K_2 i_2) - \eta_\nu} \right\| = 0 \quad (23)$$

From (21) and (22,22') we find $R^\nu_{i_0}$ and $R^\nu_{K_1 i_1, K_2 i_2}$ for each value of η_ν. The rank of the determinant (23) equals the number of one-phonon terms in the wave function (19).

It is important that eqs. (25) and (26) coincide in form with the equations given in /5,6,14/ in which only ph multipole interactions are taken into account, with the equations in /8/ in which ph multipole interactions $\lambda \mu$ = =22 and 42 are considered and with the equations in /13-15/ in which ph and pp multipole interactions $\lambda \mu$ at n_{max} =1 are taken into account. Thus, the form of equations (21) and (23) and the rank of the determinant (23) are independent of what multipole and spin-multipole interactions are taken into account and are independent of the rank n_{max} of separable interactions. This means that calculations in the QPNM can be made with any complex interactions in the separable form. The QPNM was formulated so that all complications caused by the form of interactions were concentrated in the RPA equations. It is not difficult to solve the RPA equations with complex interactions. The inclusion of ph and pp separable $n_{max} > 1$ interactions of the electric and magnetic types complicates the formulae for the two-phonon pole shift $\Delta \omega (K_1 i_1, K_2 i_2)$ and the function $U^{K_0 i_0}_{K_1 i_1, K_2 i_2}$ that is responsible for the quasiparticle-phonon interaction.

5. CONCLUSION

In the present lectures we have formulated the most general version of the QPNM. We have constructed the Hamiltonian and derived equations for ph and pp isoscalar and isovector multipole and spin-multipole finite rank separable interactions between quasiparticles. Introduction of the finite rank $n_{max} > 1$ separable interactions leads to complication of the RPA equations, which is nonessential in computer calculations. All difficulties connected with the electric and magnetic types of interactions and with the $n_{max} > 1$ separable interactions are concentrated in the RPA equations. It is important that they do not lead a noticeable complication of the QPNM equations for calculating the fragmentation of vibrational states including giant resonances. Additional difficulties caused by $n_{max} > 1$ do not arise if three-phonon terms are added to the wave function (22). They also do not arise in calculating the fragmentation of one-quasiparticle states in odd deformed nuclei.

I should like to emphasize that in solving such a complicated problem as the many-body nuclear problem one should aim at exposing the most important parts of effective interactions to be used in concrete calculations rather than at solving the problem in the most general form.

The mathematical apparatus of the QPNM constructed in this paper for deformed nuclei can serve as a basis for calculations of many characteristics of low-lying and high-lying states. We hope that the QPNM calculations will stimulate further experimental study of the structure of deformed nuclei at a new generation of accelerators and detectors.

REFERENCES.

1. Soloviev V.G. Theory of complex nuclei, Moscow, Nauka, 1971.
2. Grigoriev E.P., Soloviev V.G. Structure of even deformed nuclei, Moscow, Nauka, 1974.
3. Ivanova S.P., Komov A.L., Malov L.A., Soloviev V.G. Part.Nucl. 1976, v. 7, p. 450.
4. Soloviev V.G. Part.Nucl. 1978, v.9, p. 580.
5. Soloviev V.G. Prog.Part.Nucl.Phys., 1987, v.19, p.107.
6. Soloviev V.G.Theory of atomic nuclei. Quasiparticle and Phonons. Moscow: Energoatomizdat, 1989.
7. Soloviev V.G., Shirikova N.Yu. Z.Phys.A – Atoms and Nuclei. 1981, v. 301, p.263.
   Soloviev V.G., Shirikova N.Yu. Yad.Fys. 1982, v.36, p.1376.
8. Nesterenko V.O., Soloviev V.G., Sushkov A.V., Shirikova N.Yu. Yad.Fys. 1986, v.44, p.1443.
9. Soloviev V.G., Sushkov A.V. J.Phys. G. Nuc.Part.Phys. 1990, v.16, p.L57.
10. Alikov B.A., Badalov Kh.N., Nesterenko V.O., Sushkov A.V. Z.Phys. A – Atomic Nuclei 1980, v.331, p.265.
11. Pyatov N.I., Chernej M.I. Yad.Fiz. 1972, v.16, p.931.
12. Nojarov R., Faessler A. Nucl.Phys. 1988, v.484, p.1.
13. Soloviev V.G., Shirikova N.Yu. Izv. AN SSSR ser.fyz. 1988, v. 52, p.2005.
14. Soloviev V.G., Shirikova N.Yu. Z.Phys. – Atomic Nuclei 1989, v. 334, p. 149; Izv. AN SSSR, ser.fyz. 1990, v.54, p. 818.
15. Soloviev V.G. Z.Phys. – Atomic Nuclei 1989, v. 334, p.143.
16. Karadjov D., Soloviev V.G., Sushkov A.V. Izv.AN SSSR, ser.fyz. 1989, v. 53, p. 2150.
17. Soloviev V.G. Preprint JINR E4-90-119, Dubna, 1990.

237

Description of the Low-Lying Isovector 1+ States [*]

Amand Faessler
University of Tuebingen
Department of Physics
D-7400 Tuebingen, West-Germany

Abstract

The low-lying isovector 1^+ state in deformed nuclei is studied using phenomenological and microscopic approaches. The phenomenological approach used is the generalized Bohr-Mottelson-Model which treats the protons and the neutrons separately by allowing proton- and neutron-quadrupole deformations coupled by the symmetry energy. This approach describes also the isovector quadrupole excitations in spherical nuclei found by Denis Hamilton. In deformed nuclei it reproduces the excitation energy of the low lying isovector 1^+states correctly but it overestimates the magnetic dipole transition probability between the ground state and the 1^+ states at around 3 MeV. In this phenomenological approach the whole strength is concentrated in one state. To get a correct description one has to use a microscopic approach. We therefore develop in detail the microscopic description of excited states by the Random-Phase-Approximation (RPA). The properties of RPA solutions are discussed in the schematic model. Special emphasis is put on the spurious states connected with exact symmetries of the many-body Hamiltonian (translational and rotational invariance; proton and neutron number conservation). RPA is then applied to understand the structure of the isovector 1^+ states in deformed nuclei, including transition probabilities and inelastic electron scattering. Results are also shown for the interacting boson model 2 (IBA2), which is a phenomenological model describing proton and neutron monopole and quadrupole degrees of freedom in nuclei.

---

[*] Lectures given at the Predeal Summerschool, September 1990 in Romania. Supported by the BMFT under contract number 06 TÜ 714 and by the International Büro in Karlsruhe.

1. Introduction

In 1984 Denis Hamilton [1,2] and A. Richter [3,4] found independently of each other in spherical and in deformed nuclei low-lying isovector states, respectively. The states of Denis Hamilton and coworkers [1,2] are in spherical nuclei at around 2 MeV and have angular momenta 2^+. He found them in (n, γ) reactions mainly performed at the high flux reactor in Grenoble near the mass number $A \approx 130$. Achim Richter and coworkers [3,4] found low-lying isovector states in deformed nuclei with angular momenta 1^+ at about an excitation energy of 3 MeV with magnetic dipole transitions from the ground state into these 1^+ states with about $1\mu_N^2$.

Such states have been predicted by several theoretical groups: In generalizing the Bohr-Mottelson model to include proton and neutron degrees of freedom[5] I obtained low-lying isovector states in spherical nuclei. A generalization to deformed nuclei of this model yields the 1^+ isovector states found by A. Richter. In the 70s this model had been extensively used by Greiner, Maruhn and coworkers[6]. Specializing the generalized Bohr-Mottelson model to static axially symmetric proton and neutron deformations one obtains the two rotor-model of Lo Iudice and Palumbo[7]. A similar model has also been used by Suzuki and Rowe[8] and by R. Hilton [9].

In a similar way as the Bohr-Mottelson model can be generalized to include proton and neutron degrees of freedom the interacting boson model[10] can also be extended (IBA2)[11,12]. IBA2 predicts mixed symmetry (isovector) states where the proton and neutron bosons are not excited in phase. The usual low-lying isoscalar excitations correspond to completely symmetric proton and neutron boson states.

2 The Generalized Bohr-Mottelson Model in Spherical Nuclei.

We follow here the extension of the Bohr-Mottelson model to coupled proton and neutron quadrupole degrees of freedom given by Faessler [5] in 1966. We consider the nucleus to consist out of a proton and a neutron liquid which can perform independent quadrupole oscillations.

$$R_p(\vartheta,\varphi) = R_0\left(1 + \sum_{\mu=-2}^{2} \alpha_{p2\mu}(t) Y_{2\mu}(\vartheta,\varphi)\right)$$
$$R_n(\vartheta,\varphi) = R_0\left(1 + \sum_{\mu=-2}^{2} \alpha_{n2\mu}(t) Y_{2\mu}(\vartheta,\varphi)\right) \quad (1)$$

The dynamical variables are the quadrupole deformations $\alpha_{\tau 2\mu}$. They describe classically vibrations around the spherical equilibrium. The spherical harmonics $Y_{2\mu}$ give a complete basis set for all the quadrupole deformations. Lines of equal density are characterized by the parameter r_0.

$$r_\tau(\vartheta,\varphi) = r_0(1 + \alpha_\tau \cdot Y_2) \quad (2)$$

The subscript τ indicates the equal density lines $r_\tau(\vartheta,\varphi)$ for the protons and the neutrons. The dot between the quadrupole deformations α_τ and the spherical harmonics Y_2 indicate a scalar product including a sum over μ from -2 to +2. Since we use in the future only quadrupole deformations we omitted the subscript 2 in $\alpha_{\tau\mu}$. The Hamiltonian which describes the dynamics of the proton and neutron quadrupole deformations $\alpha_{\tau\mu}$ is given for spherical nuclei as a quadratic expression in the α's.

$$\hat{H} = \sum_{\tau=p,n}\left[\frac{1}{2}B_\tau\dot{\alpha}_\tau\cdot\dot{\alpha}_\tau + \frac{1}{2}C_\tau\alpha_\tau\cdot\alpha_\tau\right] + G(\alpha_p - \alpha_n)\cdot(\alpha_p - \alpha_n)$$
$$with: \quad \alpha_\tau \cdot \alpha_\tau \equiv \sum_{\mu=-2}^{2} \alpha_{\tau\mu}^+ \alpha_{\tau\mu} \quad (3)$$

Scholtz, Kyrchev and Faessler[13] gave recently with the help of the group SP(4,R) an exact solution of the Hamiltonian (3). They showed that (3) represents exactly two 5 dimensional uncoupled harmonic oscillators of protons and neutrons oscillating in and out of phase.

Here we want to follow the solution given in reference 5 assuming that the mass B_τ and the restoring force parameters C_τ for the protons and the neutrons are roughly equal.

The last term in the Hamiltonian (3) takes into account that the demixing of protons and neutrons costs energy as for example reflected in the symmetry energy term of the Bethe-Weizsaecker mass formula.

$$E_{sym} = a_s \frac{(N-Z)^2}{A} \qquad (4)$$

The value of the symmetry energy parameter $a_s (\approx 27$ MeV$)$ allows to determine the demixing restoring force constant G. For this we write the symmetry energy term of the Bethe-Weizsaecker mass formula as an integral over proton and neutron densities.

$$E_{sym} = \int a_s(\rho) \frac{\left(\rho_n(\mathbf{r}) - \rho_p(\mathbf{r})\right)^2}{\rho_n + \rho_p} d\tau$$
with: $\quad a_s(\rho) = C(\rho_p + \rho_n)^{2/3} \qquad (5)$
$$\rho(\mathbf{r}) = \frac{\rho_0}{1 + exp([r_0 - R_0]/a)}$$

Here r_0 is the parameter of equation (2) defining equal density surfaces and thus if eliminated with eq.(2) it introduces a dependence on the deformation parameter $\alpha_{\tau\mu}$. If one expands the symmetry energy E_{sym} into powers of α_τ one gets a constant contribution to (3), which one can neglect, and the last term of the Hamiltonian (3) with a quantitative expression for the constant G. In 1966 the parameter a_s of the symmetry energy (4) has been treated as a constant. In reality it is density dependent (5). Since the surface oscillations are mainly concentrated on the surface the density is smaller than the saturation density. Therefore the value of $a_s(\rho)$ is considerably reduced compared to a value of a_S from the Bethe-Weizsaecker mass formula. Thus the excitation energies of the isovector proton-neutron vibrations had been predicted in 1966 to lie too high. The dependence of the symmetry parameter on the density (5) can be derived in the Fermi gas model, but a numerical analysis of the energy per nucleon in nuclear matter as a function of the proton and the neutron density yields also the same dependence.

To find values for the excitations of Hamiltonian (3) we introduce the coordinates

$$\delta_\mu = \frac{1}{2}(\alpha_{p\mu} + \alpha_{n\mu})$$
$$\zeta_\mu = \frac{1}{2}(\alpha_{p\mu} - \alpha_{n\mu}) \tag{6}$$

which should be close to normal coordinates and should decouple the proton and neutron oscillations in phase (d bosons) and out of phase (z bosons).

$$\delta_\mu = \sqrt{\frac{E_d}{2C_d}}\left(d_\mu^+ + (-)^\mu d_{-\mu}\right)$$
$$\zeta_\mu = \sqrt{\frac{E_z}{2C_z}}\left(z_\mu^+ + (-)^\mu z_{-\mu}\right)$$
$$\text{with:} \quad E_d = \hbar\sqrt{\frac{C_d}{B_p + B_n}} = \hbar\sqrt{\frac{C_d}{B}} \tag{7}$$
$$E_z = E_d\sqrt{\frac{C_d + 8G}{C_d}}$$

The E2 transition probabilities can be calculated starting from the transition operator

$$M(E2,\mu) = \int \rho_p(\mathbf{r}) r^2 Y_{2\mu} d\tau$$
$$\text{with:} \quad \rho_p(\mathbf{r}) = \sum_{i=p} \delta(\mathbf{r} - \mathbf{r}_i) \tag{8}$$

If one uses for ρ_p a homogeneous charge distribution within the proton surface from equation (1) one obtains expanding the transition operator (8) in powers of α_p the expression

$$M(E2,\mu) = \frac{3z}{4\pi} R_0^2 \alpha_{p\mu}^+ \left[e\ fm^2\right]$$
$$= \frac{3z}{4\pi} R_0^2 \left[\sqrt{\frac{E_d}{2C_d}}(d_\mu + (-)^\mu d_{-\mu}^+) + \sqrt{\frac{E_z}{2C_z}}(z_\mu + (-)^\mu z_{-\mu}^+)\right] \tag{9}$$

describing one phonon transitions of d and z bosons. The low-lying d boson states allow to fit of the energy E_d to the low-lying 2^+ energy and the restoring force parameter C_d from the reduced transition probability (10).

$$B(E2; 0^+ \to 2_d^+) = 5\frac{E_d}{2C_d}\left(\frac{3z}{4\pi}R_0\right)^2 \ [e^2 fm^4]$$
$$B(E2; 0^+ \to 2_z^+) = 5\frac{E_z}{2C_z}\left(\frac{3z}{4\pi}R_0\right)^2 \ [e^2 fm^4]$$
(10)

If the demixing restoring force parameter G from the Hamiltonian (3) is determined with the help of the symmetry energy (5), the excitation energy of the isovector vibrations (z-bosons) and the transition probability from the ground state into the one boson 2_z^+ state is completely determined since

$$C_z = C_d + 8G \qquad (11)$$

as can be seen from the expressions (7).

The details will depend on the assumption which nucleons participate in the d and z vibrations to determine the density and the radial derivative of the participating nucleons. In figure 1 we give the excitation energy for a small basis (dashed-dotted line) including protons in the N=4 and neutrons in the N=5 oscillator shell only and for a large basis (solid line) including all protons and neutrons in the nucleus. The density distribution is calculated using a standard Saxon-Wood potential and including pairing correlations[14]. A detailed analysis of the mixing ratio (E2/M1 from the 2_z^+ to the 2_d^+ shows that only the nucleons in the last shell participate in the excitation (see figure 2).

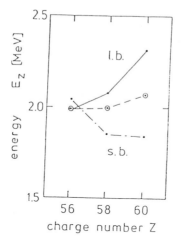

Figure 1

Excitation energy of the 2_z^+ isovector states in isotons to $^{140}_{56}Ba_{84}$. The dashed line gives the data of Hamilton and coworkers[1,2]. The solid line is calculated in a large basis including into the mass distribution all the nucleons while the dashed-dotted line includes only the protons from the N=4 and the N=5 oscillator shells for the protons and neutrons, respectively.

3. The Generalized Bohr-Mottelson Model and the Isovector 1^+ states in Deformed Nuclei.

The generalized Bohr-Mottelson model[5] can easily been applied to the isovector 1^+ states in deformed nuclei[15]. We start from a Hamiltonian

$$\hat{H} = \sum_{\tau=p,n} \left\{ \frac{1}{2} B_\tau \dot{\alpha}_\tau \cdot \dot{\alpha}_\tau + V_\tau(\alpha_\tau) \right\} + G(\alpha_p - \alpha_n) \cdot (\alpha_p - \alpha_n) \quad (12)$$

with a potential energy for the protons V_p and for the neutrons V_n which enforce axially deformed proton and neutron distributions

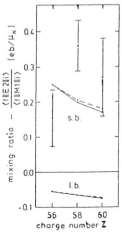

Figure 2

Mixing ratio E2/M1 in eb/μ_N for some isotones of $^{140}_{56}Ba_{84}$. The results indicate that only the small basis can describe the data. That means that only the nucleons in the last oscillator shell are participating in these isovector surface vibrations.

$$R_p(\vartheta) = R_0\bigl(1 + \beta_p Y_{20}(\vartheta)\bigr)$$
$$R_n(\vartheta) = R_0\bigl(1 + \beta_n Y_{20}(\vartheta)\bigr) \qquad (13)$$
$$\text{with:} \quad \beta_p = \beta_n = \beta$$

in their respective intrinsic systems. We again use the scalar product defined in equation (3). We assume that the potential $V_\tau(\alpha_\tau)$ is so deep that the vibrations around the equilibrium shape are very stiff and can be neglected for the following discussion. We now transform to the intrinsic system (diagonalizing the moment of inertia tensor) of the combined mass distribution of the protons and the neutrons.

$$a_{\tau\mu} = D^2_{\mu 0}(0, \tau \cdot \eta, 0)\beta = d^2_{\mu 0}(\pm\eta)\beta$$
$$d^2_{\mu 0}(\tau\eta) = <2,0|e^{iJ_y\tau\eta}|2,\mu>$$
$$\approx <2,0|1 \pm \frac{1}{2}(J_+ - J_-)\eta + \frac{1}{8}(J_+ - J_-)^2\eta^2 \pm \ldots|2\mu> \qquad (14)$$
$$\text{for protons:} \tau = 1 \quad \text{and neutrons:} \tau = -1$$

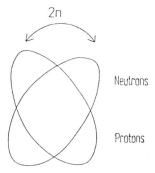

Figure 3
The axially symmetric proton and neutron mass distributions are assumed to perform scissors type oscillations with an opening angle 2η. The protons are rotated in the figure by η out of the common intrinsic system and the neutrons by the angle $-\eta$.

It is obvious that only $d^2_{1,0}(\tau \cdot \eta)$ will contribute to the last term in the Hamiltonian (12) due to the minus sign ($\tau = \pm 1$) between the proton and the neutron quadrupole deformations. Since we have

$$d^2_{10}(-\eta) = -d^2_{10}(+\eta) \qquad (15)$$

the last term of the Hamiltonian (12) can be written as a harmonic oscillator potential for the η vibrations

$$\begin{aligned}
\hat{H}_G &= G(\alpha_p - \alpha_n) \cdot (\alpha_p - \alpha_n) \\
&= G(a_p - a_n) \cdot (a_p - a_n) \\
&= 4G|d_{10}^2(\eta)|^2 \beta^2 \\
&= \frac{1}{2}(12G\beta^2)\cos^2\eta \, \sin^2\eta \\
&\approx \frac{1}{2}C\eta^2
\end{aligned} \qquad (16)$$

with: $\quad C = 12G\beta^2$

$$d_{10}^2(\eta) = \sqrt{\frac{3}{2}} \cos\eta \, \sin\eta$$

Since the potential $V_r(\alpha_r)$ in the Hamiltonian (12) enforces the axially symmetric deformation in the respective intrinsic system of the proton and the neutron mass distribution and since we neglect shape vibrations these terms are not contributing to the final Hamiltonian. We therefore must only transform the kinetic energies into the intrinsic system. This yields independent axially symmetric rotors for the protons and the neutrons.

$$\begin{aligned}
\hat{H}_{kin} &= \frac{(\mathbf{J}_p)^2}{2\Theta_p} + \frac{(\mathbf{J}_n)^2}{2\Theta_n} \\
&= \frac{(\mathbf{J}_p + \mathbf{J}_n)^2}{2\Theta} + \frac{(\mathbf{J}_p - \mathbf{J}_n)^2}{2\Theta}
\end{aligned} \qquad (17)$$

with moment of inertia: $\quad \Theta = \Theta_p + \Theta_n$

The total Hamiltonian for the scissors vibrations

$$\hat{H} = \frac{(\mathbf{J})^2}{2\Theta} + \frac{(\mathbf{J}_p - \mathbf{J}_n)^2}{2\Theta} + \frac{1}{2}C\eta^2 \qquad (18)$$

describes now independent rotations with the total angular momentum **J** and scissors vibrations of the proton and the neutron distribution with the angle η. The excitation energy is now given by the usual formula which we know from a pendulum with the moment of inertia Θ and the restoring force C.

$$E_{1^+} = \hbar\sqrt{\frac{C}{\Theta}} \qquad (20)$$

Nucleus	$E(1^+, exp)$ [MeV]	$E(1^+, th)$ [MeV]	$B(M1, exp)\uparrow$ $[\mu_N^2]$	$B(M1, th)\uparrow$ $[\mu_N^2]$
^{154}Sm	3.2	3.5	0.8±0.2	7.6; 5.1
^{156}Gd	3.1	3.5	1.3±0.2	7.0; 5.1
^{158}Gd	3.2	3.4	1.4±0.3	7.6; 4.9
^{164}Dy	3.11; 3.16	3.3	1.5±0.3	8.1; 5.1
^{168}Er	3.4	3.7	0.9±0.2	8.2; 5.7
^{174}Yb	3.35; 3.55	3.5	0.8±0.2	8.2; 5.4

Table 1

Isovector 1^+ states in rare earth nuclei. The experimental excitation energy for the state with the largest M1 strength is listed in column 2. If one finds two states with comparable strength both excitation energies are listed. Within the scissors vibrational picture such a splitting could be understood due to an asymmetric γ deformation of the nucleus. The splitting in ^{164}Dy would correspond to $\gamma = 0.8$ and in ^{174}Yb to $\gamma = 2$ degrees. The theoretical reduced magnetic dipole transition probability B (M1,Th) from the groundstate into the 1^+ state is calculated including all the protons and the neutrons (first number) and including only those protons and neutrons which contribute 98% to the moment of inertia according to the Beliaev moment of inertia formula based on the cranking model and pairing. The transition probabilities are by a factor of about 5 too large.

The magnetic dipole transition probability is by about a factor 5 too large for the scissors model. This model concentrates all the magnetic dipole strength in one state, while in reality this strength is distributed over many states. A quantitative description can therefore only be given in a microscopic approach which treats the nucleon degrees of freedom individually. An approach which can do that is the Random Phase Approximation (RPA) based on quasi-particles which take the pairing correlations between the nucleons into account. An important point in treating the nuclear many-body problem with the RPA approach is the violation of exact symmetries of the many-body Hamiltonian like translational and rotational invariance. This violations lead to so-called spurious states which have to be removed to get the physical solutions. We will therefore devote one chapter to these symmetries and the removal of the spurious states.

4. The Random Phase Approximation.

In the Random Phase Approximation (RPA) one includes already correlations into the ground state. Figure 4 shows a self consistent (Hartree-Fock) potential which is filled up to the Fermi surface and empty for the levels above. But in reality the collisions between the nucleons are smearing out the Fermi surface by an appreciable amount and induce ground state correlations. The RPA and especially the quasi-particle RPA takes a large amount of these correlations into account. It treats the difference between the correlated ground state and the excited state without having a need to know the exact nature of the correlated ground state $|g>$.

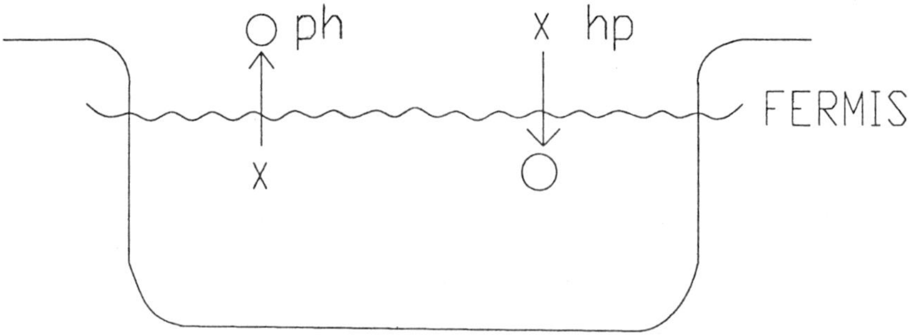

Figure 4
Self-consistent potential (e.g. Hartree-Fock potential) with levels filled up to the Fermi surface and empty above. In real nuclei the residual interaction between the nucleons which can not be put into a self-consistent field smear out the Fermi surface. Thus one can go from the ground state $|g>$ to the excited state $|\nu>$ not only by particle-hole (ph) excitations but also by annihilating a particle above the Fermi surface and creating one below (hole-particle excitations).

Including ground state correlations we can come to an excited state by bringing a particle from below into a state above the Fermi surface (particle-hole excitations), but also by bringing a particle which is due to the ground state correlations above the Fermi surface into an empty state below the Fermi surface. Naturally the last type of excitations are restricted due to the fact that the ground state correlations must have emptied at least partially a level below the Fermi surface.

$$\psi_\nu \equiv |\nu> \equiv Q_\nu^+|g> = \sum_{\substack{n>F \\ j<F}} [X_{nj}^\nu a_n^+ a_j - Y_{nj}^\nu a_j^+ a_n]|g> \qquad (20)$$

$$Q_\nu|g> = 0$$

We assume that the ground state $|g>$ is the vacuum of the operators Q_ν. Finally we want to treat them as harmonic oscillator bosons acting on the nuclear ground state as their vacuum. In the Ansatz (20) we have neglected two particle-two hole (2p-2h) excitations. They lie higher in energy and do not appreciably affect the lowest states ν. But the higher ones are strongly mixed with 2p-2h states and their strength is spread over many states. To obtain an equation for the particle-hole and the hole-particles amplitudes X_{nj}^ν and Y_{nj}^ν we used the Ritz variational principle.

$$<\delta\psi_\nu|\hat{H} - E_\nu|\psi_\nu> = 0 \qquad (21)$$

The coefficients X and Y are freely varied. We put now all equal to zero apart from one specific coefficient X_{mi}^ν and in a second equation we put all coefficients zero apart from one specific coefficient Y_{mi}^ν. The expressions multiplied with these two coefficients must be independently zero.

$$\begin{aligned} <g|[a_i^+ a_m, \hat{H}]|\nu> &= (E_\nu - E_0) <g|a_i^+ a_m|\nu> \\ <g|[a_i^+ a_m, \hat{H}]|\nu> &= (E_\nu - E_0) <g|a_i^+ a_m|\nu> \\ \text{with:} \quad \hat{H}|g> &= E_0|g> \\ E_\nu - E_0 &= E_\nu^{RPA} \end{aligned} \qquad (22)$$

Here we have used the fact that the Hamiltonian applied to the ground states yields the ground state energy E_0. In this way it is possible to introduce on the left side the commutator and on the right side the difference between the ground state energy E_0 and the energy of the excited state E_ν. Using the fact that the operator

Q_ν yields zero applied to the ground state one can with the help of equation (27) introduce on the left-hand side the double commutator and obtains:

$$\begin{aligned} <[[a_i^+ a_m, \hat{H}], Q_\nu^+]> &= E_\nu^{RPA} <[a_i^+ a_m, Q_\nu^+]> \\ <[[a_m^+ a_i, \hat{H}], Q_\nu^+]> &= E_\nu^{RPA} <[a_m^+ a_i, Q_\nu^+]> \end{aligned} \quad (23)$$

Here as also in equation (20) to (32) m, n are single nucleon states above and i, j are single nucleon states below the Fermi surface. The empty brackets in (23) indicate ground state $|g>$ expectation values. The aim is to get from (23) an equation for the coefficients X_{nj}^ν and Y_{nj}^ν defined in equation (20). This can be reached in two ways: Either we calculate the ground state expectation values in equation (23) with Hartree-Fock determinants or we assume boson commutation relations for the operators Q_ν (quasi-bosons). Both procedures yield the following result:

$$\begin{pmatrix} A & B \\ -B & -A \end{pmatrix} \begin{pmatrix} X^\nu \\ Y^\nu \end{pmatrix} = E_\nu^{RPA} \begin{pmatrix} X^\nu \\ Y^\nu \end{pmatrix}$$

with:
$$\begin{aligned} A_{mi,nj} &= (\epsilon_m - \epsilon_i)\delta_{mi,nj} + <mj|V_{eff}|in - ni> \\ B_{mi,nj} &= <mn|V_{eff}|ij - ji> \\ \epsilon_a &= <a|\frac{\mathbf{p}^2}{2m}|a> + \sum_{b<F} <ab|V_{eff}|ab - ba> \\ H &= \sum_{a=1}^A \frac{\mathbf{p}_i^2}{2m} + \sum_{i<j}^A V(i,j) \end{aligned} \quad (24)$$

The diagrams corresponding to the matrixes A and B are indicated in figure 5.

Figure 6 shows in the lower part one of the diagrams which are summed up in the RPA approach to describe the excited states. Above one sees one of the ring diagrams which are summed up to give the ground state correlations.

Equation (25) shows the iterated RPA equation (24). One sees immediately that the RPA eq. yields always two solutions of which one has the negative eigenvalue of the other. Naturally the states with negative eigenvalues are unphysical

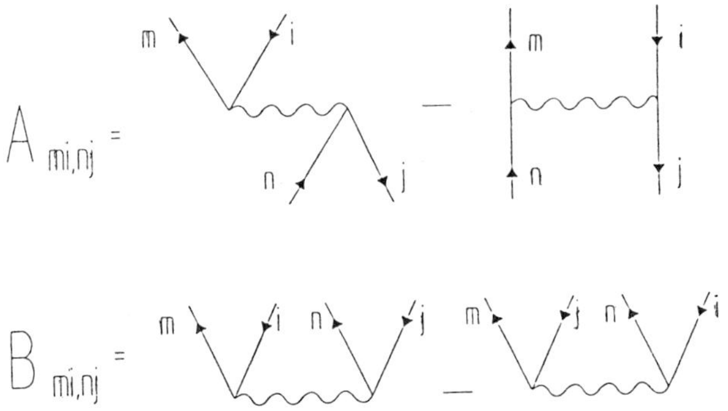

$A_{mi,nj} =$ [diagram] $-$ [diagram]

$B_{mi,nj} =$ [diagram] $-$ [diagram]

Figure 5
Diagrams of the effective interaction describing the particle-hole forces in the RPA equation (32). The second line of equation (24) contains the hermitian conjugate of these expressions.

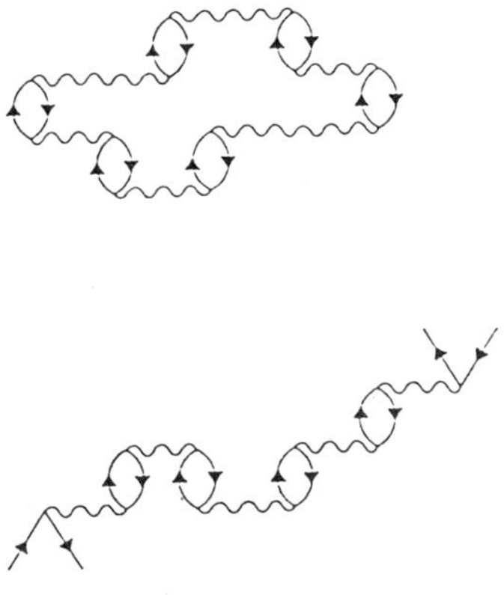

Figure 6
The lower part shows one of the RPA diagrams summed up to give the excited states. The upper part shows one of the ring diagrams summed up to describe the ground state correlations.

and have to be disregarded, since the ground state is the lowest energy state. But equation (25) also shows that the eigenvalues E_ν^{RPA} are real, although the matrix

(24) is not hermitian

$$\begin{pmatrix} (A^2 - B^2) & 0 \\ 0 & -(A^2 - B^2) \end{pmatrix} \begin{pmatrix} X^\nu \\ Y^\nu \end{pmatrix} = (E_\nu^{RPA})^2 \begin{pmatrix} X^\nu \\ Y^\nu \end{pmatrix} \qquad (25)$$

The normalization of the states is given by:

$$<\Psi_\nu|\Psi_\nu> \equiv <\nu|\nu> = <g|[Q_\nu, Q_\nu^+]|g> = \sum_{m>F, i<F} (X_{mi}^2 - Y_{mi}^2) \qquad (26)$$

In the formulation of the RPA equations we have always assumed that the phases are chosen so that the matrix elements are real. Thus we have omitted signs of complex conjugation.

In spherical nuclei the RPA quasi-bosons have a good angular momentum. That must be ensured by coupling the single nucleon angular momenta of the particles and the holes in equation (20) to the given angular momentum and parity.

$$Q_{\nu;\pi JM}^+ = \sum_{\substack{n>F \\ j<F}} [X_{nj}^{\nu;\pi JM} \{a_n^+ a_j\}_{\pi JM} - Y_{nj}^{\nu;\pi JM} \{a_j^+ a_n\}_{\pi JM}] \qquad (27)$$

In deformed nuclei the RPA solution is calculated in the intrinsic system which performs a rotation relative to the laboratory frame.

$$Q_{\nu;K\pi JM}^+ = \frac{2I+1}{16\pi^2(1+\delta_{K,0})} [D_{M,K}^J + (-)^{J-K} D_{M,-K}^J] \\ \sum_{\substack{n>F \\ j<F}} [X_{nj}^{\nu;K\pi} \{a_n^+ a_j\}_{\pi K} - Y_{nj}^{\nu;K\pi} \{a_j^+ a_n\}_{\pi K}] \qquad (28)$$

The low-lying isovector 1^+ states in deformed nuclei should be described by equation eq. (28) with J=1, K=1 and π=1.

In addition to the ground state correlations described by the diagram on the right-hand side in figure 6 we have also the more "short-ranged" pairing correlations. The nucleon-nucleon interaction is of short range and thus the particle-particle interaction is especially strong between nucleons which have the same position probability. This is the case for nucleons which are in the same orbits but have opposite angular momentum projection. Their position probability is identical, they are running around the same orbits but in opposite directions. Classically speaking they are colliding head-on after half a circulation. This head-on collision is a S-wave interaction and therefore especially strong. These correlations smear out also the ground state and are not yet contained in the RPA approach, since this includes only particle-hole interactions. The particle-particle pairing interactions are included by using quasi-particles

$$
\begin{aligned}
b_a^+ &= u_a a_a^+ - v_a a_{\bar{a}} \\
|\bar{a}> &= a_{\bar{a}}^+|0> = (-)^{j_a - m_a} a_{j_a, -m_a}^+ |0>
\end{aligned}
\quad (29)
$$

which are linear combinations over particles and holes. They take into account the smearing of the Fermi surface. v_a^2 is the probability that a state "a" is occupied, u_a^2 is the probability that the state "a" is unoccupied. The hole has to be in the time reversed state, since a missing particle yields for the many-particle state just the opposite angular momentum of the particle state in which a nucleon is missing.

The Ansatz for the RPA state with quasi particles corresponding to eq. (20) is:

$$
\begin{aligned}
|\Psi_{\nu;\pi KJM}^+> &= \frac{2I+1}{16\pi^2(1+\delta_{K,0})} [D_{M,K}^J + (-)^{J-K} D_{M,-K}^J] \, \Gamma_{\pi K}^+ \, |g> \\
\Gamma_{\pi K}^+ &= \frac{1}{2} \sum_{ab} [X_{ab}^{\nu;\pi K} b_a^+ b_b^+ - Y_{ab}^{\nu;\pi K} b_b b_a]
\end{aligned}
\quad (30)
$$

Here we have immediately written the RPA state for axially symmetric deformed nuclei as needed to study the low-lying isovector 1^+ states. For spherical nuclei the two quasi-particle creation operators at the coefficient X and the two quasi-particle annihilation operators at the coefficient Y have to be coupled to a definite angular momentum by Clebsch-Gordan coefficients. The quasi-particle RPA state (30) has not a definite proton and neutron number. One introduces therefore Lagrange multipliers λ_p and λ_n with which we subtract from the many-body Hamiltonian the operators for proton and neutron numbers. The Langrange multipliers are then chosen to have in average the correct number of protons and neutrons.

$$\hat{H}' = \hat{H} - \lambda_p \sum_{a=p} a_a^+ a_a - \lambda_n \sum_{a=n} a_a^+ a_a$$
$$< \sum_{a=\tau} a_a^+ a_a > = N_\tau = \left\{ \begin{array}{c} Z \\ N \end{array} \right\} \tag{31}$$

The RPA equations for the quasi-particle RPA (QRPA) are of the same structure as equation (24) but the matrix elements A and B are multiplied with coefficients which depend on the u_a and v_a of the quasi-particle transformation (29). They are e.g. given in the book of Ring and Schuck on page 334 and 344[16].

5. The Spurious States

The Hamiltonian of the nuclear many-body problem

$$\hat{H} = \sum_{k=1}^{A} \frac{\mathbf{p}_k^2}{2m_N} + \sum_{k<l=1}^{A} V(k,l)$$
$$[H, \hat{P}_z] = 0$$
$$[H, J_y] = 0 \tag{32}$$
$$[H, \hat{N}_\tau] = 0 \quad \text{for } \tau = p, n$$

has several exact symmetries. It is translational invariant and therefore commutes which each component of the total momentum as indicated for the z component in equation eq. (39). It also is rotational invariant and therefore commutes which each component of the total angular momentum **J**. It conserves the number of protons and neutrons and therefore commutes with the proton and neutron particle number operator $\hat{N}_r$.

Now we want to show that if we apply one of these symmetry operators to the correlated ground state of the nucleus $|g>$ in the spirit of the RPA approximation one obtains an RPA eigenstate with the excitation energy zero. We show this explicitly for the z component of the centre of mass momentum $\hat{\mathbf{P}}$ but the operator $\hat{P}_z$ could be replaced by any operator which commutes with the many-body Hamiltonian $\hat{H}$)

$$\hat{P}_z \approx \sum_{\substack{n>F \\ j<F}} \left[<n|\hat{P}_z|j> a_n^+ a_j + <j|\hat{P}_z|n> a_j^+ a_n \right]$$

$$\Psi_{\hat{P}_z} = \hat{P}_z |g>$$

(33)

From the fact that $\hat{P}_z$ commutes with the total many-body Hamiltonian we get immediately that the following ground state expectation values are identically zero.

$$< [a_i^+ a_m, [\hat{H}, \hat{P}_z]] > = 0$$
$$< [a_m^+ a_i, [\hat{H}, \hat{P}_z]] > = 0$$

(34)

We want now to show that these equations are identical with the RPA equations (23) with the eigenvalue $E_\nu^{RPA} = 0$. To do this we use the Jacobi identity for double commutators

$$[A,[B,C]] + [B,[C,A]] + [C,[A,B]] = 0$$
$$\hat{H}> \equiv \hat{H}|g> = E_0|g>$$

(35)

and the fact that the many-body Hamiltonian $\hat{H}$ applied to the ground state $|g>$ has the eigenvalue E_0.

$$< [[a_i^+ a_m, \hat{H}], \hat{P}_z] > = 0$$
$$< [[a_m^+ a_i, \hat{H}], \hat{P}_z] > = 0$$

(36)

Equation (36) is now identical with the RPA equation (23) with the eigenvalue zero. So we have proved: There exists always RPA solutions corresponding to each exact symmetry of the many-body Hamiltonian with eigenvalues zero which are unphysical (a translated or a rotated nucleus) and which have to be disregarded. Since these solutions are exactly degenerate with the ground state and since they are automatically orthogonal to all the other solutions it is no problem to separate these solutions out. This is naturally only the case if the many-body Hamiltonian used commutes with the symmetry operations which implies e.g. also that self-consistent single particle energies as defined in eq. (24) are used. But in practically all calculations one is starting from a potential. We use e.g. a deformed Saxon-Wood potential. That means that the total momentum operator is not anymore commuting with the many-body Hamiltonian since an explicit potential is not translational invariant. The same is true for a deformed potential for rotational invariance. But even if one uses a self-consistent potential but shifts the Hartree-Fock single particle energies slightly to improve the agreement with experimental single particle energies the spurious states are not anymore eigenstates and do not have the energy eigenvalue zero. They are spread over several RPA eigenstates and therefore hard to remove.

In the past a usual procedure to "remove" the spurious state was to use a free strength parameter of the residual particle-hole interaction and to determine this parameter so that one of the RPA solutions has the energy zero. This solution was then identified with the spurious state and the rest of the RPA states where assumed not to be contaminated. This is an extremely dangerous recipe to remove spurious states. For the rotational symmetry the state found in this way at energy zero starting with a deformed Saxon-Woods potential has only an overlap of about 20 % with the exact spurious state created by acting with the total angular momentum operator on the ground state. The remaining 80 % of the spurious state is contaminating the physical states which one wants to compare with the data. This procedure has for example been used by Bes and Broglia[17] by Hammamoto et al [18] and by Hilton, Ring and Mang [19]. Hammamoto [18] improved on that by requesting the Pyatov condition, that means that the commutator of the total many body Hamiltonian with the total angular momentum is zero as a ground state expectation value with RPA wave functions. This does not ensure that the spurious state is separated out as an eigenstate of the RPA equations at energy zero, but it enlarges the overlap of the spurious state with the eigenstate at energy zero to about 80 %. But still 20 % of the spurious state are distributed over the different physical states. Mainly the lowest lying RPA states are contaminated with this 20 % of the spurious rotational state. To obtain reliable results a different approach is therefore needed to remove exactly the spurious rotational state, which has the same quantum numbers as the low-lying 1^+ isovector state. (The spurious state from the total momentum has the quantum numbers 1^- and does therefore not mix with the 1^+ state.)

To ensure the orthogonality of the physical states with the spurious rotational state we introduce a Lagrange multiplier which allows to enforce this orthogonality.

$$\begin{aligned} \hat{H}' &= \hat{H} - \sum_\nu \lambda_\nu J_- \left(\Gamma^\nu_{K\pi=1+}\right)^+ \\ <S|\nu> &= N <g|J_-\left(\Gamma^\nu_{1+}\right)^+|g> = 0 \\ |S> &= N \, J_+|g> \end{aligned} \qquad (37)$$

The added term contains therefore an interaction between the rotation and the microscopic vibrations. Therefore we sometimes call it the rotation-vibration interaction term. This term introduces a self-consistency into the solution. The Hamiltonian depends on the solution and therefore the solution has to be iterated. Since the Hamiltonian depends on the Langrange multipliers λ_ν of all the solutions the RPA states $|\nu>$ are still orthogonal to each other and they are also orthogonal to the spurious state $|S>$. The proof for the orthogonality of the RPA states is not modified by introducing this rotation-vibration term. A test of the orthogonality of the numerical RPA solutions indicate that this orthogonality is fulfilled to better than 10^{-8} that means within the accuracy of the computer calculation.

6. RPA in the Schematic Model

It is obvious that the particle-hole matrix elements in expressions A and B in the RPA equation (24) can be written in a separable way if one neglects the exchange terms indicated in figure 5.

$$A_{mi,nj} = \epsilon_{mi}\,\delta_{mi,nj} + \kappa D_{mi}D_{nj}$$
$$B_{mi,nj} = \kappa D_{mi}D_{nj} \qquad (38)$$

Here ϵ_{mi} is the difference between the particle ϵ_m and the hole energy ϵ_i. If one puts the expression (38) into the RPA equation (31) one obtains a system of equations which can be solved analytically.

$$(E - \epsilon_{mi})X_{mi} = \kappa D_{mi}\sum_{nj}D_{nj}(X_{nj}+Y_{nj})$$
$$(E + \epsilon_{mi})X_{mi} = -\kappa D_{mi}\sum_{nj}D_{nj}(X_{nj}+Y_{nj})$$
$$X_{mi} = N\,\frac{D_{mi}}{E-\epsilon_{mi}} \qquad (39)$$
$$Y_{mi} = -N\,\frac{D_{mi}}{E+\epsilon_{mi}}$$
$$\text{with:}\qquad N = \kappa\sum_{nj}D_{nj}(X_{nj}+Y_{nj})$$

Here again m, n indicate particle states above the Fermi surface and i, j hole states below the Fermi surface. N is a normalization constant which is determined by the norm (25). To get the secular equation which determines the RPA energy E we divide the first two equations of (39) by the energy brackets on the left-hand side, multiply with D_{mi} and sum over all particle m and hole i states. Adding up both equations yields also on the left-hand side the expression for "N" defined in the last equation of (46). Dividing both sides by N yields the secular equation for the RPA energy E.

$$\frac{1}{\kappa} = \sum_{mi}\left[\frac{D_{mi}^2}{E-\epsilon_{mi}} - \frac{D_{mi}^2}{E+\epsilon_{mi}}\right]$$
$$\frac{1}{\kappa} = \sum_{mi}2\epsilon_{mi}\frac{D_{mi}^2}{E^2-\epsilon_{mi}^2} \qquad (40)$$

The secular equation (40) can be solved graphically (see figure 7). The particle-hole interaction is attractive ($\kappa < 0$) for isoscalar states. The spurious states corresponding to a rotation of the whole nucleus around some specific angle is such an isoscalar state. The usual recipe to "remove" the spurious states $|S>$ defined in eq. (37) is to adjust the strength of the particle-hole force κ so that the energy E lies at zero. As explained above this is no guarantee that the RPA state obtained in this way at energy zero is identical with the spurious state (37). In figure 26 the poles of equation (40) are indicated by vertical dashed lines. They are separated in $0\hbar\omega$ and $2\hbar\omega$ particle-hole excitations within an oscillator shell and across two oscillator shells. The low-lying isovector 1^+ state is then the one indicated by the intersection with the positive $1/\kappa$ line above the highest $0\hbar\omega$ particle-hole pole.

7. Numerical Results

As a model Hamiltonian we use a deformed Saxon-Woods potential with the parametrisation of Soloviev and coworkers[30]. The phase of the non-diagonal matrix element of the spin-orbit coupling is correct [21]. The pairing force strength $G_{p/n}$ are taken from the odd-even mass differences of the nuclei which yield gap parameters Δ of the order of 1 MeV. The deformation parameter β is taken from the electric quadrupole transition probabilities between the first excited 2^+ state and the ground state.

$$\hat{H}' = \hat{H} - \sum_a \lambda_{\tau_a} a_a^+ a_a = \sum_a (\epsilon_a - \lambda_{\tau_a}) a_a^+ a_a \qquad (41)$$

The effective particle-hole force V_{eff}^{ph} for the particle-hole interaction must include contributions where the particles and the holes couple to the angular momentum projection K=1 in the intrinsic system with positive parity. These are isoscalar and isovector quadrupole-quadrupole, spin-spin and spin-quadrupole forces.

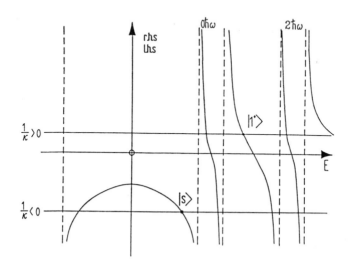

Figure 7
Graphical solution of the RPA secular equation (40) in the schematic model for an attractive ($1/\kappa < 0$) and a repulsive ($1/\kappa > 0$) force. Along the ordinate we plot the right- and the left-hand side of equation (40) as a function of the energy E. The particle-hole poles are indicated by dashed lines. They are separated into particle-hole excitations within an oscillator shell ($0\hbar\omega$) and across two oscillator shells ($2\hbar\omega$). The lowest energy solution $|S>$ would be identical with the rotational spurious state if the many-body Hamiltonian of the nucleus would exactly commute with the total angular momentum. In this case the line with $1/\kappa < 0$ would intersect the solid line (right-hand side of the secular equation (40)) at energy E=0. If one uses a Hamiltonian with a deformed potential the total angular momentum does not commute with the many-body Hamiltonian of the nucleus and thus even by adjusting κ so that the lowest energy RPA solution lies at E=0 this state is not identical with the spurious state $|S>$ defined in equation (37).

$$V_{eff}^{ph} = \sum_{i<j}[\kappa_0 + \kappa_1 \tau(i) \cdot \tau(j)]\hat{q}(i)\hat{q}(j)$$
$$+ \sum_{i<j}[c_0 + c_1 \tau(i) \cdot \tau(j)]S(i)S(j)$$
$$+ \sum_{i<j}[d_0 + d_1 \tau(i) \cdot \tau(j)][S\hat{q}(i)]_1 [S\hat{q}(j)]_1 \qquad (42)$$

with: $\hat{q}_\mu(i) = r_i^2 Y_{2\mu}(\mathbf{r}_i)$

$S(i) = \frac{1}{2}\sigma(i)$

The isoscalar and isovector pieces of the spin-spin force (c_0 and c_1) are taken from the work of Dickhoff, Faessler, Muether and Wu [22] who extracted this strength from a reaction matrix calculated from the Reid soft core potential. The strength of the spin-quadrupole is taken from the work of Soloviev and coworkers[23]. It turns out that the spin-quadrupole force is not very essential for describing the data.

The quadrupole force constants κ_0 and κ_1 are determined by trying to restore at least partially the rotational symmetry by requesting that the Hamiltonian (41) commutes with the total angular momentum.

$$[\hat{H}_0 + \hat{H}_I, J_+] \stackrel{!}{=} 0 \qquad (43)$$
$$[\hat{H}_0, J_+] = -[\hat{H}_I, J_+]$$

We call this the Pyatov condition [24] since this was first requested by Pyatov. Naturally the condition (43) can not be fulfilled exactly. The unperturbed Hamiltonian $\hat{H}_0$ contains the deformed Saxon-Woods potential and $\hat{H}_I$ all the particle-hole interactions. For $\hat{H}_I$ we make now an Ansatz which is for a deformed harmonic oscillator potential identical with a quadrupole-quadrupole force. For a deformed Saxon-Woods potential this is only approximately fulfilled. But since we are using the Pytov condition only to determine the quantity k this approximate equality should not matter.

$$\hat{H}_I = \kappa [\hat{H}_0, J_+]^+ [\hat{H}_0, J_+] \approx \kappa Q \cdot Q \qquad (44)$$

If we now introduce expression (44) into the Pyatov condition we obtain on the right-hand side a anti-commutator bracket.

$$[\hat{H}_0, J_+] = \kappa \left\{ [\hat{H}_0, J_+], [J_+, [\hat{H}_0, J_+]] \right\}_+$$
$$[J_+, [\hat{H}_0, J_+]] = \frac{1}{2\kappa} \qquad (45)$$

One realizes that the first equation (45) which is identical with the Pyatov condition is solved exactly if the second equation is fulfilled. Since the second equation can not be exactly fulfilled we determine κ from the ground state expectation value of the second equation (45). This ground state expectation value

$$< [J_+, [\hat{H}_0, J_+]] > = \frac{1}{2\kappa} \qquad (46)$$

can be written down for the proton, the neutron and the proton-neutron part yielding three different coupling constants κ_p, κ_n and κ_{pn} as explained in reference 25 und 26. The reason that the proton and the neutron quadrupole force constant is different is connected with the fact that the Saxon-Woods potential for the protons and the neutrons is not identical. Thus the corresponding expectation values (53) yield different results.

The Saxon-Woods potential is not always optimally reproducing the experimental single particle states [27]. We have therefore taken the liberty to adjust a few of the single particle levels of the Saxon-Woods potential to single particle energies derived from neighbouring odd mass nuclei.

Figure 27 gives the results for ^{56}Gd. The excitation energy (3.1 MeV) of the strongest 1^+ state agrees very favourably with the data measured by the Darmstadt group [3]. The reduced magnetic dipole transition probability $B(M1; 0 \to 1^0) [\mu_N^2]$ agrees also nicely with the data. The low-lying states between 2 and 5 MeV are mostly of orbital nature. The ratio of the matrix elements squared of the orbital contribution to the spin contribution varies between 5 and 100. The repulsive spin-spin force helps to push the spin-flip degrees of freedom to higher energies. They are concentrated between 9 and 12 MeV excitation energy.

The upper part of figure 8 shows the overlap squared of the scissors state

$$|R> = [\alpha_p J_{yp} - \alpha_n J_{yn}]|g>$$
$$<R|R> = 1 \quad ; \quad <R|S> = 0 \tag{47}$$

with the different RPA solutions. The most collective 1^+ which has been called the scissors mode by the experimentalists has only an overlap of 14% with the microscopically constructed scissors state of equation (47). The two parameters of the scissors mode α_p and α_n are determined by the normalization and the orthogonality to the spurious state (37).

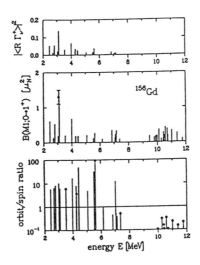

Figure 8
Spectrum of magnetic dipole excitations in ^{156}Gd plotted as the reduced magnetic dipole transition probabilities $B(M1; 0 \to 1^+)$ $[\mu_N^2]$ as a function of the excitation energy. The lower part shows the ratio of the orbit-to-spin of the M1 transition matrix element [25]. The upper part gives the overlaps squared [28] with the scissors state R defined in equation (47).

Figure 9 shows the transverse M1 form factors for inelastic electron scattering calculated from the transition density. The left-hand side gives the form factor

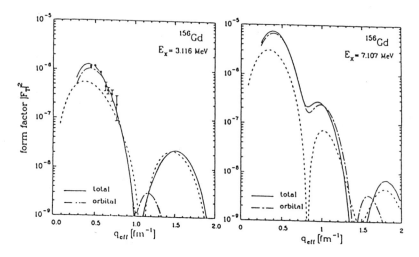

Figure 9
Transverse M1 form factors for inelastic electron scattering obtained from transition densities and compared with experiment [3]. The dashed line is the plane wave Born approximation (PWBA). The solid line represents the DWBA form factor. The dashed-dotted line shows the orbital part only of the distorted wave Born approximation (DWBA). The right-hand side gives the same results for the scissors mode (47). The excitation energy given is not an eigenvalue but the expectation value ($E_x = 7.107 MeV$).

squared in the 1^+ at 3.1 MeV, while the right-hand side gives the form factor for the scissors state.

Figure 10 shows the distribution of the M1 strength for ^{164}Dy. There are three very close lying states between 3.1 and 3.22 MeV. This range is blown up on the right-hand side of the figure. The lower part of figure 10 shows the orbit-to-spin ratio of the matrix elements indicating that the measured three states are essentially orbital type excitations.

Figure 11 shows the transversal form factors squared calculated with the destorted wave Born approximation using as in figure 9 the DWBA code of J.

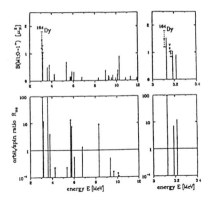

Figure 10

$B(M1; 0^+ \to 1^+) [\mu_N^2]$ distribution as a function of the excitation energy in MeV. On the right-hand side the energy range between 3 and 3.4 MeV is blown up to show the three experimentally measured states and to compare them with the RPA calculation. The experimental data are from the Darmstadt group [20,47] and from resonance fluorescence work from Giessen and Koeln [48,49]. The theoretical RPA results are from Tuebingen [46].

Heisenberg [33].

Figure 14 shows the transition density for the M1 transition from the ground state into the 1^+ in ^{164}Dy at 3.1 MeV [25]. It must be compared with the transition density extracted with the help of a Fourier-Bessel analysis [34] from Darmstadt data.

Figure 14, 15 and 16 show the transition density, the form-factors squared and the transition density calculated with the hole states in the interacting boson approximation (IBA2)[11,12] for ^{164}Dy. To perform this calculation one has to calculate microscopic S and D bosons of two fermion creation operators coupled to angular momentum 0^+ and 2^+. The approach is described in detail in reference 44. The transition density of figure 14 of IBA2 agrees quite well[34] with the data shown in figure 13. The differential cross section shown in figure 15 comes out a little bit

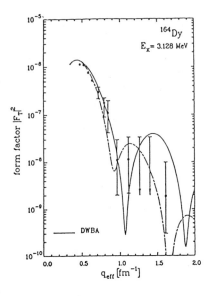

Figure 11
Transversal form factor of the 3.1 MeV 1^+ state in ^{164}Dy calculated in the DWBA approach [33] (solid line) compared with data from Darmstadt [3]. The dashed-dotted line gives the pure orbital contribution, arising from the convection current of the protons, as a function of the effective momentum transfer q.

too large compared to the data. Here we did not use the usual renormalization applied normally for all the IBA2 cross sections which multiply the theoretical cross section by a factor to adjust theory to the experiment at the maximum. The number of bosons is counted relative to the proton shell Z=50 and to the neutron shell N=82.

At least in theory it should not matter if we use 16 protons above the Z=50 shell or 16 proton holes relative to the Z=82 shell. The transition distribution calculated with 16 proton holes (8 proton bosons) relative to the Z=81 shell is very different from the transition density calculated with 16 proton particles relative to Z=50. The transition density shown in figure 35 does also not agree with the data of figure 13.

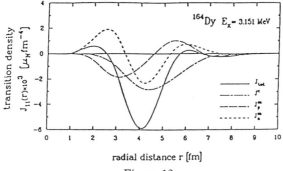

Figure 12

Transition density for the transition from the ground state into the isovector 1^+ state at 3.1 MeV in units $[19^{-3}\mu_N fm^{-4}]$ as a function of the radial distance from the centre of the nucleus in fm. The solid line is the total transition density the long dashed - short dashed line is the transition density arising from the proton convection current. This transition density could arise from a scissors mode. The long dashed curve is the magnetization current arising from the spin of the protons and the short dashed represents the magnetization current due to the spin of the neutrons. At low momentum transfer only the orbital current (long dashed - short dashed) contributes, while at higher momentum transfer (shorter wave length) the spin contributions get dominant [26].

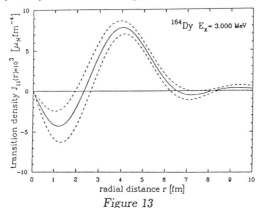

Figure 13

Transition density in units $[10^{-3}\mu_N fm^{-4}]$ as a function of the radial distance from the centre of the nucleus extracted by a Fourier-Bessel analysis from Darmstadt data at an excitation energy of 3.0 MeV [34]. Comparing the experimental data of this figure with the transition density in figure 31 one should keep in mind that the sign of the transition density can not be extracted from the differential cross section. One should multiply these data with a factor (-1) to compare with the theory of figure 12.

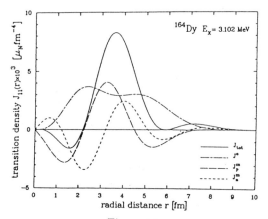

Figure 14
Transition density calculated [34] in IBA2. The solid line is the total transition density which has to be compared with the experimental transition density given in figure 13. The orbital contribution arising from the convection current of the protons is given as the long dashed - short dashed curve. The magnetisation current of the proton spins is indicated by the long dashed curve, while the short dashed curve shows the magnetisation current of the neutron spins.

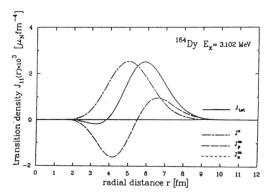

Figure 16
Transition density calculated [15] in IBA2 relative to the closed shell $Z=82$ for the protons (16 proton hole states) and the neutron shell $N=82$. The notation is the same as in figure 14. Although in theory this transition density should be the same if one considers 16 proton particles relative to $Z=50$ or 16 proton holes relative to $Z=82$, one finds a very different transition density which does not agree with the experimental result shown in figure 13.

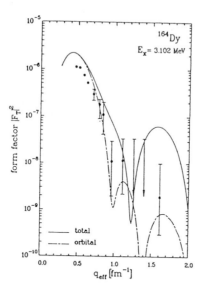

Figure 15

Transversal M1 form factor squared as a function of the effective momentum transfer in fm^{-1} for ^{164}Dy calculated in IBA2 [15] using $Z=50$ and $N=82$ as closed shells. The solid line is the total transversal form factor squared while the long dashed - short dashed curve is the orbital contribution arising from the proton convection current. Both curves are calculated in the DWBA approach using a program of Heisenberg [33].

8. Summary

First we were discussing excitation within one oscillator shell isovector states in spherical and deformed nuclei in the generalized Bohr-Mottelson model treating the quadrupole deformation degrees of freedom of the protons and the neutrons separately [5]. In spherical nuclei one can show [13] that this leads to two independent five dimensional quadrupole harmonic oscillators of d and z bosons.

All the parameters of the z isovector boson excitations in spherical nuclei can be determined from the excitation energy of the isoscalar low-lying 2^+ state and the $B(E2, 0^+ \to 2_d^+)$ transition probability. An additional parameter can be determined from the symmetry energy of the Bethe-Weiszaecker mass formula. The agreement with the available isovector data in spherical nuclei is satisfactory.

In deformed nuclei the collective generalized Bohr-Mottelson model yields good results for the excitation energies of the 1^+ states found at Darmstadt and by resonance fluorescence measurments from Giessen and Koeln. But the transition probabilities are by about a factor five too large compared with the data. This is due to the fact that a generalized Bohr-Mottelson model [5] concentrates all the M1 strength in one state. To have a reliable description of the M1 transition strength one needs a microscopic description. Therefore we introduce the Quasi-particle Random Phase Approximation (QRPA). We put special emphasis on removing the spurious state of rotations which has the same quantum numbers as the 1^+ state. It is not enough to determine the quadrupole force strength κ so that one RPA solution is degenerate with the ground state. This state is only identical with the spurious state if the nuclear many-body Hamiltonian commutes with the total angular momentum and if the single particle energies are determined self-consistently. Since one starts from a deformed Saxon-Woods potential both conditions are not fulfilled. The state obtained by the above procedure can have an overlap as low as 20 % with the spurious state (37). We therefore introduce a constraint which enforces the orthogonality of all physical states with the spurious state $|S>$. By requesting that the ground state expectation value of the commutator of the many-body Hamiltonian and the total angular momentum is zero we get equations to determine the quadrupole force strength for the protons, for the neutrons and for the proton-neutron interaction. Deriving the repulsive spin-spin force from Brueckner reaction matrix elements of the Reid soft core potential and taking the parameters of the Saxon-Woods potential and for the spin-quadrupole force from Soloviev and collaborators [20,23] one has practically a parameter free Hamiltonian which gives surprisingly good agreement with the experimental data.

Finally we can also answer the question: Are the isovector 1^+ states scissors modes? Our results indicate that the states found in Darmstadt and Giessen have about an overlap squared of 10 % to 20 % with the scissors mode (47) constructed by rotating the protons opposite to the neutrons and by requiring orthogonality to the spurious state (37). The reason for this small overlap with the scissors mode is that only a few two-quasi- particle states (2-3 if one requests a contribution of more than 10%) do contribute to the 1^+ isovector states found experimentally[36].

References

1. W. D. Hamilton, P. Hungerford, G. Jung, P. Pfeiffer, S. M. Scott, J. Phys. **G9** (1983) 763.
2. W. D. Hamilton, A. Irbäck, J. P. Elliott, Phys. Rev. Lett. **53** (1984) 2469.
3. D. Bohle, A. Richter, A. E. L. Dieperink, N. Lo Indice, F. Palumbo, O. Scholten, Phys. Lett. **B137** (1984) 27.
4. D. Bohle, G. Küchler, A. Richter, W. Steffen, Phys. Lett. **B148** (1984) 260 and U. E. P. Berg, C. Bläsing, J. Drexler, R. D. Heil, U. Kneissl, W. Naatz, R. Ratzek, S. Schennach, R. Stock, T. Weber, H. Wickert, B. Fischer, H. Hollick, D. Kollewe, Phys. Lett. **B149** (1984) 59.
5. A. Faessler, Nucl. Phys. **85** (1966) 653.
6. W. Greiner et al., Phys. Lett. **57B** (1975) 109.
7. N. Lo Iudice, F. Palumbo, Phys. Rev. Lett. **41** (1978) 1532.
8. T. Suzuki, D. J. Rowe, Nucl. Phys. **A289** (1977) 461.
9. R. Hilton, Z. Phys. **A316** (1984) 121.
10. D. Jansen, R. V. Jolos, F. Dönau, Nucl. Phys. **A224** (1974) 93.
11. F. Iachello, Nucl. Phys. **A358** (81) 89c.
12. A. E. L. Dieperink, Progr. Part. Nucl. Phys. **9** (1983) 121.
13. F. Scholtz, G. Kyrchev, A. Faessler, Nucl. Phys. **A491** (1989) 91.
14. A. Faessler, R. Nojarov, Phys. Lett. **166B** (1986) 367.
15. A. Faessler, R. Nojarov, Phys. Lett. **166B** (1986) 367, R. Nojarov, Z. Bochnacki, A. Faessler, Z. Phys. **A324** (1986) 289 and J. Phys. **G12** (1986) L47.
16. R. Ring, P. Schuck, The Nuclear Many Body Problem p. 301 - 345; Springer Verlag, Heidelberg
17. D. R. Bes, R. A. Broglia, Phys. Lett. **137B** (1984) 141.
18. I. Hamamoto, S. Aberg, Phys. Lett. **B145** (1984) 163 and Physica Scripta **34** (1986) 697.
19. R. R. Hilton, S. Iwasaki, H. J. Mang, R. Ring, M. Faber, preprint 1986.
20. F. A. Gareev, S. P. Ivanova, V. G. Soloviev, S. I. Fedotov, Sov. J. Part. Nucl. **4** (9174) 148.
21. J. Speth, D. Zawischa, Phys. Lett. **B211** (1988) 247; Phys. Lett. **B219**

(1989) 529.
22. W. H. Dickhoff, A. Faessler, H. Müther, S. S. Wu, Nucl. Phys. **A405** (1983) 534.
23. V. G. Soloviev, Ch. Stoyanov, V. V. Voronov, Nucl. Phys. **A304** (1978) 503 and V. Yu, Ponomarev, V. G. Soloviev, Ch. Stoyanov, A. I. Vdovin, Nucl. Phys. **A323** (1979) 446.
24. N. I. Pyatov, M. I. Cerney, Yad. Fiz. **16** (1972) 931 and V. V. Paltchik, N. I. Pyatov, Yad. Fiz. **32** (1980) 924.
25. R. Nojarov, A. Faessler, Nucl. Phys. **A484** (1988) 1.
26. A. Faessler, R. Nojarov, F. G. Scholtz, Nucl. Phys. (1990) to be published.
27. W. Ogle, S. Wahlborn, R. Piepenbring, S. Fredrikson, Rev. Mod. Phys. **43** (1971) 424.
28. A. Faessler, R. Nojarov, T. Taigel, Nucl. Phys. **A492** (1989) 105.
29. F. G. Scholtz, R. Nojarov, A. Faessler, Phys. Rev. Lett. **63** (1989) 1356.
30. U. Hartmann, D. Bohle, T. Guhr, K. D. Hummel, G. Kilgus, U. Milkau, A. Richter, Nucl. Phys. **A465** (1987) 25.
31. U. E. P. Berg, C. Bläsing, J. Drexler, R. D. Heil, U. Kneissl, W. Naatz, R. Ratzek, S. Schennach, R. Stock, T. Weber, H. Wickert, B. Fischer, H. Hollick, D. Kollewe, Phys. Lett. **B149** (1984) 59.
32. C. Wesselborg, P. von Brentano, K. O. Zell, R. D. Heil, H. H. Pitz, U. E. P. Berg, U. Kneissl, S. Lindenstruth, U. Seemenn, R. Stock, Phys. Lett. **B207** (1988) 22.
33. J. Heisenberg, H. P. Block, Ann. Rev. Nucl. Sci. **33** (1983) 569.
34. P. O. Lipas, M. Koskinen, H. Harter, R. Nojarov, A. Faessler, Phys. Lett. **B230** (1989) 1 and to be published in Nucl. Phys.A (1990).
35. Davis, R., Phys. Rev. **97** (1955) 766.
36. Lee, T.D. and Yang, C.N., Phys. Rev. **104** (1956) 254.
37. Wu, C.S., Ambler, E., Hayward, R.W., Hopper, D.D. and Hudson, R.P., Phys. Rev. **105** (1957) 1413.
38. Langacker, P., Phys. Rep. **72** (1981) 185.
39. Fritzsch, H. and Minkowski, R., Phys. Rep. **73** (1981) 67.
40. W. C. Haxton, G. J. Stephenson, Progr.Part.Nucl.Phys. **12** (1984) 409.

41. Tomoda, T., Faessler, A., Schmid, K.W. and Grümmer, F., Nucl. Phys. **A452** (1986) 591.
42. Avignone III, F.T., Brodzinski, R.L., Evans, J.C., Ir., Hensley, K., Miley, H.S. and Reeves, J.H., Phys. Rev. **C34** (1986) 666.
43. Caldwell, D.O., Eisberg, R.M., Grumm, D.M., Hale, D.L., Witherell, M.S., Goulding, F.S., Laudis, D.A., Madden, N.W., Malone, D.F., Pehl, R.H. and Smith, A.R., Phys: Rev. **D33** (1986) 2737.
44. Caldwell, D.O., Osaka Conference, June 1986 $(\tau_{1/2}^{0\nu}(^{76}Ge) > 3.9 \cdot 10^{23}$ years); Berkeley Conference, July 1986 $(\tau_{1/2}^{0\nu}(^{76}Ge) > 4.7 \cdot 10^{23}$ years).
45. Kirsten, R., in Proceedings of the International Symposium on Nuclear Beta Decay and Neutrinos, Osaka, Japan 1986, Edited by T. Kotani, H. Ejiri, E. Takasugi (World Scientific, Singapore, 1986, p. 81)
46. Manuel, O.K., Proc. Int. Symp. on Nuclear Beta Decay and Neutrinos, Osaka, Japan, June 1986, eds. T. Kotani, H. Ejiri, E. Takasugi (World Scientific, Singapore, 1986, p. 71)
47. Elliott, S.R., Hahn, A.A. and Moe, M.K., Phys. Rev. Lett. **59** (1987) 2020.
48. Eliott, S.R. Hahn, A.A. and Moe, M.K., Proc. Int. Symp. on Weak and Electromagnetic Interactions in Nuclei, July 1986, ed. H.V. Klapdor (Springer, Berlin 1986) p. 692
49. Civitarese, O., Faessler, A. and Tomoda, T., Phys. Lett. **B194** (1987) 11.
50. Nolte, E. et al., Z. Physik **A306** (1985) 223.
51. Kleinheinz, P. et al., Phys. Rev. Lett. **55** (1985) 2664.
52. Bohr, A. and Mottelson, B.R., Nucl. Structure, Vol. 1 (Benjamin, New York, 1969).
53. Bertsch, G., The Practitioners Shell Model (American Elsevier, New York, 1972)
54. Tomoda, T. and Faessler, A., Phys. Lett. **B199** (1987) 2383.
55. Vogel, P., and Zirnbauer, M.R., Phys. Rev. Lett. **57** (1986) 3148.
56. Muto, K. and Klapdor, H.V.,
57. Engel, J., Vogel, P., Civitarese, O. and Zirnbauer, M.R., Phys.Lett. **B208** (1988) 187.

LOW LYING DIPOLE MODES IN THE RARE EARTH REGION

P. von Brentano and A. Zilges
Institut für Kernphysik, Universität zu Köln, D-5000 Köln 41, Germany
R.D. Heil, U. Kneissl, H.H. Pitz and C. Wesselborg *
Institut für Kernphysik, Justus-Liebig-Universität Giessen, D-6300 Giessen, Germany

Abstract

High resolution Nuclear Resonance Fluorescence (NRF) experiments have been carried out to study low lying dipole excitations in rare earth nuclei. In particular, we identified the so called M1 scissors mode near 3 MeV excitation energy. The sensitivity of the NRF method makes it possible to observe weaker transitions and a fragmentation of the dipole strength. We summarize our results both on 1^+ and 1^- states and give a short survey of recent theoretical investigations on the M1 excitations.

Introduction

Magnetic dipole modes of nuclei have been investigated intensively in terms of various models from the sixties to the eighties [1-6]. One particular feature of these studies has been the prediction of a new low lying collective orbital magnetic dipole mode [3-6]. In 1983 the Darmstadt group around Richter and Bohle identified this mode by observing a strong M1 transition at 3.075 MeV in ^{156}Gd using electron scattering [7, 8]. This exciting discovery started an intensive work both in the experimental and in the theoretical field. It was found by the Giessen group around Kneissl and Berg that the M1-mode can also be seen in high resolution (γ,γ')-experiments [9]. Systematic NRF experiments were performed in different regions of the isotopic table ranging from fp-shell nuclei to the actinides by the Giessen–Cologne–Stuttgart collaboration [10]. We will focus on the well deformed nuclei in the rare earth region. Whereas the dipole strength in the Dy-isotopes is concentrated in a few transitions it is considerably fragmented in the examined Gd-isotopes, and even more so in the Yb-isotopes. We will first briefly summarize the information that can be obtained from photon scattering experiments. The next section will present the experimental results for the systematics of the K=1 and K=0 states in the rare earth nuclei. Finally we will discuss the results with respect to recent theoretical work on the nature of the magnetic dipole transitions.

---
*Feodor-Lynen-Fellow of the A.v. Humboldt Foundation at Brookhaven National Laboratory

The NRF technique

The NRF-experiments have been performed at the bremsstrahlung facility of the Stuttgart Dynamitron described in detail in refs. [10-12]. The electromagnetic excitation mechanism is well understood and therefore the photon scattering experiments allow the determination of the following quantities in a completely model independent way:

- The excitation energies,
- the spins,
- the absolute transition strengths,
- the parities, and
- (with adequate assumptions) the K-quantum numbers.

In the past it was only possible to extract the parities by comparing the (γ, γ') data with electron scattering (e,e') experiments of the Darmstadt group [13, 14] or by using the polarized off-axis bremsstrahlung available at the Giessen LINAC [15]. In our present experiments in collaboration with the Goettingen group at the Stuttgart Dynamitron we use two different Compton polarimeter set-ups for parity determination: A classical 5 Ge-detector set-up and a sectored single crystal Ge-detector. First results for dipole states in ^{142}Nd, ^{150}Nd and ^{232}Th have been published in refs. [16, 17]. The data evaluation of a recent polarimeter measurement on ^{162}Dy is under way.

Figure 1 shows a typical NRF spectrum of ^{176}Yb between 2 and 3 MeV measured with a photon endpoint energy of 4 MeV. The good energy resolution (FWHM$\simeq$3 keV at 3 MeV) makes it possible to resolve even very close lying transitions. For each excited level one observes groundstate transition and the corresponding transition to the 2_1^+-state. From the strength ratios of these transitions the K-quantum numbers are suggested by the Alaga rules. A recent study shows that in most of the cases the K-quantum number is still rather good in the examined energy region [18]. The (γ, γ') spectrum of ^{174}Yb between 2.0 and 2.6 MeV is shown in fig 2. In this experiment the bremsstrahlung endpoint energy was reduced to $\simeq$3.2 MeV.

Experimental Results

Figure 3 shows the distribution of $\Delta K=1$ strength in the examined Gd-[12], Dy-[19] and Yb-[20] isotopes. For J=1, K=1 states both positive and negative parity is possible. An elegant way in which the data can be plotted without knowing the parity is to give the parity independent decay width Γ_0 or better the reduced decay width $\Gamma_0^{red} = \Gamma_0/E_\gamma^3$. Therefore we have plotted this quantity times an adequate scale factor in figs. 3-5. As we expect that a significant fraction of the $\Delta K=1$ is magnetic strength we have chosen for these transitions a scale factor b in such a way that $b \cdot \Gamma_0^{red}(1^+ \to 0^+) = B(M1, 0^+ \to 1^+)$ in μ_N^2. This amounts to choice b=0.2598 μ_N^2 (MeV)^3(meV)$^{-1}$ if Γ_0^{red} is given in mev/(MeV)^3.

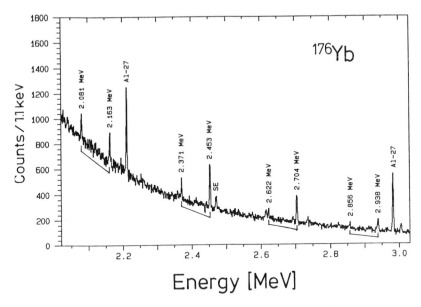

Fig. 1: NRF spectrum of ^{176}Yb measured with 4 MeV endpoint energy.

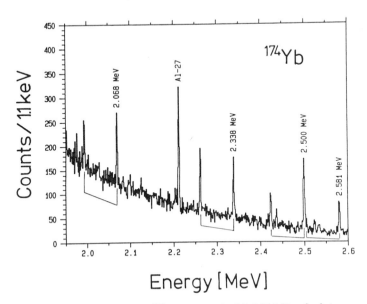

Fig. 2: NRF spectrum of ^{174}Yb measured with 3.2 MeV endpoint energy.

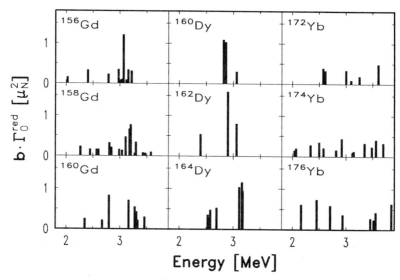

Fig. 3: Distribution of $\Delta K=1$ strength in the examined Gd-, Dy- and Yb-isotopes. In case of positive parity the ordinate value can be identified with the $B(M1)\uparrow$-strength in μ_N^2, i.e.
$b \cdot \Gamma_0^{red}(1^+ \to 0^+) = B(M1, 0^+ \to 1^+)$ in μ_N^2.

The positive parity of several states could be assigned by comparing the (γ,γ') data with electron scattering results of the Darmstadt group or with Compton polarimeter measurements of the Giessen–Goettingen–Cologne collaboration. The orbital magnetic scissors-mode is expected around an excitation energy of 3 MeV and in fact, we identified $\Delta K=1$ dipole strength around this energy in all examined nuclei. But in contradiction to the theoretical approaches based on a simple geometric collective picture [4], the M1 strength is fragmented into several transitions. In addition we found some K=1 states at lower energies of $\simeq 2.5$ MeV. A recent ^{165}Ho(t,α)^{164}Dy experiment by Freeman et al. [21] suggested a two quasiproton 7/2$^-$ [523] × 5/2$^-$ [532] structure for one of these $K^\pi=1^+$, J=1 states around this energy. Within a quasiparticle description of the Nilsson model it seems to be possible to reproduce most of the detected M1 transitions [22]. For a comparison with the predictions for the scissors mode in various models with the experimental data we sum up the strengths of all J=1, K=1 states between 2.7 and 3.6 MeV. Figure 4 shows the behaviour of this sum in the mass region A=156-176. We included the data points for ^{166,168,170}Er from an earlier photon scattering experiment by Metzger [23]. The data evaluation of a recent (γ,γ')-experiment on ^{166}Er by the Giessen-Cologne collaboration is under way and will improve the value for A=166. Unless dipole strength is shifted above 4 MeV, we observe a dramatic re-

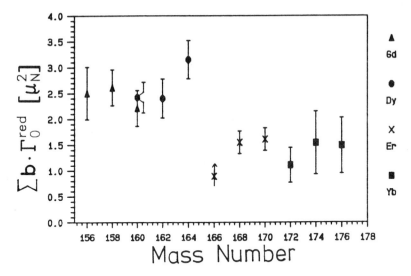

Fig. 4: Systematics of the summed $\Delta K=1$ strength between 2.7 and 3.6 MeV in the mass region A=156–176. We note that $b \cdot \Gamma_0^{red}(1^+ \to 0^+) = B(M1, 0^+ \to 1^+)$ in μ_N^2.

duction of the strength for mass numbers $A \geq 166$. This anomaly is not yet understood and clearly warrants a NRF-measurement on ^{164}Er which is in preparation.

For the remaining J=1, K=0 states negative parity is highly suggested both from the experimental data and from the IBA-2 model. Figure 5 shows the $\Delta K=0$ dipole strength found in Gd-, Dy- and Yb-nuclei. The ordinate gives the dipole strength $c \cdot \Gamma_0^{red}$ where Γ_0^{red} is the parity independent groundstate decay width. The scale factor c is chosen in such a way that $c \cdot \Gamma_0^{red}(1^- \to 0^+) = B(E1, 0^+ \to 1^-)$ in $10^{-3}e^2\text{fm}^2$. One finds $c=2.866 \cdot 10^{-3}e^2\text{fm}^2(\text{MeV})^3(\text{meV})^{-1}$ if Γ_0^{red} is given in meV/(MeV)^3. In all nuclei except in ^{176}Yb we could detect a very strong E1-transition around 1.5 MeV which is suggested to be the bandhead of a $K^\pi=0^-$ band, originating from an octupole vibration around a quadrupole deformed shape. The transition strengths from the groundstate to these 1^--states are $B(E1) \uparrow \simeq 10 - 20 \cdot 10^{-3}e^2\text{fm}^2$ in almost all nuclei. The origin of the structure of the weaker $\Delta K=0$ transitions at higher energies is unknown and needs further investigation, see also the discussion in ref. [24].

Discussion

The existence of a collective magnetic dipole mode in deformed nuclei has been predicted in the late seventies by Lo Iudice and Palumbo in the Two Rotor Model [4]. Within this geometrical model the deformed proton and neutron fluids perform a scissors-like rotation. Palumbo and Richter tried to explain the experimentally observed

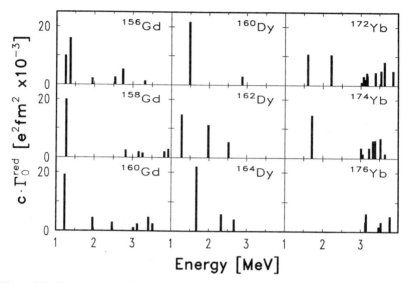

Fig. 5: Distribution of $\Delta K=0$ strength in the examined Gd-, Dy- and Yb-isotopes. In case of negative parity the ordinate value can be identified with the $B(E1)\uparrow$-strength in $e^2 fm^2 \cdot 10^{-3}$, i.e. $c \cdot \Gamma_0^{red}(1^- \to 0^+) = B(E1, 0^+ \to 1^-)$.

fragmentation by taking into acount a triaxial deformation of the nucleus which leads to a splitting of the M1-mode into two states [25]. Recently Lo Iudice et al. published an improved description of the scissors mode within the Two Rotor Model [26]. They predict an excitation energy of $E_x \simeq 2.8$ MeV and a $B(M1)\uparrow$-strength of $6.2\,\mu_N^2$ for ^{164}Dy. This should be compared with the experimental value $B(M1)\uparrow = 3.15 \pm 0.37\,\mu_N^2$ measured in (γ, γ') experiments of the Giessen-Cologne collaboration [19].

The algebraic approach of the Interacting Boson Model was extended by Arima et al. to the IBA-2 by incorporating an additional p-n degree of freedom [27]. In the IBA-2 it is possible to describe states of non-maximum symmetry with respect to the interchange of proton and neutron bosons, called mixed symmetry states. The lowest lying mixed symmetry state in heavy deformed nuclei is expected to be a $J^\pi=1^+$ state which is the bandhead of a K=1 band around 3 MeV [5, 28]. The expected $B(M1)\uparrow$ strength depends on the number of valence bosons N_π and N_ν and on the difference of the boson g-factors $(g_\pi - g_\nu)^2$. The standard values for the g-factors chosen on qualitative microscopic grounds are $g_\pi=1$ and $g_\nu=0$ whereas a systematic study of measured g-factors for the 2_1^+-states by Wolf et al. yielded $g_\pi=0.65$ and $g_\nu=0.08$ in the rare earth nuclides [29]. With the latter parameter set the measured M1-strength is underestimated by the IBA-2 [20]. It has been proposed that it is necessary to include g-bosons in the IBA to obtain a correct description [30, 31]. A fragmentation of the M1 strength into several states can be reproduced partly in the framework of the IBA-2 by using a complete Majorana operator [32].

Otsuka evaluated the M1-strength in terms of the particle-number-conserved Nilsson+BCS formalism [33]. He suggested that there is more than one M1 excitation mode, namely the unique-parity spin modes, the normal-parity spin modes and the scissors mode. The latter carries most of the orbital strength. Otsuka's results for the scissors mode in ^{156}Gd, ^{164}Dy and ^{166}Er overestimate the experimental strengths located around 3 MeV by a factor two. A systematic study of M1 strength in the Rare Earth Region within the Nilsson model given by De Coster and Heyde [34] pointed to a relation between the transition strengths and the ground-state quadrupole deformation. Other random-phase-approximation or quasirandom-phase-approximation calculations reproduced the fragmentation of the M1-mode, but the detailed results presented in different papers still differ considerably [35-38].

In conclusion the high resolution photon scattering experiments of the Giessen-Cologne collaboration identified the M1 scissors mode in all examined well deformed even-even rare earth nuclei. The $\Delta K=1$ strength is fragmented over a broad energy interval between 2 and 4 MeV excitation energy. Microscopic calculations seem to be more effective for the reproduction of this fragmentation than the simpler but more illustrating picture of a scissors-like motion in the collective models. Nevertheless a complete theoretical understanding of both the $\Delta K=1$ and the remaining $\Delta K=0$ dipole excitations is still warranted.

We gratefully acknowledge the experimental support by A. Jung, H. Friedrichs, S. Lindenstruth, J. Margraf, H. Schacht, B. Schlitt, U. Seemann, R. Stock and T. Weber from Giessen, R.D. Herzberg from Cologne, B. Kasten, G. Müller, K.W. Rose and W. Scharfe from Goettingen, B. Fischer, H. Hollick, E. Kutz and J. Lefevre from Stuttgart and valuable discussions with R.F. Casten, C. De Coster, W. Frank, A. Gelberg, I. Hamamoto, K. Heyde, S.D. Hoblit, R. Jolos, P.O. Lipas, B. Mottelson, A.M. Nathan, A. Richter, M. Schumacher, and P. Vogel. This work was supported by the Deutsche Forschungsgemeinschaft under contract Br 799-32 and Kn 154-18.

References

[1] W. Greiner, Nucl. Phys. **80**, 417 (1966).

[2] A. Faessler, Nucl. Phys. **85**, 653 (1966).

[3] S.I. Gabrakov, A.A. Kuliev, N.I. Pyatov, D.I. Salamov and, H. Schultz, Nucl. Phys. **A182**, 625 (1972).

[4] N. Lo Iudice and F. Palumbo, Phys. Rev. Lett. **41**, 1532 (1978).

[5] F. Iachello, Nucl. Phys. **A358**, 89c (1981).

[6] R.R. Hilton, Zeit f. Phys. **A316**, 121 (1984) and references therein.

[7] A. Richter, Proc. Int. Nuclear Physics Conf., vol.2, ed. P. Blasi and R.A. Ricci, (Florence,1983) p.189.

[8] D. Bohle, A. Richter, W. Steffen, A.E.L. Dieperink, N. Lo Iudice, F. Palumbo, and O. Scholten, Phys. Lett. **B137**, 27 (1984).

[9] U.E.P. Berg, C. Bläsing, J. Drexler, R.D. Heil, U. Kneissl, W. Naatz, R. Ratzek, S. Schennach, R. Stock, T. Weber, H. Wickert, B. Fischer, H. Hollick, and D. Kollewe, Phys. Lett. **B149**, 59 (1984).

[10] U. Kneissl, Part. and Nucl. Phys. **24**, 41 (1989).

[11] P. von Brentano, C. Wesselborg and A. Zilges, in Recent Advances in Nuclear Physics, edited by M.Petrovici and N.V.Zamfir (World Scientific, Singapore,1989), p.77.

[12] H.H. Pitz, U.E.P. Berg, R.D. Heil, U. Kneissl, R. Stock, C. Wesselborg, and P. von Brentano, Nucl. Phys. **A492**, 411 (1989).

[13] D. Bohle, G. Küchler, A. Richter, and W. Steffen, Phys. Lett. **B148**, 260 (1984).

[14] D. Bohle, A. Richter, U. E. Berg, J. Drexler, R. D. Heil, U. Kneissl, H. Metzger, R. Stock, B. Fischer, H. Hollick, and D. Kollewe, Nucl. Phys. **A458**, 205 (1986).

[15] U. E. Berg and U. Kneissl, Ann. Rev. Nucl. Part. Sci. **37**, 27 (1987).

[16] B. Kasten, R.D. Heil, P. von Brentano, P.A. Butler, S.D. Hoblit, U. Kneissl, S. Lindenstruth, G. Müller, H.H. Pitz, K.W. Rose, W. Scharfe, M. Schumacher, U. Seemann, Th. Weber, C. Wesselborg, and A. Zilges, Phys. Rev. Lett. **63**, 609 (1989).

[17] R.D. Heil, B. Kasten, W. Scharfe, P.A. Butler, H. Friedrichs, S.D. Hoblit, U. Kneissl, S. Lindenstruth, M. Ludwig, G. Müller, H.H. Pitz, K.W. Rose, M. Schumacher, U. Seemann, J. Simpson, P. von Brentano, Th. Weber, C. Wesselborg, and A. Zilges, Nucl. Phys. **A506**, 223 (1990).

[18] A. Zilges, P. von Brentano, A. Richter, R.D. Heil, U. Kneissl, H.H. Pitz, and C. Wesselborg, Phys. Rev. **C42**, 1945 (1990).

[19] C. Wesselborg, P. von Brentano, K.O. Zell, R.D. Heil, H.H. Pitz, U.E.P. Berg, U. Kneissl, S. Lindenstruth, U. Seemann, and R. Stock, Phys. Lett. **B207**, 22 (1988).

[20] A. Zilges, P. von Brentano, C. Wesselborg, R.D. Heil, U. Kneissl, S. Lindenstruth, H.H. Pitz, U. Seemann, and R. Stock, Nucl. Phys. **A507**, 399 (1990).

[21] S. J. Freeman, R. Chapman, J. L. Durell, M. A. Hotchkis, F. Khazaie, J. C. Lisle, J. N. Mo, A. M. Bruce, R. A. Cunningham, P. V. Drumm, D. D. Warner and J. D. Garrett, Phys. Lett. **B222**, 347 (1989).

[22] C. De Coster, private communication.

[23] F.R. Metzger, Phys. Rev. **C13**, 626 (1976).

[24] H.H. Pitz, R.D. Heil, U. Kneissl, S. Lindenstruth, U. Seemann, R. Stock, C. Wesselborg, A. Zilges, P. von Brentano, S.D. Hoblit, and A.M. Nathan, Nucl. Phys. **A509**, 587 (1990).

[25] F. Palumbo and A, Richter, Phys. Lett. **B158**, 101 (1985).

[26] N. Lo Iudice, F. Palumbo, A. Richter, and H. J. Wörtche, Phys. Rev. **C42**, 241 (1990).

[27] A. Arima, T. Otsuka, F. Iachello, and I. Talmi, Phys. Lett. **B66**, 205 (1977).

[28] A. E. L. Dieperink, Prog. Part. and Nucl. Phys. **9**, 121 (1988).

[29] A. Wolf, R. F. Casten, and D. D. Warner, Phys. Lett. **B190**, 19 (1987).

[30] Y. Akiyama, Symmetries and Nuclear Structure, eds. R. A. Meyer and V. Paar, (New York, 1987) p.368.

[31] I. Morrison, P. von Brentano, and A. Gelberg, J. Phys. **G15**, 801 (1989).

[32] W. Frank, private communication.

[33] T. Otsuka, Nucl. Phys. **A507**, 129c (1990).

[34] C. De Coster and K. Heyde, Phys. Rev. Lett. **63**, 2797 (1989).

[35] R. Nojarov and A. Faessler, Nucl. Phys. **A484**, 1 (1988).

[36] K. Sugawara-Tanabe and A. Arima, Phys. Lett. **B206**, 573 (1988).

[37] J. Speth and D. Zawischa, Phys. Lett. **B211**, 247 (1988); **B219**, 529 (1989).

[38] D. Zawischa, M. Macfarlane, and J. Speth, Phys. Rev. **C42**, 1461 (1990).

DIRECT AND MULTIPLE EXCITATIONS IN MAGIC ^{96}Zr FROM 22 MeV $(\vec{d},d')$ *)

D. HOFER, M. BISENBERGER, G. GRAW,
R. HERTENBERGER, H. KADER, P. SCHIEMENZ
Sektion Physik der Universität München, D-8046 Garching, FRG

G. MOLNÀR
Institute of Isotopes, Budapest, Hungary

*) Work supported in part by the BMFT

ABSTRACT

In a high energy resolution, 22 MeV polarized $(\vec{d},d')$ measurement on doubly closed subshell ^{96}Zr at the Munich Q3D spectrograph, we are able to resolve all excited states of ^{96}Zr up to 4.3 MeV excitation energy. The excitation strength is concentrated on a few low lying states, especially the first 2^+ and the first 3^-, suggesting the existence of a $3^- \otimes 3^-$ quadruplett and a $2^+ \otimes 3^-$ quintuplett of states. A preliminary coupled channel analysis of angular distributions gives informations about one step (direct) and two step (multiple) parts of the excitation mechanism.

1. Introduction

The low lying excitation spectrum ($E_x \leq 4.5$ MeV) of ^{96}Zr is of considerable interest due to features of double shell closure occuring at the $2p_{1/2}$ proton and the $2d_{5/2}$ neutron subshell[1]. Most outstanding is the B(E3)= 69 ± 20 WU collectivity of the 3_1^- state at 1897 keV, obtained from a recent lifetime measurement[2]. Similar to ^{40}Ca one has in the low lying spectrum excitations with predominantly two quasiparticle character and intruder bands, based on multiparticle excitations like the first excited 0^+ at 1582 keV. With respect to the search for double octupole excitations, ^{96}Zr is of special interest due to the large B(E3) value, the low excitation energy of the 3_1^- and the relatively low level density at twice this energy. Because E3 transitions can hardly be observed, double octupole excitations can be identified by analyzing inelastic scattering data with the coupled channel approach.

2. Experiment

In addition to unpolarized, recent ^{96}Zr (p, p') at 65 MeV[3] and ^{96}Zr (d, d') at 52 MeV[4] studies, we measured ^{96}Zr ($\vec{d}$, d') at 22 MeV with a 200 nA polarized deuteron beam (p=0.6) and a position resolving focal plane detector with anode and periodic cathode readout[5] for the Munich Q3D spectrograph. Using a 190μg/cm^2 thin ZrO$_2$ target, enriched to 95.6 % in ^{96}Zr, we get an energy resolution from 5 to 7 keV, resolving all known excited states up to 4.3 MeV excitation energy. Fig. 1 shows the spectrum.

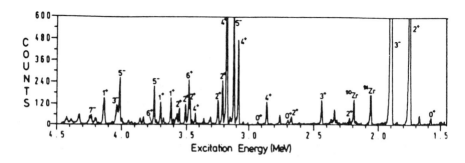

Fig.1 Excitation spectrum of ^{96}Zr at $\theta_{Lab} = 35°$.

3. Data Analysis

A preliminary theoretical analysis of the most interesting states was made in full second order coupled channel approach for vibrational nuclei[6] using the computercode ECIS[7] and standard values for the optical potential[8]. The deformation parameters for quadrupole and octupole vibration, β_2 and β_3 and the surface absorption W_D were adjusted to fit simultaneously the ground state, the 2_1^+ and the 3_1^-. With these values fixed, calculations for two phonon states were performed. Strong states which seem to be mainly directly excited (RPA particle hole excitations) are calculated in this approach by assuming direct β_λ excitation. Fig. 2 shows the coupling schemes, used for the various cases.

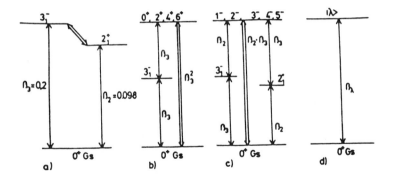

Fig.2 Coupling schemes for calculation of: a) the 2_1^+ and the 3_1^-, b) octupole two phonon states, c) quadrupole octupole phonon states and d) direct excitation.

3.1 Direct Excitations

The strongest states in the measured energy range are the 2_1^+ at 1750 keV and the 3_1^- at 1897 keV, shown in Fig.3. With a β_3 of 0.2 we yield a B(E3) of 46 WU that agrees within the error bars with the value of the lifetime measurement. The second order coupling between 2_1^+ and 3_1^- was necessary to get a reasonable good description of the data. This may indicate the existence of a $2^+\otimes 3^-$ multiplett of states. The description of the 2_1^+ is less good than for the 3_1^-, in this case the microscopic formfactor seems to be quite different from the simple collective one. The strongest 2^+, 4^+ and 5^- excitation at 3212 keV, 3176 keV and 3120 keV shown in fig.3, are assumed to be direct excitations from the ground state, corresponding to the lowest lying particle hole excitations that come out in an preliminary RPA calculation[9].

3.2 Multiple Excitations

Candidates to identify double octupole excitations are 0^+ and 6^+ states[10], lying in the range of twice the 3^- energy at 3.8 MeV. For the 2^+ and 4^+ states other mechanisms provide much larger cross sections. Below 4 MeV there are three known 0^+ and two known 6^+ states. For the 0^+ states, the observed cross sections are about a factor of ten below double octupole predictions, calculated as in fig.2 b). The cross section of the first 6^+ state (fig. 4) (being 10 times larger than for the second 6^+ state) of 100 µb/sr has the same order of magnitude as a full second order calculation for two phonon octupole excitation but not the right shape (fig. 4 dashed line). The angular dependence of the cross section is better reproduced assuming direct collective β_6 excitation (fig. 4 full line). However the analyzing power shows that the situation is surely more complicated.

In the case of octupole quadrupole excitations expected at about 3.65 MeV excitation energy only a 5^- candidate at 3.749 MeV (fig.4) will be discussed here (the other two known 5^- states show direct excitation character), because the spin assignment of a 3^- candidate is not yet shure and other members of the $2^+ \otimes 3^-$ multiplett are not observed. The dashed line in fig. 4 represents a two phonon calculation as in fig.1 d), the full line a direct β_5 calculation. One observes the same discrepancy concerning the shape as in the case of the double octupole 6^+.

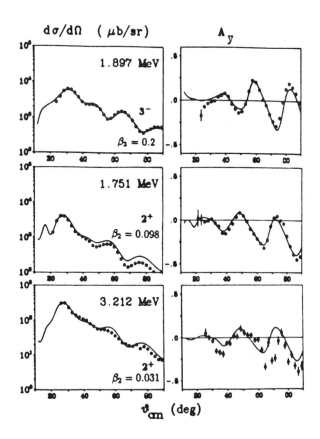

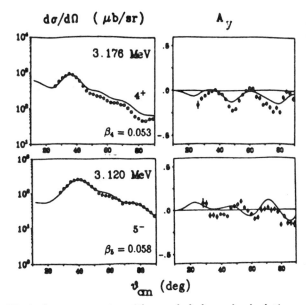

Fig.3 Strongest states with coupled channel calculations.

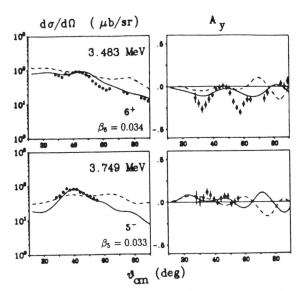

Fig.4 Candidates for double octupole 6^+ and octupole quadrupole 5^- with calculations for direct β_λ (full lines) and pure two phonon (dashed lines) excitation

4. Discussion

Most of the strong, low lying states can be described by a direct β_λ excitation, this confirms the magic structure of ^{96}Zr. Considering the data of double octupole candidates, the observations seem not to be in favour of the presence of such kind of excitations below 4 MeV. One has, however to study in more detail the interference between two step octupole excitation and direct two phonon excitation of a $3^- \otimes 3^-$ state, resulting as a second order effect in the expansion of the collective formfactor (fig.2 b)). The contribution of both terms to the cross section are calculated for the 6^+ in fig. 5.

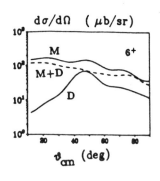

The second order direct term (fig. 5 curve D) interferes negativ with the first order two step term (fig.5 curve M) and is very large. In the collective model however, the direct second order matrix element is very restricted to going with β_3^2 times the second derivative of the collective formfactor[6]. It remains the question whether this is the full truth. Further investigations of the octupole two phonon vibration (that has never been observed yet) will consider the microscopic natur of the second order direct term and his interaction with other excitation mechanismes.

Fig.5 Contribution of first order multistep (M) and second order direct (D) term to the double octupole cross section (D+M) of the 6^+

References

1. G. Molnàr et al. *Nucl.Phys.* A500 (1989), 43
2. H. Ohm et al., preprint Jülich, submitted to *Phys.Lett.*
3. M. Fujiwara et al., RCNP Osaka *Univ.Ann.Rep.* (1989), 59
4. G. Molnar, KFA Jülich *Ann.Rep.* (1989)
5. H. Wessner et al., *NIM* A286 (1990), 175
6. T. Tamura, *Rev.Mod.Phys.* 37 Vol.4 (1965), 679
7. J. Raynal, C.E.N. Saclay, code ECIS79
8. W. Daehnick, J. Childs, Z. Vrcelj, *Phys.Rev.* C21 (1980), 2253
9. W. Unkelbach, KFA Jülich *Ann.Rep.* (1989)
10. D. Kusnezov et al., *Phys.Lett.* B228 (1989), 11

ID
LINEARISED COLLECTIVE SCHRÖDINGER EQUATION FOR NUCLEAR QUADRUPOLE SURFACE VIBRATIONS

Martin Greiner[1+2], Dirk Heumann[1] and Werner Scheid[1]

[1] Institut für Theoretische Physik der Justus Liebig Universität,
Heinrich-Buff-Ring 16, 6300 Giessen, Germany
[2] Gesellschaft für Schwerionenforschung, Planckstrasse 1,
Postfach 110552, 6100 Darmstadt, Germany

ABSTRACT:
The linearisation of the Schrödinger equation for nuclear quadrupole surface vibrations yields a new spin degree of freedom, which is called collective spin and has a value of 3/2. With the introduction of collective spin dependent potentials, this linearised Schrödinger equation is then used for the description of low energy spectra and electromagnetic transition probabilities of some even-odd Xe, Ir and Au nuclei which have a spin 3/2 in their groundstate.

INTRODUCTION

It is well known that the usual Schrödinger equation does not incorporate spin degrees of freedom. Phenomenologically spin degrees of freedom can be introduced by adding spin-dependent terms to the Schrödinger Hamiltonian, as, for example, a spin-orbit coupling or a Pauli term. A more fundamental way to proceed is to linearise the free Schrödinger equation [1]. In this case one constructs an equivalent free equation where the energy and momentum operators only appear in first order. The linearised free Schrödinger equation then describes a particle with spin 1/2. A much more familiar application of the linearisation scheme is the derivation of the free Dirac equation from the free Klein-Gordon equation [2]. In the free Klein-Gordon equation the energy and momentum operators appear in second order whereas in the free Dirac equation they only enter in first order. As a consequence of the linearisation scheme a spin degree of freedom appears, which is not due to the theory of relativity as usually assumed, but is due to the linearisation of the equations of motions.

In nuclear physics Schrödinger equations for multipole degrees of freedom have a wide range of application for the description of nuclear spectra and collective phenomena, e.g., collective surface or density vibrations. In this contribution we linearise the corresponding free collective Schrödinger equations; that means we construct equivalent equations with energy and momentum appearing only in first order. The new equations contain spin degrees of freedom, which we denote as collective spins [3].

In section 2 we sketch the linearisation procedure for a free Schrödinger equation of an arbitrary multipole degree of freedom and deduce the collective spin from the linearised equation. For nuclear quadrupole surface degrees of freedom we explicitly introduce collective spin-dependent potentials, which simmulate the interaction between the valence nucleon and the core, into the linearised Schrödinger equation, which is then transformed into an effective Schrödinger equation with collective spin-dependent potentials; this is the subject of section 3. Applications with respect to the description of collective aspects of some even-odd nuclei like Xe, Ir and Au are also presented in section 3. Concluding remarks are given in section 4.

LINEARISATION OF THE FREE $(2\lambda+1)$-DIMENSIONAL SCHRÖDINGER EQUATION

Schrödinger equations describing $2\lambda+1$ multipole degrees of freedom are used for example for the treatment of nuclear collective surface vibrations. In this case the vibrations of spherical nuclei are described by collective surface coordinates $\alpha_{\lambda\mu}$ defined by the expansion of the nuclear surface as follows [4]

$$R(\vartheta,\varphi) = R_0 \left\{ 1 + \sum_{\lambda,\mu}(-1)^{\mu}\alpha_{\lambda-\mu}Y_{\lambda\mu}(\vartheta,\varphi) \right\}. \tag{1}$$

The Hamiltonian for surface vibrations of multipolarity λ is rotationally invariant and has the following structure in lowest order in these coordinates,

$$H_\lambda = \frac{1}{2B_\lambda}\sum_\mu (-1)^\mu \pi_{\lambda\mu}\pi_{\lambda-\mu} + \frac{C_\lambda}{2}\sum_\mu (-1)^\mu \alpha_{\lambda\mu}\alpha_{\lambda-\mu}. \tag{2}$$

The quantities $\pi_{\lambda\mu}$ are the canonically conjugate momenta. Since the coordinates $\alpha_{\lambda\mu}$ are complex, the introduction of real coordinates

$$\begin{aligned}x^{(\lambda)}_{\lambda+1-\mu} &= \tfrac{1}{\sqrt{2}}[\alpha_{\lambda\mu}+(-1)^\mu \alpha_{\lambda-\mu}], & x^{(\lambda)}_{\lambda+1} &= \alpha_{\lambda 0},\\ x^{(\lambda)}_{\lambda+1+\mu} &= \tfrac{i}{\sqrt{2}}[\alpha_{\lambda\mu}-(-1)^\mu \alpha_{\lambda-\mu}], & (\mu &= 1,\ldots,\lambda),\end{aligned} \tag{3}$$

and the corresponding momenta $p_i^{(\lambda)}$ fulfilling the usual commutation relations

$\left[p_i^{(\lambda)}, x_j^{(\lambda)}\right] = -i\hbar\delta_{ij}$ yields a familiar expression for the Hamiltonian (2):

$$H_\lambda = \frac{1}{2B_\lambda} \sum_{i=1}^{2\lambda+1} p_i^{(\lambda)^2} + \frac{C_\lambda}{2} \sum_{i=1}^{2\lambda+1} x_i^{(\lambda)^2}. \tag{4}$$

In a completely analogous manner as the three-dimensional Schrödinger equation is linearised [1] or the Dirac equation is derived from the Klein-Gordon equation [2] we linearise the $(2\lambda + 1)$-dimensional free Schrödinger equation

$$\left(i\hbar\frac{\partial}{\partial t} - \frac{1}{2B_\lambda}\sum_{i=1}^{2\lambda+1} p_i^{(\lambda)^2}\right)\psi = 0. \tag{5}$$

The linearised free Schrödinger equation must have the following structure:

$$\Theta_\lambda \psi \equiv \left[P^{(\lambda)}\left(i\hbar\frac{\partial}{\partial t}\right) + \sum_{i=1}^{2\lambda+1} Q_i^{(\lambda)} p_i^{(\lambda)} + R^{(\lambda)}\right]\psi = 0. \tag{6}$$

In this equation the time derivative and momenta appear only in first order. We assume that the wave function ψ solves the free Schrödinger equation (5) and the linearised free Schrödinger equation (6). This is achieved by the definition of a second linear operator

$$\Theta'_\lambda \equiv \left[P^{(\lambda)\prime}\left(i\hbar\frac{\partial}{\partial t}\right) + \sum_{i=1}^{2\lambda+1} Q_i^{(\lambda)\prime} p_i^{(\lambda)} + R^{(\lambda)\prime}\right] \tag{7}$$

with the property

$$\Theta'_\lambda \Theta_\lambda = 2B_\lambda\left(i\hbar\frac{\partial}{\partial t} - \frac{1}{2B_\lambda}\sum_{i=1}^{2\lambda+1} p_i^{(\lambda)^2}\right). \tag{8}$$

A comparison of the terms on the left and right hand side of eq. (8) determines the unknown operators $P^{(\lambda)}, Q_i^{(\lambda)}$ and $R^{(\lambda)}$, so that the linearised $(2\lambda + 1)$-dimensional free Schrödinger equation (6) finally reads

$$\begin{pmatrix} I & 0 \\ 0 & 0 \end{pmatrix}\left(i\hbar\frac{\partial}{\partial t}\right)\psi = H_L^{(\lambda)}\psi, \tag{9}$$

where

$$\psi = \begin{pmatrix} \varphi \\ \chi \end{pmatrix},$$

$$H_L^{(\lambda)} = -\begin{pmatrix} 0 & 0 \\ 0 & I \end{pmatrix} 2B_\lambda - \sum_{j=1}^{2\lambda+1}\begin{pmatrix} 0 & i\gamma_j^{(\lambda-1)} \\ -i\gamma_j^{(\lambda-1)} & 0 \end{pmatrix} p_j^{(\lambda)}. \tag{10}$$

The matrices $\gamma_i^{(\lambda)}$ have to fulfill the Clifford algebra

$$\gamma_i^{(\lambda)}\gamma_j^{(\lambda)} + \gamma_j^{(\lambda)}\gamma_i^{(\lambda)} = 2\delta_{ij}I, \qquad (11)$$

$i,j = 1,\ldots,2\lambda+1$. Both functions φ and χ satisfy the free Schrödinger equation (5). It is observed that the linearised $(2\lambda + 1)$-dimensional free Schrödinger equation can be regarded as the "nonrelativistic limit" of a $(2\lambda+1)$-dimensional free Dirac equation, i.e. $E_{Dirac} = B_\lambda + E$ with $E \ll 2B_\lambda$. This gives an anschaulich argument for eq. (11).

The linearised free Schrödinger equation (9) contains spin degrees of freedom which we denote as collective spin degrees of freedom. This notation is justified since Schrödinger equations for multipole degrees of freedom generally descrice collective degrees of freedom, e.g. the nuclear surface vibrations. In order to deduce the collective spin we investigate the commutator between the angular momentum operator and the linearised free Hamiltonian. The Cartesian components of the angular momentum operator [4], expressed within the new coordinates $x_i^{(\lambda)}$ and momenta $p_i^{(\lambda)}$, are specific linear combinations of the angular momentum tensor

$$L_{mn}^{(\lambda)} = x_m^{(\lambda)}p_n^{(\lambda)} - x_n^{(\lambda)}p_m^{(\lambda)}. \qquad (12)$$

For example, for quadrupole degrees of freedom the angular momentum operators are

$$\begin{aligned} L_x^{(2)} &= L_{41}^{(2)} + L_{52}^{(2)} + \sqrt{3}L_{43}^{(2)}, \\ L_y^{(2)} &= L_{12}^{(2)} + L_{54}^{(2)} + \sqrt{3}L_{23}^{(2)}, \\ L_z^{(2)} &= 2L_{51}^{(2)} + L_{42}^{(2)}. \end{aligned} \qquad (13)$$

Each element of the angular momentum tensor $L_{mn}^{(\lambda)}$ is of course hermitean. The following commutator algebra between two elements is valid:

$$\left[L_{ij}^{(\lambda)}, L_{mn}^{(\lambda)}\right] = -i\hbar\left(\delta_{jm}L_{in}^{(\lambda)} + \delta_{jn}L_{mi}^{(\lambda)} + \delta_{im}L_{nj}^{(\lambda)} + \delta_{in}L_{jm}^{(\lambda)}\right). \qquad (14)$$

The commutator between the free linearised Hamiltonian $H_L^{(\lambda)}$ of (10) and $L_{mn}^{(\lambda)}$ yields

$$\left[H_L^{(\lambda)}, L_{mn}^{(\lambda)}\right] = i\hbar\left[\begin{pmatrix} 0 & i\gamma_m^{(\lambda-1)} \\ -i\gamma_m^{(\lambda-1)} & 0 \end{pmatrix}p_n^{(\lambda)} - \begin{pmatrix} 0 & i\gamma_n^{(\lambda-1)} \\ -i\gamma_n^{(\lambda-1)} & 0 \end{pmatrix}p_m^{(\lambda)}\right]. \qquad (15)$$

Introducing a hermitean spin tensor

$$S_{mn}^{(\lambda)} = -\frac{1}{4}i\hbar\left[\gamma_m^{(\lambda)}, \gamma_n^{(\lambda)}\right], \qquad (16)$$

$\lambda =$	1	2	3	4
$s =$	$\frac{1}{2}$	$\frac{3}{2}$	3 0	5 2

Table 1: The collective spin s for different multipolarities λ.

we find

$$\left[H_L^{(\lambda)}, L_{mn}^{(\lambda)} + S_{mn}^{(\lambda)}\right] = 0. \quad (17)$$

In addition the commutator between two elements of the spin tensor yields the same algebra as in the case of the "orbital" angular momentum tensor (see eq. (14)):

$$\left[S_{ij}^{(\lambda)}, S_{mn}^{(\lambda)}\right] = -i\hbar \left(\delta_{jm} S_{in}^{(\lambda)} + \delta_{jn} S_{mi}^{(\lambda)} + \delta_{im} S_{nj}^{(\lambda)} + \delta_{in} S_{jm}^{(\lambda)}\right). \quad (18)$$

As a consequence one can immediately define the Cartesian components of the spin operator analogously to the angular momentum operator. For quadrupole degrees of freedom the spin operators are then

$$\begin{aligned} S_x^{(2)} &= S_{41}^{(2)} + S_{52}^{(2)} + \sqrt{3} S_{43}^{(2)}, \\ S_y^{(2)} &= S_{12}^{(2)} + S_{54}^{(2)} + \sqrt{3} S_{23}^{(2)}, \\ S_z^{(2)} &= 2 S_{51}^{(2)} + S_{42}^{(2)}. \end{aligned} \quad (19)$$

There are at least also two other possibilities how to derive the expressions of the spin operators. Here we have investigated the commutator between the linearised Hamiltonian (10) and the angular momentum operator (11,12). But one could also investigate the transformational behaviour of the wavefunction solving the linearised Schrödinger equation (9) under spatial rotations or require time reversal symmetry in order to deduce the spin operator [3].

The spins as functions of the multipolarity λ of the coordinates can be calculated by diagonalising the corresponding irreducible spin operators. The results are shown for $\lambda \leq 4$ in tab.1.

THE LINEARISED SCHRÖDINGER EQUATION (λ=2) WITH SPIN--DEPENDENT POTENTIALS FOR EVEN-ODD Xe, Ir AND Au NUCLEI

In the following we will only concentrate on the quadrupole degrees of freedom, because they are the most important collective degrees of freedom for nuclear structure. In this case the collective spin turns out to be 3/2, which is a halfinteger value. This observation alone immediately leads to the assumption, that the linearised Schrödinger equation might be used to describe some

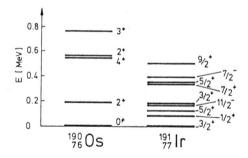

Figure 1: Experimental energy spectra of ^{190}Os and ^{191}Ir [5].

even-odd nuclei, whereas the usual Schrödinger equation describes the neighbouring even-even nuclei. The even-odd nuclei should then have a total angular momentum of $3/2^+$ in their groundstate and the first excited states should have angular momenta $1/2^+, 3/2^+, 5/2^+$ and $7/2^+$, resulting from the coupling of the collective spin to the first excited 2^+ state of the neighbouring even-even nucleus. In figure 1 this point is depicted again for a possible canditate, namely ^{191}Ir; the negative parity states shown can not be included into the present theory. Other possible canditates are for example other even-odd Ir isotopes as well as some even-odd Xe and Au nuclei.

It is now the task to introduce spin-dependent potentials into the linearised Schrödinger equation (9) in order to describe some of the stated even-odd nuclei. These potentials depend explicitly on the collective spin degree of freedom. The linearised Schrödinger equation then reads

$$i\hbar \frac{\partial}{\partial t}\psi = H_{LA}\psi$$
$$= (H_L + A_{spin} \cdot f(x_i, p_i))\psi, \qquad (20)$$

where the second operator on the right hand side represents the spin-dependent potentials. The spin-dependence of these potentials is contained in a coupling matrix denoted as A_{spin}, which is 8×8 dimensional. It can be represented as a linear combination of 64 linearly independent, hermitean 8×8 dimensional Γ-matrices. These Γ-matrices are products of the six γ-matrices

$$\gamma_j = \begin{pmatrix} 0 & i\tilde{\gamma}_j \\ -i\tilde{\gamma}_j & 0 \end{pmatrix}, \gamma_6 = \begin{pmatrix} I & 0 \\ 0 & -I \end{pmatrix}, \qquad (21)$$

which are given in terms of the 4×4 dimensional matrices $\tilde{\gamma}_j$ known from the

Dirac equation,

$$\tilde{\gamma}_1 = \begin{pmatrix} 0 & \sigma_x \\ \sigma_x & 0 \end{pmatrix}, \quad \tilde{\gamma}_2 = \begin{pmatrix} 0 & \sigma_z \\ \sigma_z & 0 \end{pmatrix}, \quad \tilde{\gamma}_3 = \begin{pmatrix} I & 0 \\ 0 & -I \end{pmatrix},$$

$$\tilde{\gamma}_4 = i \begin{pmatrix} 0 & I \\ -I & 0 \end{pmatrix}, \quad \tilde{\gamma}_5 = \begin{pmatrix} 0 & \sigma_y \\ \sigma_y & 0 \end{pmatrix}, \quad (22)$$

where the σ_i are the well known Pauli matrices. The Γ-matrices are then defined as follows:

$$\begin{aligned} \Gamma^S &= I, \\ \Gamma^V_\mu &= \gamma_\mu, & \mu &= 1,\ldots,6, \\ \Gamma^{T2}_{\mu\nu} &= i\gamma_\mu\gamma_\nu = \tfrac{i}{2}[\gamma_\mu,\gamma_\nu] = \tfrac{-2}{\hbar}S_{\mu\nu}, & \nu > \mu &= 1,\ldots,6, \\ \Gamma^{T3}_{\mu\nu\tau} &= i\gamma_\mu\gamma_\nu\gamma_\tau, & \tau > \nu > \mu &= 1,\ldots,6, \quad (23) \\ \Gamma^{PS} &= i\gamma_1\gamma_2\gamma_3\gamma_4\gamma_5\gamma_6 \equiv \gamma_7, \\ \Gamma^{PV}_\mu &= i\gamma_7\gamma_\mu, & \mu &= 1,\ldots,6, \\ \Gamma^{PT2}_{\mu\nu} &= i\gamma_7\gamma_\mu\gamma_\nu, & \nu > \mu &= 1,\ldots,6. \end{aligned}$$

In close analogy to the conventions used in the case of the Dirac equation, the index S stands for scalar, V for vector, T2 and T3 for tensor of second and third rank, PS for pseudoscalar, PV for pseudovector and PT2 for pseudotensor of second rank. The Γ-matrices have to be multiplied with coordinate- and momentum-dependent functions $f(x_i, p_i)$ under the condition that the required symmetries for the linearised Schrödinger equation (20) like hermiticity, SO(3)-, parity- and time reversal-invariance are fulfilled. The interaction operator $A_{spin} \cdot f(x_i, p_i)$ of eq. (20) can now be written in the following manner:

$$A_{spin} \cdot f(x_i, p_i)$$

$$= \Gamma^S \cdot f^S(x_i, p_i) + \sum_{\mu=1}^{6} \Gamma^V_\mu \cdot f^V_\mu(x_i, p_i) + \sum_{\nu>\mu=1}^{6} \Gamma^{T2}_{\mu\nu} \cdot f^{T2}_{\mu\nu}(x_i, p_i)$$

$$+ \sum_{\tau>\nu>\mu=1}^{6} \Gamma^{T3}_{\mu\nu\tau} \cdot f^{T3}_{\mu\nu\tau}(x_i, p_i) + \Gamma^{PS} \cdot f^{PS}(x_i, p_i)$$

$$+ \sum_{\mu=1}^{6} \Gamma^{PV}_\mu \cdot f^{PV}_\mu(x_i, p_i) + \sum_{\nu>\mu=1}^{6} \Gamma^{PT2}_{\mu\nu} \cdot f^{PT2}_{\mu\nu}(x_i, p_i). \quad (24)$$

As a first illustrative example let us discuss the following, very restrictive coupling

$$\left(\frac{(1+\alpha)}{2}\Gamma^S + \frac{(1-\alpha)}{2}\Gamma^V_6\right) \cdot V(\beta) = \begin{pmatrix} I & 0 \\ 0 & \alpha I \end{pmatrix} \cdot V(\beta), \quad (25)$$

which introduces only the collective potential of the neighbouring even-even nucleus of the even-odd nucleus of interest. Here, α is a real coupling constant.

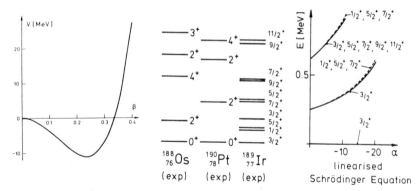

Figure 2: The experimental spectra of ^{188}Os, ^{190}Pt and ^{189}Ir and the energy levels of ^{189}Ir calculated with the linearised Schrödinger equation (20) and (25). The potential used is shown on the left hand side.

The collective potential depends only on the radial quadrupole coordinate $\beta = \left(\sum_{i=1}^{5} x_i^2\right)^{\frac{1}{2}}$. The expression (25) represents a scalar coupling of a potential which has been extensivly studied in ref. [6]. Using the ansatz

$$\psi = \begin{pmatrix} \varphi \\ \chi \end{pmatrix} exp\left(-\frac{i}{\hbar}Et\right) \qquad (26)$$

and the inequality

$$\alpha V(\beta) \ll 2B \qquad (27)$$

the linearised Schrödinger equation (20) with the coupling (25) included can be approximated up to first order in α by

$$E\varphi = \left\{ \frac{1}{2B}\sum_{i=1}^{5}p_i^2 + V(\beta) + \frac{\alpha}{4B^2}\frac{1}{\beta}\frac{\partial V(\beta)}{\partial \beta}\sum_{i,j=1}^{5}\tilde{S}_{ij}L_{ij} \right.$$
$$\left. +\frac{\alpha}{4B^2}V(\beta)\sum_{i=1}^{5}p_i^2 - \frac{i\hbar\alpha}{4B^2}\frac{1}{\beta}\frac{\partial V(\beta)}{\partial \beta}\sum_{i=1}^{5}x_ip_i \right\}\varphi. \qquad (28)$$

The SO(5) spin-orbit coupling (third term) and the correction to the kinetic energy (fourth and fifth term) both give rise to a level shift, which is not sufficient to describe the first excited levels of the even-odd Ir nuclei with respect to their relative positions as well as to magnitude; see fig.2. The level shift is a kind of finestructure splitting. Therefore, it is important to consider more general couplings as given in (24).

Using eqs. (21), (23) and (26), we can write the linearised Schrödinger equation (20) as a set of two coupled differential equations:

$$E\varphi = \left[(f^S + f_6^V) + \sum_{\mu=1}^{5} \tilde{\gamma}_\mu \left(f_\mu^{PV} + f_{\mu 6}^{PT2}\right) + i \sum_{\nu>\mu=1}^{5} \tilde{\gamma}_\mu \tilde{\gamma}_\nu \left(f_{\mu\nu}^{T2} + f_{\mu\nu 6}^{T3}\right)\right]\varphi$$

$$+ \left[(f^{PS} - if_6^{PV}) - i\sum_{\mu=1}^{5} \tilde{\gamma}_\mu \left(p_\mu - f_\mu^V + if_{\mu 6}^{T2}\right)\right.$$

$$\left. +i\sum_{\nu>\mu=1}^{5} \tilde{\gamma}_\mu \tilde{\gamma}_\nu f_{\mu\nu}^{PT2} - \sum_{\tau>\nu>\mu=1}^{5} \tilde{\gamma}_\mu \tilde{\gamma}_\nu \tilde{\gamma}_\tau f_{\mu\nu\tau}^{T3}\right]\chi,$$

$$2B\chi = \left[(f^{PS} + if_6^{PV}) + i\sum_{\mu=1}^{5} \tilde{\gamma}_\mu \left(p_\mu - f_\mu^V - if_{\mu 6}^{T2}\right)\right. \tag{29}$$

$$\left. +i\sum_{\nu>\mu=1}^{5} \tilde{\gamma}_\mu \tilde{\gamma}_\nu f_{\mu\nu}^{PT2} + \sum_{\tau>\nu>\mu=1}^{5} \tilde{\gamma}_\mu \tilde{\gamma}_\nu \tilde{\gamma}_\tau f_{\mu\nu\tau}^{T3}\right]\varphi$$

$$+ \left[(f^S - f_6^V) - \sum_{\mu=1}^{5} \tilde{\gamma}_\mu \left(f_\mu^{PV} - f_{\mu 6}^{PT2}\right) + i \sum_{\nu>\mu=1}^{5} \tilde{\gamma}_\mu \tilde{\gamma}_\nu \left(f_{\mu\nu}^{T2} - f_{\mu\nu 6}^{T3}\right)\right]\chi.$$

In this general form the coupled equations are not easy to handle. It is conveniant to solve first the second equation of (29) by assuming that we can use the inequality

$$\left[(f^S - f_6^V) - \sum_{\mu=1}^{5} \tilde{\gamma}_\mu \left(f_\mu^{PV} - f_{\mu 6}^{PT2}\right) + i \sum_{\nu>\mu=1}^{5} \tilde{\gamma}_\mu \tilde{\gamma}_\nu \left(f_{\mu\nu}^{T2} - f_{\mu\nu 6}^{T3}\right)\right] \ll 2B \tag{30}$$

and expand this equation in terms of $\frac{1}{2B}$. This inequality as well as inequality (27) are fulfilled in all cases of interest, because the spin-dependent potentials lead to a much smaller splitting of the energy levels than the value of the mass parameter (measured in MeV).

Finally, the second equation of (29) can be solved for χ by using (30). The result is inserted into the first equation, which then leads to an effective Schrödinger equation:

$$E\varphi = \left[\frac{1}{2B}\sum_{i=1}^{5} p_i^2 + W(x_i, p_i) + \sum_{\mu=1}^{5} \tilde{\gamma}_\mu A_\mu (x_i, p_i) + \sum_{\nu>\mu=1}^{5} \tilde{\gamma}_\mu \tilde{\gamma}_\nu F_{\mu\nu} (x_i, p_i)\right]\varphi$$

$$= \left[\frac{1}{2B}\sum_{i=1}^{5} p_i^2 + V(x_i) + T_{corr}(x_i, p_i) + \sum_{\mu=1}^{5} \tilde{\gamma}_\mu A_\mu (x_i, p_i)\right.$$

$$\left. +\frac{2i}{\hbar} \sum_{\nu>\mu=1}^{5} \tilde{S}_{\mu\nu} F_{\mu\nu} (x_i, p_i)\right]\varphi. \tag{31}$$

The functions W, A_μ and $F_{\mu\nu}$ depend on the various functions f and their derivatives. In the second step, we split up the function $W(x_i, p_i)$ into a potential part $V(x_i)$ and a correction $T_{corr}(x_i, p_i)$ to the kinetic energy and use the transformation

$$\tilde{\gamma}_\mu \tilde{\gamma}_\nu = \frac{1}{2}[\tilde{\gamma}_\mu, \tilde{\gamma}_\nu] = \frac{2i}{\hbar}\tilde{S}_{\mu\nu}. \tag{32}$$

The two last terms of the right hand side of (30) depend on 15 linearly independent 4×4 matrices $\tilde{\gamma}_\mu$ and $\tilde{S}_{\mu\nu}$. They are referred as effective collective spin-dependent potentials from now on.

Let us first consider an easy and demonstrating version of eq.(31) [7],

$$E\varphi = \left(\frac{1}{2B}\sum_{i=1}^{5}p_i^2 + V(\beta) + \epsilon_1 \sum_{\mu,\nu=1}^{5}\tilde{S}_{\mu\nu}L_{\mu\nu} + \epsilon_2 \sum_{i=1}^{3}\tilde{S}_i L_i\right)\varphi. \tag{33}$$

Here we choose the collective potential $V(\beta)$ to depend only on the radial quadrupole coordinate β; calculations of the spectra of the neighbouring even-even nuclei Os and Pt of the even-odd Ir nuclei for example indicate that the collective potential is γ-soft and does almost not depend on the other intrinsic quadrupole coordinate γ [8]. The third and fourth terms in (33) represent collective SO(5) and SO(3) spin-orbit couplings, respectively; ϵ_1 and ϵ_2 are real coupling constants. Both spin orbit terms are of lowest order in the coordinates and momenta for a tensor coupling fulfilling SO(3), parity and time-reversal invariance. For a moment we disregard a correction term to the kinetic energy and a vector coupling term. - In order to solve (33) analytically for the lowest excited states, we introduce the following ansatz for the wavefunction φ,

$$\varphi = \varphi_{n\lambda\mu L J M_J} = g_{n\lambda}(\beta)\sum_{M_L}(L\frac{3}{2}J|M_L M_J - M_L M_J)Y_{LM_L}^{\lambda\mu}\left(\frac{x_i}{\beta}\right)\chi_{\frac{3}{2},M_J-M_L}. \tag{34}$$

The radial wave function has the labels n and λ denoting the number of knots and the SO(5) quantum number characterizing the SO(5) spherical harmonic $Y_{LM_L}^{\lambda\mu}$. L and M_L stand for the "core" angular momentum quantum number and its projection and μ is an additional label in order to specify the group reduction $SO(5) \supset SO(3)$ completely. The SO(5) spherical harmonics can also be expressed through the intrinsic quadrupole coordinate γ and the Euler angles; for details see ref. [4]. The spinors $\chi_{\frac{3}{2}M_S}$ have four components and correspond to the collective spin 3/2. - The SO(3) spin orbit coupling of (33) is of course always diagonal for the wavefunction (34) and so is the SO(5) spin orbit coupling, too, but only for $\lambda = 0$ or $\lambda = 1$. Nevertheless this is no restriction for the lowest excited states, because $\lambda = 0$ corresponds to a "core" angular momentum L=0 and $\lambda = 1$ to L=2. For these lowest states the energy eigenvalues of the effective Schrödinger equation (33) then result to be (λ=0 or 1)

$$E = E_{n\lambda} + \epsilon_1\left(\pm(\lambda + \frac{3}{2}) - \frac{3}{2}\right) + \frac{\epsilon_2}{2}\left(J(J+1) - L(L+1) - \frac{15}{4}\right), \tag{35}$$

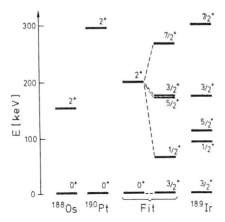

Figure 3: Comparison of experimental levels of ^{189}Ir [9] with theoretical energy eigenvalues obtained from eq. (35). In addition the low energy spectra of the neighbouring nuclei ^{188}Os and ^{190}Pt are shown.

where $E_{n\lambda}$ is the eigenvalue of the Schrödinger equation

$$\left(\frac{1}{2B}\sum_{i=1}^{5} p_i^2 + V(\beta)\right)\varphi_{n\lambda\mu LJM_J} = E_{n\lambda}\varphi_{n\lambda\mu LJM_J}. \quad (36)$$

We apply this formula for the description of the energies of the lowest excited levels of even-odd Ir nuclei. In fig.3 we compare the theoretical result (35) with the lowest experimental energy levels of ^{189}Ir. The quantities E_{01}, ϵ_1 and ϵ_2 of (35) have been assumed as free parameters and determined by a mean square fit to the shown experimental energy levels. Strictly taken, E_{01} is no free parameter and should result from a specific γ-soft potential. The fitted value of E_{01} is $E_{01} = 201 keV$ (E_{00} is set to zero) and agrees very satisfactorily with the mean value $E_{2^+} = 226 keV$ of the first excited 2^+-state of the neighbouring even-even nuclei ^{188}Os and ^{190}Pt and with the weighted mean value of 199 keV of the first excited $1/2^+, 3/2^+, 5/2^+$ and $7/2^+$ states of ^{189}Ir. The theoretical splitting of the first excited levels agrees quite well with the corresponding experimental one. For the other even-odd Ir nuclei we get a similiar good agreement.

Figure 4 shows that the coupling constant ϵ_1 of the SO(5) spin-orbit coupling is nearly constant for the various Ir nuclei. The value of the coupling constant ϵ_2 of the SO(3) spin-orbit coupling increases with increasing neutron number of the Ir nuclei. Also the energy eigenvalue E_{01} shows the same rise. Because these three parameters show only a soft and moderate variation for the different even-odd Ir isotopes, this simple treatment already shows that spin-dependent potentials in the linearised Schrödinger equation are necessary to describe successfully at least the lowest excited states of the even-odd Ir nuclei. In order to obtain the energies of higher excited states or even electromagnetic transition probabilities, a more complete quantitative treatment is needed, to which we

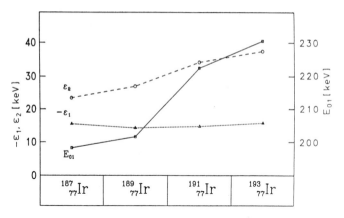

Figure 4: The parameters E_{01}, ϵ_1 and ϵ_2 for various even-odd Ir nuclei.

turn now.

For a more general treatment of the effective Schrödinger equation (31), we will turn back to the spherical notation,

$$\tilde{\gamma}_{2\pm 2} = \tfrac{1}{\sqrt{2}}(\tilde{\gamma}_1 \mp i\tilde{\gamma}_5), \quad \tilde{\gamma}_{2\pm 1} = \pm\tfrac{1}{\sqrt{2}}(\tilde{\gamma}_2 \mp i\tilde{\gamma}_4), \quad \tilde{\gamma}_{20} = \tilde{\gamma}_3 ; \qquad (37)$$

for the coordinates and momenta confer eq. (3). For the effective Hamiltonian we make the following ansatz,

$$\begin{aligned}
H \;=\;& \frac{\sqrt{5}}{2B}\left[\pi^{[2]}\otimes\pi^{[2]}\right]^{[0]} + c_T\left\{\left[\left[\alpha^{[2]}\otimes\pi^{[2]}\right]^{[2]}\otimes\pi^{[2]}\right]^{[0]}\right\} \\
& c_1\left[\alpha^{[2]}\otimes\alpha^{[2]}\right]^{[0]} + c_2\left[\left[\alpha^{[2]}\otimes\alpha^{[2]}\right]^{[2]}\otimes\alpha^{[2]}\right]^{[0]} + c_3\left(\left[\alpha^{[2]}\otimes\alpha^{[2]}\right]^{[0]}\right)^2 \\
& + c_4\left(\left[\alpha^{[2]}\otimes\alpha^{[2]}\right]^{[0]}\right)\left(\left[\left[\alpha^{[2]}\otimes\alpha^{[2]}\right]^{[2]}\otimes\alpha^{[2]}\right]^{[0]}\right) \\
& + c_5\left(\left[\alpha^{[2]}\otimes\alpha^{[2]}\right]^{[0]}\right)^3 + c_6\left(\left[\left[\alpha^{[2]}\otimes\alpha^{[2]}\right]^{[2]}\otimes\alpha^{[2]}\right]^{[0]}\right)^2 \\
& + \epsilon_1\left[\tilde{\gamma}^{[2]}\otimes\alpha^{[2]}\right]^{[0]} + \epsilon_2\left\{\left[\tilde{\gamma}^{[2]}\otimes\left[\alpha^{[2]}\otimes\alpha^{[2]}\right]^{[2]}\right]^{[0]}\right\} \\
& + \epsilon_3\left\{\left[\tilde{\gamma}^{[2]}\otimes\left[\pi^{[2]}\otimes\pi^{[2]}\right]^{[2]}\right]^{[0]}\right\} \\
& + \epsilon_4\left\{\left[\left[\tilde{\gamma}^{[2]}\otimes\tilde{\gamma}^{[2]}\right]^{[1]}\otimes\left[\alpha^{[2]}\otimes\pi^{[2]}\right]^{[1]}\right]^{[0]}\right\}
\end{aligned}$$

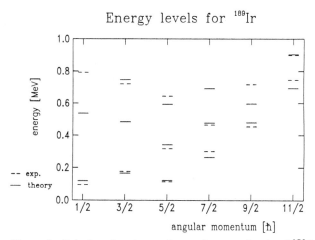

Figure 5: Calculated and experimental energy levels of ^{189}Ir.

$$+\epsilon_5 \left\{ \left[\left[\tilde{\gamma}^{[2]} \otimes \tilde{\gamma}^{[2]} \right]^{[3]} \otimes \left[\alpha^{[2]} \otimes \pi^{[2]} \right]^{[3]} \right]^{[0]} \right\}. \qquad (38)$$

The curly brackets appearing in the Hamiltonian mean symmetrisation with respect to permutations. The first term represents the kinetic energy, whereas the second one is a correction to it. The terms with coefficients $c_1, \ldots, c_6$ stand for an expansion of the collective potential up to sixth order in the coordinates. This part of the Hamiltonian, i.e. the first eight terms, will be identified with the Hamiltonian describing the neighbouring even-even nucleus of the specific even-odd nucleus of interest. As a consequence the eight parameters $B, c_T, c_1, \ldots, c_6$ are fixed from what we know about the stated even-even nucleus; in fact, however, we allow for a small change in order to simmulate the difference of the parameters between the two existing neighbouring even-even nuclei of the chosen even-odd nucleus. - The last five terms of the Hamiltonian (38) are the effective spin dependent potentials up to second order in the coordinates and momenta; the terms with ϵ_1, ϵ_2 and ϵ_3 represent a vector coupling, whereas the terms with ϵ_4 and ϵ_5 represent a tensor coupling. In fact the term with ϵ_4 is once again the SO(3) spin-orbit coupling; the term with ϵ_5 is a linear combination of the SO(3) and SO(5) spin-orbit coupling. The parameters $\epsilon_1, \ldots, \epsilon_5$ are taken as free parameters.

In order to extract the corresponding energy eigenvalues and eigenfunctions, the Hamiltonian (38) is diagonalised within a fivedimensional oscillator basis [4], reflecting the group chain $U(5) \supset SO(5) \supset SO(3)$, coupled to the collective spin on a $SO(3)$ level. This basis has the structure of the wavefunction of eq. (34), where now $g_{n\lambda}(\beta)$ are the fivedimensional oscillator radial wavefunctions.

The calculated energy eigenvalues for ^{189}Ir are shown in fig.5 together with the known experimental energy levels [9]. The five free parameters $\epsilon_1, \ldots, \epsilon_5$ of the effective spin-dependent potentials have been chosen in such a way to get a reasonable and good agreement with the lowest experimental levels of the

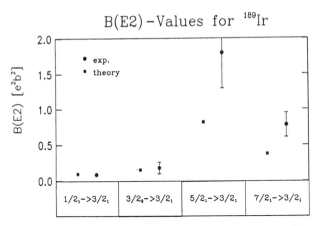

Figure 6: Calculated and experimental B(E2)-values of ^{189}Ir.

groundstate- and first excited band. It turns out that the two terms of the Hamiltonian (38) with coupling constants ϵ_1 and ϵ_3 are only of minor importance; changing the value of these two coupling constants in a reasonable range does almost not influence the calculated energy eigenvalues. As a consequence this approach effectively has only three free parameters, namely $\epsilon_2, \epsilon_4, \epsilon_5$. There is some small staggering of the calculated levels of the two lowest bands around the experimental levels, but nevertheless the overall agreement is remarkably good. For those levels not belonging to the lowest bands we lack some information about the experimental situation, which might not be complete. - For the other even-odd Ir isotopes and also the even-odd Xe and Au isotopes we find a similar good agreement between theory and experiment.

Now it is always a relatively easy thing for a model to reproduce the experimental energy levels. A far more sensitive test would be to look also on the electromagnetic transition probabilities. For E2-transitions we use the quadrupole operator [4]

$$Q_{2\mu} = \frac{3ZR_0^2}{4\pi}\left\{\alpha_{2\mu}^* - \frac{10}{\sqrt{70\pi}}\left[\alpha^{[2]} \otimes \alpha^{[2]}\right]_{2-\mu}\right\} ; \qquad (39)$$

Z is the charge and R_0 the mean radius (see also eq. (1)) of the nucleus. Here no additional free parameter comes into play. The corresponding B(E2)-values of transitions between the first excited $1/2^+, 3/2^+, 5/2^+, 7/2^+$ states to the groundstate $3/2^+$ of ^{189}Ir are shown in fig.6. The B(E2)-values of the interband transitions $(1/2^+)_1 \rightarrow (3/2^+)_1$ and $(3/2^+)_2 \rightarrow (3/2^+)_1$ are small compared to the ones of the intraband transitions $(5/2^+)_1 \rightarrow (3/2^+)_1$ and $(7/2^+)_1 \rightarrow (3/2^+)_1$. The absolute values of the B(E2)-values of the last two stated transitions are a little bit smaller than the experimental values. But nevertheless the trend is correct and for this somehow extraordinary pure collective approach, i.e. without use of an additional single particle degree of freedom, the agreement is astonishingly good.

CONCLUDING REMARKS

At this very end, since it turns out, that the concept of the linearisation of collective Schrödinger equations is even more fruitful for quantitative treatments than expected, a few more comments should be made.

The choice of the effective spin-dependent potentials entering the effective Schrödinger equation has been limited only to the lowest order terms in the collective coordinate and momenta. Taking higher order terms into account, which could be necessary to refine the quantitative treatment, one does not know from the very beginning, which of them are really important and which are not. Therefore it might be reasonable to translate, for example, the $U(6) \otimes U(4)$ Interacting Boson Fermion Model (IBFA) [10], a model which is known to work in the Ir- and Au-region, with a coherent state method. This approach on the one hand would reveal the interrelationship of different models and on the other hand would yield those higher order effective potentials which should be considered in an extension of the present approach.

Up to now only those nuclei have been considered, which have a total angular momentum of 3/2 in their groundstate. It would be interesting to extend the present approach in order to describe other even-odd nuclei. The corresponding wave equations necessary for that investigation can be constructed from the linearised Schrödinger equation in a procedure which is similar to the construction of the Bargmann-Wigner equations [11] in relativistic quantum mechanics.

A question still to raise is to ask for a microscopical interpretation of the collective spin. Since the linearised collective Schrödinger equation turns out to be amazingly successful, finding the answer for this question should now be a major goal. In this context we want to mention a similar application of the linearisation scheme to some algebraic nuclear structure models [12], which points to the indication, that the collective spin might be interpreted as a pseudospin.

REFERENCES

1. J.M. Lévy-Leblond, Commun.Math.Phys. 6 (1967) 286.

2. P.A.M. Dirac, Proc.Roy.Soc. A117 (1928) 610.

3. M. Greiner, W. Scheid and R. Herrmann, Mod.Phys.Lett. A3 (1988) 859; J.Phys.A:Math.Gen. 21 (1988) 3227.

4. J.M. Eisenberg and W. Greiner, Nuclear Theory: Nuclear Models, vol.1 (third edition), Amsterdam: North-Holland (1987).

5. C.M. Lederer and V.S. Shirley, Table of Isotopes (seventh edition), New York: John Wiley & Sons (1978).

6. M. Greiner and W. Scheid, Phys.Rev. C42 (1990) 262.

7. M. Greiner, D. Heumann and W. Scheid, Z.Phys.A-Atomic Nuclei 336 (1990) 139.

8. P.O. Hess, J. Maruhn and W. Greiner, J.Phys.G:Nucl. Phys. 7 (1981) 737.

9. S. André et.al., Nucl.Phys. A243 (1975) 229.

10. A.B. Balantekin, I. Bars and F. Iachello, Nucl.Phys. A376 (1981) 284.

11. U. Bargmann and E. Wigner, Proc.Nat.Acad.Sci.(USA) 34 (1948) 211.

12. H. Wu, M. Greiner and D.H. Feng, Phys.Rev. C39 (1989) 1059.

MICROSCOPIC INSIGHTS INTO THE SHAPE COEXISTENCE IN THE A=70 REGION

A.Petrovici

Institute of Atomic Physics, Bucharest,
P.O.Box MG-6, Romania

ABSTRACT

The shape coexistence phenomena dominating the structure of the doubly-even nuclei in the A=70 mass region are explained in the frame of the EXCITED VAMPIR model. The influence of the additional correlations on top of the symmetry projected quasi-particle mean fields of the EXCITED VAMPIR type is investigated by a new variational procedure, which increases the confidence level of the resulting wave functions introducing the dominant correlations for each particular configuration in a systematic way. Special predictions concerning the general trends in this mass region are confirmed by recent experimental data.

1. INTRODUCTION

According to both the experimental systematics and theoretical calculations, the mass A$\sim$70 nuclei have a transitional character and display a rich pattern of shape coexistence and shape transition. The multiplicity of structures, the rapid and also drastic changes of their behaviour for a variation of the number of nucleons by only two units make the A=70-80 region an important testing ground of nuclear models[1]. Of course a parameter-free theoretical description of the complex behaviour of these nuclei is highly desirable. Unfortunately, a microscopic insight into the shape coexistence phenomena dominating the structure of the mass 70 region by a complete diagonalization of a suitable chosen effective many-nucleon Hamiltonian, as it is done in the Shell-model Configuration Mixing (SCM) approach is severely hindered, because a large number of valence nucleons are distributed among many single particle levels.

The coexistence of overlaping bands built on different nuclear shapes, and also the observation of many low spin states, e.g. 0^+ or 2^+ states, at very low excitation energy is related to the competition between various large gaps at different deformations displayed by the Nilsson single particle energy diagrams. Furthermore, the nuclei in this mass region have usually both active protons as well as neutrons in the $0g_{9/2}$ shell-model orbit. Since there are many different ways to couple these particles to intermediate and high spin values, a competition of many configurations at these spin values is rather likely, too.

For a deeper theoretical understanding of the complex experimental situation encountered in the mass $A\sim70$ nuclei, one needs a microscopic model, in which all the essential degrees of freedom of collective as well as single particle nature, coexisting and competing are not introduced " by hand ", but accounted for in a completely microscopic fashion. Our way to construct such a model is based on varia - tional techniques used in all the approaches developed in the frame of the so called VAMPIR [2] (Variation After Mean Field Projection in Realistic model spaces) family. The two most sophisticated procedures belonging to this model family continuously extended and improved during the last couple of years, EXCITED VAMPIR and EXCITED FED VAMPIR approximations, are completely adequate to describe the shape coexistence phenomena.

The EXCITED VAMPIR model is essentially a mean field theory, which solves the Hartree-Fock-Bogoliubov problem with spin and number projection before the variation for yrast and non-yrast states. In this approximation an optimized mean field is obtained for each state of a given symmetry by a chain of independent variational calculations up to finally the residual interaction in the resulting A-nucleon configuration space is diagonalized. In the first stage we studied the low spin excitation spectra for a couple of doubly-even Ge as well as Se isotopes[3] in the frame of the EXCITED VAMPIR model. We found that a variable mixing of prolate and oblate more or less deformed projected quasi-particle determinants is able to produce the experimental picture for the low spin states of the considered Ge and Se isotopes, like the presence of 4-5 coexisting 0^+ or 2^+ states in 3-4 MeV excitation energy, as well as the general trends in the quadrupole moments, B(E0) and B(E2) values, and proton and neutron occupations of the spherical single particle orbitals. Our interpretation strongly support the shape coexistence as a dominant feature of the low spin states in even Ge and Se nuclei. On the other hand, this new microscopic understanding of the shape coexistence motivated the extension of our EXCITED VAMPIR investigations to high spin states in the ^{68}Ge, ^{70}Se and ^{72}Se nuclei[4]. It turned out that the shape coexistence persists and manifests specifically at high spin states. We predicted a strong bunching of states of a given spin and parity in a small excitation energy interval and a variable, sometimes very strong, mixing in between these states which creates a complex feeding pattern for the yrast band, including competing M1, ΔI=0 transitions. Nevertheless, it came out that the high spin states are grouping into multiple bands based on different structures, some of them connected by E2 crossing transitions.

This complex situation asked for the extension of the theoretical approximation in order to understand how reliable is the EXCITED VAMPIR description of the considered states. A new stage was achieved[5] going beyond the symmetry projected quasi-particle mean field approximations by the so called EXCITED FED (from FEw Determinants) VAMPIR method. In this approach each state is not described by only one, but by a linear combination of few projected determinants obtained successively in a chain of variational calculations, asking in each step for maximum residual interaction and by this for maximum additional contribution to that state. The additional configurations on top of each main underlying mean field can at least partly compensate for the missing correlations originating in the symmetry restrictions imposed on the Hartree-Fock-Bogoliubov transformation in order to get a numerically feasible approximation. By this procedure the remaining residual interaction between the resulting correlated few lowest states of a given symmetry becomes much smaller than in the case of the uncorrelated EXCITED VAMPIR solutions, and thus increases the reliability of the wave functions considerably. The nature of each high spin band is more precisely distinguished since the dominant correlations for each particular configuration are included in a systematic way. Reinvestigating the low and high spin states of ^{68}Ge with this new method we found that the qualitative features of the EXCITED VAMPIR description persist. But now it was possible to identify between the many overlaping bands one having considerably larger deformation, $\beta_2 \sim 0.42$, than the other bands in ^{68}Ge and in other Ge nuclei. This is a strikingly large deformation for such a light nucleus, however, as we shall see, recent experimental data confirm our predictions. Recently[7] we proposed a new, significant test of the EXCITED FED VAMPIR description of the shape coexistence in the A=70 region calculating charge and transition charge densities for a chain of even Ge nuclei. As will be shown, the experimental data obtained from (e,e') reactions[8] give support to our results.

2. THE THEORETICAL BACKGROUND

A brief presentation of the basic ideas behind the approaches used to describe the complex experimental situation encountered in the mass A$\sim$70 nuclei, including the treatment of the charge and transition charge densities will be given in the following.

In the various approaches of the so called VAMPIR family, the variational techniques are used to solve the Hartree-Fock-Bogoliubov problem with number and spin projection before variation for an effective many-body Hamiltonian, based on a nucleon-nucleon interaction renormalized for a finite M-dimensional spherical single particle basis adequate for the A=70 mass region, consisting of general one- and two-body terms.

The variational procedure starts with the calculation of the first projected HFB wave function for the yrast state of a given symmetry, s=NZI$^\pi$. Only positive parity, even spin states can be treated, since we enforce conservation of parity and isospin projection on the HFB transformation, assume time-reversal and axial symmetry and admit only real transformations. The underlying quasiparticle vacuum $|F^1>$ is calculated directly from the chosen effective many-body Hamiltonian minimizing the energy functional

$$E_1^s \equiv <\phi^1;sM|H|\phi^1;sM> \qquad (1)$$

with respect to general variations of the HFB transformation, and so one gets the VAMPIR approach to the considered yrast state, representing the optimal approximation to this state, which can be constructed from a single symmetry-projected HFB determinant

$$|\phi^1;sM> = \hat{(H)}^s_{M0} | F_1 > \eta^1_1 \qquad (2)$$

Here $\hat{(H)}^s_{M0}$ projects on the desired symmetry s, $|F_1>$ stays for the HFB-type vacuum and η^1_1 ensures the normality of the wave function. Then, in complete analogy can be calculated the excited states with the same symmetry, with the only difference being that one has to ensure the orthogonality of the current trial wave function with respect to all the solutions already obtained, and this is the so called EXCITED VAMPIR approximation. In this approach an optimal mean field is obtained for each one of the lowest m states of symmetry s and then finally the wave functions are given diagonalizing the residual interaction between them. Before the final diagonalization the EXCITED VAMPIR solution for the i-th state of symmetry s gets the form

$$|\phi_i;sM> \equiv \hat{(H)}^s_{M0} \sum_{j=1}^{i} |F_j > \eta^i_j \qquad (3)$$

However, introducing in the A-nucleon configuration basis for a given symmetry only one additional determinant, optimal for each new excited state to be considered, the residual interaction between the lowest m such configurations does not necessarily account for the dominant correlations on top of each EXCITED VAMPIR solution. An improved variational procedure, the EXCITED FED VAMPIR approach, allows to go beyond this mean field approximation and to include the most important missing configurations in a systematic way.

In this approximation the i-th state of a given symmetry s is represented as a linear combination of few projected quasiparticle determinants constructed successively in a chain of variational calculations, the state being Schmidt orthogonalized with respect to all the i-1 already found solutions. By this procedure the main additional correlations on top of the underlying mean field for each state are obtained in a systematic way, asking in each step of the variational chain for maximum additional contribution to the energy of the previous approximation to that wave function. So a properly normalized linear combination of n_i different HFB-type determinants $|F_k>$ (k = $1,\ldots,n_i$) is created for the i-th state of symmetry s

$$|\phi_i^{(n_i)};sM> \equiv \hat{T}^{(i)} \hat{\textcircled{H}}_{MO}^s \sum_{\nu=1}^{n_i} | F_{q+\nu} > f_\nu^{in_i}$$

$$\equiv \hat{\textcircled{H}}_{MO}^s \sum_{j=1}^{\omega(i)} | F_j > n_j^i \qquad (4)$$

where $\hat{T}^{(i)}$ eliminates the first i-1 solutions from the model space with q indicating the total number of building determinants, $q \equiv \sum_{j=1}^{i-1} n_j$, and $\omega(i) \equiv \sum_{\ell=1}^{i} n_\ell$. In each step of the variational chain for the i-th "correlated" wave function one varies only the last added mean field and the configuration mixing coefficients which ensures the orthonormality of the $1 \leq n_i$ resulting linear independent solutions. Finally as already in the EXCITED VAMPIR approximation, the residual interaction between the m energetically lowest linear combinations is diagonalized and the lowest m physical states of symmetry s

$$|\psi_\alpha^{(m)}> = \sum_{i=1}^{m} |\phi_i^{(n_i)};sM> g_{i\alpha}^{(m)} , \quad \alpha = 1,\ldots,m \qquad (5)$$

and the corresponding energies E_α ($\alpha=1,\ldots,m$) are obtained

$$(H - E\mathbf{1})g = 0$$
$$g^+g = \mathbf{1} \tag{6}$$

The EXCITED FED VAMPIR variational procedure, going beyond the symmetry-projected quasiparticle mean field approximations, automatically selects the relevant degrees of freedom for the main mean field underlying the structure of a particular state, and also for the most important additional correlations with respect to the approximation already obtained. In this way arbitrary drastic changes of the structure with increasing excitation energy or spin are accessible and furthermore various types of shape coexistence could be distinguished. The particular mentioned features of this model are essential for an adequate description of the structure of nuclei displaying a dynamical shape coexistence. With respect to the EXCITED VAMPIR approximation this procedure has the big advantage that the dominant correlations on top of the optimal projected mean field solutions are accounted for in each state separately. As a consequence each of the still neglected configurations could bring to the energy for the already obtained states only gains comparable with that from the last correlating configuration included for any of them. Also the residual interaction will be at most at this level and of course much smaller. Thus the confidence level of the resulting wave functions is improved.

Using these EXCITED VAMPIR and EXCITED FED VAMPIR approaches we can calculate the matrix elements of tensor operators in between arbitrary solutions. Once we obtain the spectroscopic amplitudes, the electromagnetic properties can be calculated easily.

Few details will be given concerning the description of the charge and transition charge densities obtained from electron scattering to discrete levels. For the elastic electron scattering on a 0^+ nuclear target the transverse form factor is zero and in the Born approximation, ignoring the distorsion effects on the electron wave functions the longitudinal form factor can be directly obtained

$$F_L^2(q) = f_{rec} \left(\frac{q}{q_\mu}\right)^4 \left(\frac{d\sigma(\theta)/d\Omega}{d\sigma_M(\theta)/d\Omega}\right), \qquad (7)$$

where the Mott cross section is

$$d\sigma_M(\theta)/d\Omega = \frac{Z^2 \alpha^2 \cos^2(\theta/2)}{4(E/\hbar c)^2 \sin^4(\theta/2)}, \qquad (8)$$

the recoil factor is $f_{rec} = 1 + (2E\sin^2(\theta/2)/M)$, and q and q_μ are three- and four-momentum transfers, respectively. As in this case only the Coulomb monopole contributes to the cross section, one can obtain directly the charge distributions of the nuclei in their ground state. Since in the electroexcitation of a collective state of an even-even nucleus (2^+, 4^+, 6^+,...) the Coulomb multipoles (C2,C4,...) are much larger in magnitude than the corresponding electric multipoles (E2,E4,...) only the longitudinal form factors are taking into account in the analysis of the data.

For the calculation of the transition density matrix elements we assume that the involved charge density operator is represented by the one-body operator

$$\hat{\rho}(\vec{r}) = \sum_{i=1}^{A} e(i) \delta(\vec{r} - \vec{r}_i). \qquad (9)$$

The transition charge densities are obtained as the Fourier-Bessel transforms of the form factors into the coordinate space

$$\hat{\rho}_{fi}(\vec{r}) = \frac{1}{2\pi^2} \int d^3q \, e^{i\vec{q}\cdot\vec{r}} \sum_{L(M)} (-i)^L Y^*_{LM}(\hat{q}) \cdot$$
$$(-1)^{I_f - M_f} \begin{pmatrix} I_f & L & I_i \\ -M_f & M & M_i \end{pmatrix} < f \| \hat{M}^{coul}_L (q) \| i > \qquad (10)$$

where the Coulomb multipole operator is defined by

$$\hat{M}^{coul}_{JM}(q) = \int d^3r \, j_J(qr) Y_{JM}(\Omega_x) \rho(\vec{r}) , \qquad (11)$$

with j_J the spherical Bessel functions and Y_{JM} the spherical harmonics. Two standard corrections are introduced: the finite size of the nucleons is taken into account folding the Fourier transform of the charge density operator with a Gaussian form factor $f_{SN}(q) = \exp(-q^2 a^2/6)$, where a is the root-mean-square radius of the nucleon, and to correct for the spurious center-of-mass motion an additional factor is used $f_{CM}(q) = \exp(b^2 q^2/4A)$, where b is the oscillator length. Since we use as model space for the EXCITED FED VAMPIR approach an oscillator basis the transition densities are calculated analytically.

3. RESULTS AND DISCUSSIONS

For the description of the even-even nuclei in the A=70 mass region a ^{40}Ca core was used and the model space consisted out of the $1p_{1/2}$, $1p_{3/2}$, $0f_{5/2}$, $0f_{7/2}$, $2s_{1/2}$, $1d_{5/2}$, $0g_{7/2}$ and $0g_{9/2}$ oscillator orbits for both protons and neutrons and included for the latter furthermore the $0h_{11/2}$ level. As effective two-body interaction a Brueckner nuclear matter G-matrix derived from the Bonn one-boson- exchange potential was taken. The renormalization of this G-matrix consisted of: two isospin dependent short range Gaussians (0.707fm) with strengths of -50MeV for T=1 proton-

proton and -40MeV for T=1 neutron-neutron matrix were introduced to enhance the pairing components. An isospin-independent spin-orbit Gaussian was added, and, finally the onset of deformation was influenced by a monopole shift in all the diagonal T=0 matrix elements of the form $<0g_{9/2}0f$; $IT=0|G|0g_{9/2}0f$; $IT=0>$ with $0f$ denoting either the $0f_{5/2}$ or the $0f_{7/2}$ orbit.

The first insights into the complex structures of the nuclei in the A$\sim$70 region came out from the investigation of the low spin states in some chains of even Ge and Se nuclei in the frame of the EXCITED VAMPIR model[3]. This approach yield indeed in all the considered doubly-even Ge and Se isotopes always some low lying 0^+ and 2^+ states in the first few MeV excitation energy as observed experimentally, and also the other low spin properties like the general trends in the quadrupole moments as well as in the B(E0) and B(E2) values are in fair agreement with the experimental data. The EXCITED VAMPIR wave functions for the low spin states in this mass region were in general complicated mixtures of several configurations corresponding to rather different deformations, with a special interplay of the prolate and oblate deformed mean fields and dynamical changes of the hexadecapole deformations and thus the label "shape coexistence" seems to be well justified. These microscopic results and the experimental triple forking at 8^+ into three bands with no crossing transitions observed earlier[9] gave us motivation to extend our investigations to high spin values for these nuclei. The EXCITED VAMPIR method was extended to spins up to 18^+ in ^{68}Ge and 22^+ in ^{72}Se nuclei[4]. In these calculations it turned out that the shape coexistence phenomena persist to rather high angular momenta. So, in both nuclei we found several coexisting 8^+, 10^+, 12^+ and even higher spin states in rather small intervals of excitation energy. Because of the strong mixing in between some of these states a very complicated decaying mode was obtained for each state: many decaying branches displaying a strong competition between E2, $\Delta I=2$ and M1, $\Delta I=0$ transitions. Consequently, the resulting decay pattern becomes very complicated. Nevertheless, still a grouping into several bands is possible and the fastest ways to enter the yrast line can be identified. Recent experimental data[6] in the nucleus ^{68}Ge indicate a much richer band structure and a very complicated decaying mode for this spin region than observed earlier[9] and thus confirm the theoretical predictions at least qualitatively. However, question is how reliable is our theoretical description as the EXCITED VAMPIR approach is essentially a sort of mean field approximation. It is restricted to mainly only one symmetry-projected quasiparticle determinant for each of the states considered and takes only the residual interaction between always the lowest few configurations for each symmetry into account. It may not always be sufficient, since this residual interaction does not necessarily account for the dominant correlations on top of each underlying EXCITED VAMPIR mean field for a given state, and even more there is no a priori reason to neglect contributions from configurations of higher excitation energy which are not included in the truncated chain of variational calculations built for the lowest first states of a given symmetry. Especially in regions with a high level density, where this additional correlations could be introduced within the EXCITED VAMPIR approximation only if a rather high number m of A-nucleon configurations would be constructed, this could lead to considerable changes of the EXCITED VAMPIR picture for the considered states. Thus it seems advisable to device an improved variational scheme which incorporates such correlations in a systematic way, no matter where in energy they occur, for each considered state. This new variational procedure, EXCITED FED VAMPIR approach, created to answer to such requests, was used to reinvestigate the spectrum of

^{68}Ge.

Using the same model space and effective Hamiltonian for both, EXCITED VAMPIR and EXCITED FED VAMPIR approximations a direct comparison of the two descriptions obtained for ^{68}Ge nucleus was possible. It turned out that the qualitative features of the results are not influenced by the additional correlations. Comparing the EXCITED FED VAMPIR spectrum with the EXCITED VAMPIR one it becomes evident that the essential properties like the occurence of many coexisting bands at comparable excitation energies and the complexity of the resulting decay pattern characterize both theoretical spectra. Furthermore, all various bands indicated in the EXCITED FED VAMPIR spectrum in Fig.1 appear in the EXCITED VAMPIR results, too. This fact can be inferred from a direct comparison of the electromagnetic properties, like spectroscopic quadrupole moments and B(E2) values of these bands which show a good agreement. Furthermore, the fact that in both calculations essentially the same structures are obtained is supported by the amount of angular momentum which the valence protons and neutrons in the $0g_{9/2}$ shell-model orbit contribute to the total spin in various bands. These so-called "alignment plots" are presented in Fig.2. As can be seen the inclusion of the additional correlations causes here again no essential qualitative modifications with respect to the EXCITED VAMPIR picture. Since, last but not least, also the transition energies within the various bands were in most cases not very much affected we may conclude that at least as far as the qualitative features are concerned the more sophisticated EXCITED FED VAMPIR method confirms the description obtained for the ^{68}Ge nucleus within the EXCITED VAMPIR approximation. Concerning quantitative details, however, the two approaches display some distinct differences. Of course our results depend on the renormalization of the nuclear matter

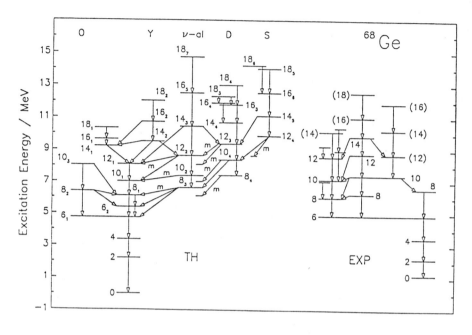

Fig.1. The theoretical spectrum as obtained within the EXCITED FED VAMPIR approach on the basis of real HFB transformations (left side) is compared to the recent experimental results [6] (right side) for the even spin positive parity states of the nucleus ^{68}Ge. The labels "m" at some of the decay-links indicate strong $\Delta I = 0$, M1 transitions. The various labels on top of the theoretical bands are explained in the text.

G-matrix adequate for the model space and the considered mass region. Even if recent experimental data on ^{68}Ge nucleus[6] indicate a rich band structure at high spins confirming our predictions, the available experimental information concerning the electromagnetic properties is rather sparse and can not help to get the final answer neither for the effective interactions nor for the effects of the additional correlations.

In order to get some information concerning the general

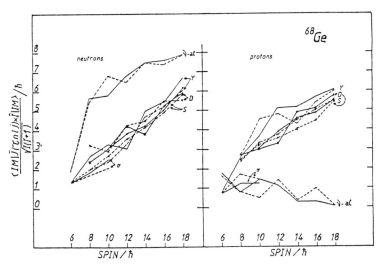

Fig.2. The contribution of the neutrons (left side) and the protons (right side) in the $0g_{9/2}$ shell-model orbit to the total angular momentum in the nucleus ^{68}Ge as obtained for some of the bands presented in Fig.1. The full lines refer to the results of the correlated EXCITED FED VAMPIR approach, the dashed ones to those of the more restricted EXCITED VAMPIR calculations.

trends in this mass region and the special changes determined by a variation of the number of nucleons by two units we investigated other two nuclei ^{70}Se and ^{72}Se. Because of the nice qualitative agreement of the EXCITED FED VAMPIR and the more restricted EXCITED VAMPIR results in ^{68}Ge, here only the much less time consuming latter approach was used. As we shall show in the following, as theoretically expected, the ^{68}Ge, ^{70}Se and ^{72}Se nuclei manifest similar features, but in the same time peculiar differences appear in these three nuclei. A direct comparison of the theoretical spectrum with the available experimental data for medium and high spin states in these three nuclei can be realized from Figs. 1,3,4.

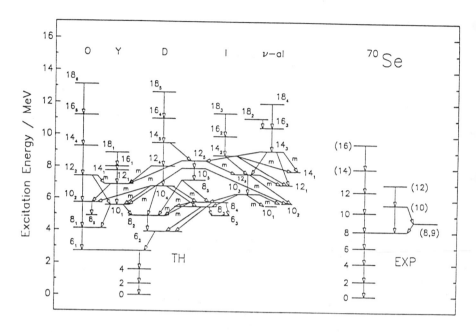

Fig.3. The theoretical spectrum of the nucleus ^{70}Se, here obtained in the EXCITED VAMPIR approximation, is compared to the available experimental data[10].

In these figures the label for each spin corresponds to the energetical order in the calculated set of states for a given angular momentum.

Concerning ^{68}Ge nucleus the EXCITED FED VAMPIR results[5] indicate, as the previous EXCITED VAMPIR calculations[4] that up to angular momentum 6^+ the calculated yrast levels are almost pure oblate states. A strong oblate-prolate mixing characterizes the wave functions for the 8^+ states. The oblate and (labelled "O" in the figure) is almost pure starting with the third 10^+, but the oblate states are at rather high excitation energy and therefore probably only

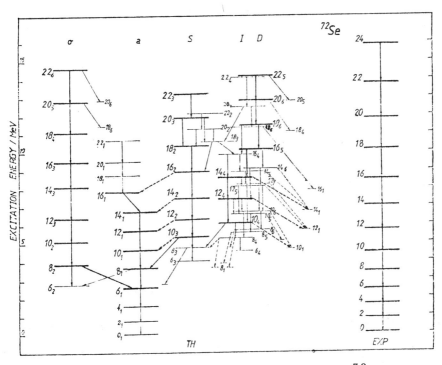

Fig.4. The theoretical spectrum of the nucleus ^{72}Se, calculated in the EXCITED VAMPIR approximation, is compared to the available experimental data[10].

weakly populated. Besides this oblate structure four other pronounced bands are obtained in the calculations. They are all prolate deformed but differ in the magnitude of their quadrupole and hexadecapole moments as well as in their pairing properties, and display different alignments, too.

The band labelled as "ν-a1" is characterized by an almost empty $0g_{9/2}$ proton level while the neutrons in the same shell-model orbit contribute here a considerable portion of the total angular momentum. This strong neutron alignment is reflected in the small g-factors of the members of this band.

The g-factor for the 8_3^+ state belonging to this band is -0.03, and therefore rather likely corresponds to the experimental 8_2^+ state, for which a g-factor of -0.28 ± 0.14 has been measured.

The two bands labelled as "D" and "S" have almost the same quadrupole deformation of the neutron side, however, on the proton side the latter band is considerably more deformed. This is evident from the intrinsic quadrupole moments ($\beta_2 \sim 0.42$ as compared to ~ 0.34) of the leading configuration as well as from the B(E2)-values, which above spin 12 are about twice as large in the "S" - than in the "D"-band. Our predictions concerning the "S"-band, displaying a strikingly large deformation for such a light nucleus, seem to be confirmed by the new experimental data[6] revealing a new band with larger moment of inertia. In fact, comparing the new experimental and theoretical calculations, one is struck by the marked agreement between the multiple band structures observed and predicted. All the predicted prolate bands have one common feature: their members become strongly mixed as soon as spin values as low as 12^+, 10^+, 8^+ are reached. Consequently, there are many competing decay branches for the various 14^+, 12^+, 10^+ and even 8^+ states. Thus the decay does not run via streched E2's within the various bands but also via some strong $\Delta I=2$ B(E2)'s as well as a couple of strong $\Delta I=0$ M1-transitions (indicated by "m" in the figure) mainly feeding the yrast band at these angular momenta. The resulting complex decay pattern may explain at least some of the irregularities seen in the experimental data, and obviously the theoretical results are strongly supported by the observation of not only several 8^+, but also a few energetically bunched 10^+, 12^+ and 14^+ states, as can be seen in Fig.1. Concerning the B(E2) values, here the data from several experiments all display large error bars, but nevertheless do still disagree in some cases. Therefore, the agreement (in some cases) or disagreement (in some other cases) of the theoretical predictions with the data is not yet relevant and needs a lot of further investigation. Even more, as the EXCITED FED VAMPIR and EXCITED VAMPIR approaches give rather similar B(E2)'s the inclusion of the additional correlations yields no new aspects.

Similar results have been obtained for the ^{70}Se and ^{72}Se nuclei investigated within the EXCITED VAMPIR approximation.

Also here a rather complicated decay pattern with a high level density in certain spin regions is obtained, as can be seen from Figs.3 and 4. Again the shape coexistence dominates the structure of the low- and high-spin states. For the low spin yrast states a dominant oblate character emerges from the structure of the wave functions and the spectroscopic quadrupole moments. At high spins an almost pure oblate band coexists with the other prolate bands.

Similar with the band calculated in ^{68}Ge a neutron aligned band "ν-al" is obtained in ^{70}Se, too. These bands are both moderately prolate deformed and the $0g_{9/2}$ neutrons are responsible for a considerable portion of the total angular momentum, as can be seen from Figs.2 and 5. As compared to the case of ^{68}Ge, however, the neutron alignment in the ^{70}Se is less pronounced. This is due to the fact that in the ^{68}Ge band there are just two $0g_{9/2}$ neutrons while de proton $0g_{9/2}$ orbit is almost empty. Thus the neutron alignment is very much favored. In ^{70}Se, however, we have two additional protons which partly occupy the $0g_{9/2}$ and also lead to a somewhat larger neutron occupation of this level.

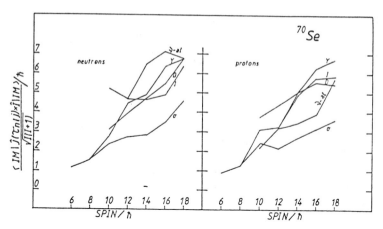

Fig.5. The neutron and proton alignments of the $0g_{9/2}$ orbit as in Fig.2, however, for the nucleus ^{70}Se which has been studied here within the EXCITED VAMPIR approximation only.

Thus here the alignment does not happen as rapidly as in the former case. Adding two more neutrons the occupation of the $g_{9/2}$ orbit is increased even more and thus the alignment severely hindered. This is the reason why no equivalent band was found in the case of ^{72}Se (Fig.4).

On the other hand the strongly deformed band "S" out of the ^{68}Ge spectrum in Fig.1 which to some extent is also seen in ^{72}Se has no counterpart in the ^{70}Se nucleus, at least in the present EXCITED VAMPIR description. Obviously, it can not be excluded that such a band may be found at higher excitation energy in this nucleus, too.

It is worth mentioning that strong magnetic dipole transitions linking the various bands with each other in the intermediate spin region are obtained in all the three considered nuclei ^{68}Ge, ^{70}Se and ^{72}Se. As already mentioned above in the introduction one expects in this spin region a competition of several configurations resulting from the various possible couplings of the $0g_{9/2}$ valence protons and neutrons, and indeed an analysis of the strong B(M1) values reveals that they are in most cases dominated by amplitudes corresponding to the reordering of nucleons in this shell - model orbit. As a consequence these transitions usually contain considerable (about 50 percent) orbital components.

Recently, we performed a new test of our description of the shape coexistence for low spin states comparing the theoretically calculated charge and transition charge densities in a chain of even Ge nuclei with the available model-independent densities obtained from electron scattering data.

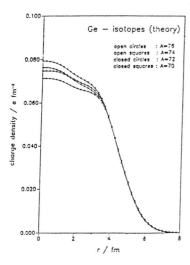

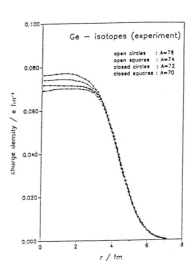

Fig.6. The theoretical charge density for $0^+_{g.s.}$ of the $70,72,74,76$Ge nuclei. The unique effective charges used for the valence nucleons were $e_n=0.17$ and $e_p=1.17$.

Fig.7. The experimental charge density for the $0^+_{g.s.}$ of the $70,72,74,76$Ge nuclei obtained from a model-independent analysis of the elastic electron scattering data[8].

New electron scattering experiments[8] provide information concerning elastic electron scattering and also electroexcitation of the first two 2^+ states for the $^{70,72,74,76}Ge$ isotopes.

We calculated the wave functions for few lowest 0^+ and 2^+ states within the EXCITED FED VAMPIR approximation. The maximum number of projected quasiparticle determinants introduced in the linear combination for each correlated wave function was obtained truncating the chain of variational calculations as soon as the energy gains due to subsequently added configurations became of the order of about 100 keV. The same effective extracharge $e_{eff}=0.17$ was used for neutrons and protons for all the investigated nuclei. A fairly good agreement between the experimental and the theoretical pictures obtained for the charge densities comes out from a comparison of the plots given in Figs.6 and 7. The variation with the mass number is correctly reproduced.

In Figs.8,9,10 are presented the transition charge densities for the first 2^+ state and also the pictorial representations for the available experimental results. The agreement theory/experiment reflects the accuracy in reproducing the experimental B(E2) values.

In Figs.11,12 are displayed some results for transition charge densities corresponding to the excitation of the second 2^+ state. The discrepancies which still persist ask for some modifications of the effective Hamiltonian and motivate our future effort to investigate the effects which can be obtained admitting general unitary Hartree - Fock - Bogoliubov transformations.

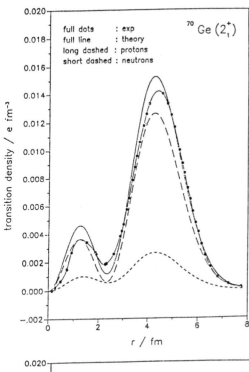

Fig.8. The comparison of the calculated transition charge density for the 2_1^+ state of ^{70}Ge with the corresponding pictorial representation obtained from a model-independent analysis of the inelastic electron scattering data. The effective charges used in the calculations are $e_n = 0.17$ and $e_p = 1.17$.

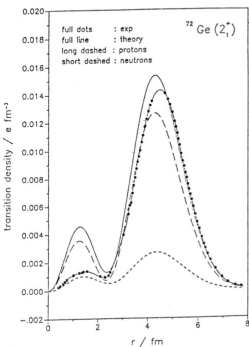

Fig.9. Same as Fig.8, but for the nucleus ^{72}Ge.

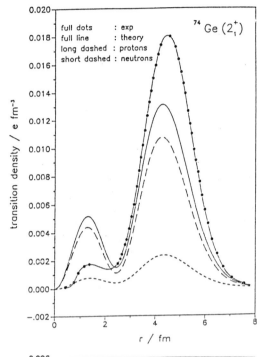

Fig. 10. Same as Fig. 8, but for the nucleus ^{74}Ge.

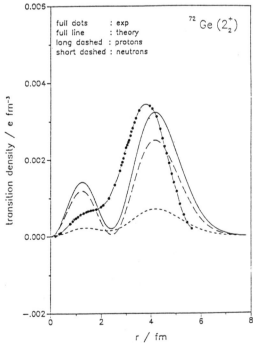

Fig. 11. Same as Fig. 9, but for the 2_2^+ state.

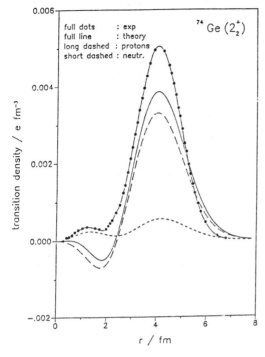

Fig.12. Same as Fig.10 but the for 2_2^+ state.

4. CONCLUSIONS

In the present lectures are discussed the two most sophisticated approaches belonging to the VAMPIR family, EXCITED VAMPIR and EXCITED FED VAMPIR approximations, and the new insights into the shape coexistence dominating the structure of the mass number $A\sim 70$ nuclei coming from this type of microscopic description.

The results presented here indicate the influence of the additional correlations on top of the symmetry projected quasiparticle mean fields of the EXCITED VAMPIR description. Such correlations introduced in a systematic way by the EXCITED FED VAMPIR procedure increase the reliability of the resulting wave functions. The qualitative features of the results obtained for the ^{68}Ge nucleus concerning the complexity of the resulting decay pattern and the competition of various configurations corresponding to different "shapes", and even the structure of the various bands are not influenced by the additional correlations.The theoretical predictions are confirmed by new experimental data for high spin states[6]. Concerning the low spin states this theoretical description has a special support from the experimentally available charge and transition charge densities[8].

More information about the general trends in this mass region was inferred studying a chain of nuclei where the number of nucleons is changed by two units : ^{68}Ge, ^{70}Se, ^{72}Se. One obtains a couple of coexisting bands being strongly mixed in some angular momentum regions and linked there by strong magnetic dipole transitions.

Obviously, neither the theoretical nor the experimental spectra are yet the final answer to the structure of these nuclei. More experimental data than available up to now,especially for the electromagnetic properties of the various states are required to make possible the still needed readjustment of the effective two-body interaction and careful investigation of the effects of the additional correlations. Thus a lot remains to be done, both theoretically as well as experimentally. Nevertheless, the results presented here provide some interesting information on the way towards a more microscopic interpretation of the complex situation encountered in the nuclei of the A∼70 mass region.

These lectures are essentially based on studies performed in collaboration with K.W.Schmid, A.Faessler (Tübingen-RFG) and F.Grümmer (Bochum-RFG). The author acknowledges the financial support from the German - Romanian collaboration.

REFERENCES

1. J.H.Hamilton, Structures of Nuclei far from Stability, in Treatise on Heavy-Ion Science, vol.8, (D.A.Bromley, ed.), Plenum Press, New York-London, 1989
2. K.W.Schmid and F.Grümmer, Rep.Prog.Phys. 50 (1987) 731
3. A.Petrovici, K.W.Schmid, F.Grümmer, A.Faessler and T. Horibata, Nucl.Phys. A434 (1988) 317
4. A.Petrovici, K.W.Schmid, F.Grümmer and A.Faessler, Nucl.Phys. A504 (1989) 277
5. a) K.W.Schmid, Zhen Ren-Rong, F.Grümmer and A.Faessler, Nucl.Phys. A499 (1989) 63
   b) A.Petrovici, K.W.Schmid, F.Grümmer and A,Faessler, Nucl.Phys. A 517 (1990) 108
6. L.Chaturvedi, X.Zhao, A.V.Ramayya, J.H.Hamilton, J. Kormicki, S.Zhu, G.Girit, H.Xie, W.-B.Gao, J.-R.Jiang, A.Petrovici, K.W.Schmid, A.Faessler, N.R.Johnson, C. Baktash, I.Y.Lee, F.K.McGowan, M.L.Halbert, M.A.Riley, J.H.McNeill, M.O.Kortelahtig, J.D.Cole, R.B.Piercey, H.Q.Jin, submitted for publication.
7. A.Petrovici, K.W.Schmid, F.Grümmer and A.Faessler,accepted for publication in Zeit. f.Physik A (1991)
8. D.Goutte, Ph.D.Thesis, Orsay, 1984
9. A.P.deLima, A.V.Ramayya, B.vanNooijen, R.M.Roningen, H.Kawakami, R.B.Piercey, E.deLima, R.L.Robinson, H.J. Kim, L.K.Peker, F.A.Rickey, R.Popli, A.J.Caffrey and J.C.Wells, Phys.Rev. C23 (1981) 213
10. T.Mylaeus, J.Busch, J.Eberth, M.Liebchen, R.Sefzig, S. Skoda, W,Teichert, M.Wiosna, P.von Brentano,K.Schiffer, K.O.Zell, A.V.Ramayya, K.H.Maier, H.Grawe, A.Kluge and W.Nazarewicz, J.Phys. G: Nucl.Phys. 15 (1989) L135.

//333

Axially Deformed Nuclei in the Relativistic Mean Field Theory

V. Blum, J. Fink, B. Waldhauser,
Institut für Theoretische Physik der Universität
6000 Frankfurt 11, West Germany
J. A. Maruhn[a,b], P. G. Reinhard[c], and W. Greiner[b]
Joint Institute for Heavy Ion Research
Holifield Heavy Ion Research Facility
Oak Ridge, TN 87831, USA

1 The model

Relativistic mean-field theories have attracted much attention because they allow to describe qualitatively nuclear saturation and spin orbit splitting within a simple model consisting of Dirac nucleons interacting via an attractive scalar (σ) and a repulsive vector (ω) meson field [1, 2, 3]. The model can be brought up to a quantitative description of nuclear bulk properties by introducing nonlinear selfcouplings of the σ-field [4], and by including an isovector-vector (ρ) meson and the photon. The parameters of the nonlinear model can be optimized such that bulk properties of spherical nuclei, such as binding energy, radius, surface thickness and spin-orbit splitting, are reproduced with high precision [5, 6].

There is still much to be done to apply and to test the model within a wider range of phenomena. The obvious step is to examine properties of (axially symmetric) deformed nuclei within that model. First studies have been concerned with the deformation minima of light nuclei [7, 8, 9]. More information about the deformation dynamics is accessed if one displays the whole potential energy surface (PES). We have extended the techniques of

---

[a]Invited Speaker
[b]Permanent address: Institut für Theoretische Physik der Universität, 6000 Frankfurt 11, West Germany.
[c]Permanent address: Institut für Theoretische Physik der Universität, 8520 Erlangen, West Germany.

ref. [7, 10] to include pairing and to allow for constrained relativistic mean-field calculations using a quadrupole constraint. A first application of these techniques was a systematic study of the PES of light nuclei [11]. As a test case in the region of the heavy nuclei we also examined the isotopes of Gd which have a transition from spherical to deformed nuclei around ^{152}Gd. We expect that such transitional nuclei have particularly sensitive deformation dynamics.

Another motivation for these studies is the use of the resulting nuclear matter equation of state for high-energy heavy ion collision simulations. Here the advantage is that the theory is formulated covariantly, so that an extension to high energies is no problem, whereas conventional heuristic expressions for the equation of state have to be restricted in their density dependence to ensure causality. Also the model has a more complicated temperature dependence than the ususally assumed separation of compressional and thermal energy, and it is relatively straightforward to include Δ-particles. Of course, it is not at all guaranteed that a model fitted to nuclear structure data will work at all for high-energy collisions, but it is certainly worth while to investigate this application.

We start from the action given by the standard ansatz [5, 6]:

$$\begin{aligned}S = \int d^4x \, \{&\overline{\psi}(\mathrm{i}\gamma^\mu\partial_\mu - m_N - g_\sigma\Phi - g_\omega V^\mu\gamma_\mu \\ &-g_\rho\vec{R}^\mu\cdot\vec{\tau}\gamma_\mu - eA^\mu\frac{1+\tau_0}{2}\gamma_\mu)\psi \\ &+\tfrac{1}{2}(\partial^\mu\Phi\partial_\mu\Phi - m_\sigma^2\Phi^2) - \frac{b_2}{3}\Phi^3 - \frac{b_3}{4}\Phi^4 - \tfrac{1}{2}(\tfrac{1}{2}G_{\mu\nu}G^{\mu\nu} - m_\omega^2 V^\mu V_\mu) \\ &-\tfrac{1}{2}(\tfrac{1}{2}\vec{B}_{\mu\nu}\cdot\vec{B}^{\mu\nu} - m_\rho^2\vec{R}^\mu\cdot\vec{R}_\mu) - \frac{1}{4}F_{\mu\nu}F^{\mu\nu}\},\end{aligned}$$

where $G_{\mu\nu} = \partial_\mu V_\nu - \partial_\nu V_\mu$, and similary for $\vec{B}_{\mu\nu}$ and $F_{\mu\nu}$ derived from $\vec{R}_\mu$ and A_μ, respectively. It includes the nucleons, a scalar field Φ, a vector field V_μ, an isovector-vector field $\vec{R}_\mu$ and the photon A_μ. This is understood to be an effective action in connection with the mean field and no sea approximation. The mean field approximation consists in treating the meson field operators as simple classical c-number fields; as a consequence, one can expand the field operators of the nucleons in terms of single particle wave functions. The no sea approximation means that the contribution of the anti-nucleon states of the so called Dirac-sea is neglected and only the occupied nucleon bound states are taken into account. Furthermore, we consider the stationary case. Thus all fields are time-independent, the space components of the vector fields and four-currents vanish, and the time-dependence of the single particle states separates as $\exp(-i\varepsilon_\alpha t)$.

For the numerical solution of the problem we first discretize the action on an axially symmetric space lattice and then derive the coupled field equations by variation of this action, similar as in comparable nonrelativistic calculations [12, 13]. The four components of the Dirac spinors for the nucleons are stored on shifted grids to avoid the problem of the so-called Fermion doubling [14]. The observables of interest are the energy, which is computed from the action by standard techniques, and the cartesian quadrupole moment

$$Q = \int d^3r\, \rho_{pr}(x,y,z)(2z^2 - x^2 - y^2),$$

calculated from the nucleon, which is the zero-component of the four-current. The Dirac equation is transformed into an effective Schrödinger equation and then solved with the damped gradient iteration method which is taken over from classical calculations [15, 16]. The meson field equations are solved with the sucessive overrelaxed iteration [17].

In order to calculate the whole PES we introduce a quadrupole constraint in the Dirac equation. A prescribed expectation value of the quadrupole moment is achieved by iterating the corresponding Lagrange multiplier of the quadrupole constraint together with the iterative solution of the Dirac equation; again the technique is taken over from nonrelativistic calculations [18]. A pure quadrupole operator causes problems because it is unbound ($\propto x^2$, y^2, z^2). This unphysical effect is avoided by damping the quadrupole operator outside the nucleus.

All calculations have been performed with a pairing scheme within the constant gap approximation using for the gap $\Delta = 11.2\,\mathrm{MeV}/A$ [20]. Approximately one shell above the Fermi level is taken into account, where the selection of the levels has been done at the initialisation stage (which corresponds to a selection in a deformed harmonic oscillator).

For the model parameters we use the precision fits to properties of spherical nuclei. An unrestricted fit within the full nonlinear model is taken from ref. [5], called set NL1 in the following. Like all nonlinear sets it has a rather low effective mass of $m^*/m_n = 0.58$. This leads to a low level density, and this will influence the deformation properties. (In fact, deformation properties are a very clean analyzing instrument for nuclear level schemes, even better than the spectra of odd nuclei which mask the single particle energies by polarisation effects.) In order to examine the influence of the effective mass m^*/m_n on the PES systematically we also consider nonlinear sets from a fit with a constraint on m^*/m_n [6]. The set NL06 corresponds to an effective mass of $m^*/m_n = 0.6$, set NL065 to $m^*/m_n = 0.65$ and accordingly for the other sets. The parameters for the sets NL1 and NL06 through NL075 are given in table 1.

set	g_σ	g_ω	g_ρ	b_2	b_3 (fm^{-1})	m_σ (MeV)	m_ω (MeV)	m_ρ (MeV)
NL1	10.138	13.285	4.976	-12.172	-36.265	492.250	795.36	763.0
NL06	9.996	12.729	4.730	-13.197	-37.978	492.363	780.0	763.0
NL065	9.322	11.732	4.752	-11.300	-25.085	491.871	780.0	763.0
NL07	8.510	10.702	4.810	-10.449	-13.175	478.975	780.0	763.0
NL075	7.511	9.515	4.926	-9.576	1.971	454.871	780.0	763.0

Table 1: The parameter sets used: coupling constants and masses of the mesons. The resulting effective masses, m^*/m, are 0.58 for the set NL1, 0.6 for the set NL06, 0.65 for NL065 and accordingly for the others.

2 Light Nuclei

We present calculations for all light even-even nuclei with equal proton and neutron number from ^{12}C to ^{40}Ca. The full potential energy surfaces are shown in fig. 1. There some parts of the PES's for the parameter sets NL1 and NL06 are missing because the calculations did not converge. The explanation for this is the following: due to shell fluctuations some parts of those nuclei have scalar densities greater then 0.22 fm^{-3}, which is considerably larger than nuclear matter scalar density. Above this critical density there is no stable solution of the scalar meson field in connection with the sets NL1 and NL06 [21].

From fig. 1 we see that since ^{16}O is a closed shell nucleus it comes out to be spherical, the same holds for ^{40}Ca. In all other cases there appears an oblate minimum, a barrier at zero deformation, and a prolate minimum or at least a saddle point. The position of the minima is almost independent of the parameter set. Compared to experimental data generally all resulting ground state deformations have the correct sign and they are always a bit too small. This is a rather desirable feature because the measured β_2 also contain the vibration correction whereas our calculated values represent purely static deformations.

The quadrupole moments are significantly smaller than those published in ref. [7]. This is purely due to the inclusion of pairing which tends to restore spherical symmetry.

Even though the minima occur at almost the same deformation for all forces the relative energies in the PES's depend sensitively on the parameter set. Increasing m^* leads to more pronounced minima and barriers, decreasing the effective mass softens the PES. There is also some effect on the total binding.

Now we discuss some nuclei in more detail. Carbon is the most problematic nucleus. The

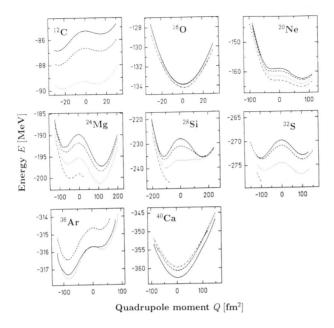

Figure 1: PES's of the light even-even nuclei from ^{12}C to ^{40}Ca. The different lines represent different parameter sets with different m^*: NL075 (full) with $m^*/m = 0.75$, NL07 (dashed) with $m^*/m = 0.70$, NL065 (dotted) with $m^*/m = 0.65$, and NL06 (dash-dotted) with $m^*/m = 0.60$.

calculated quadrupole moment is about 40% too small and the radius is 10% too large. The same difficulties occurred already in non-relativistic self-consistent calculations. It indicates that strong ground state correlations occur in the structure of ^{12}C. For example, it could be that ^{12}C consists of three α-particles sitting in a triangle. This configuration would have a large negative quadrupole moment but would not be found in an axially symmetric description.

In sulfur the prolate and the oblate minima are of nearly equal depth and at the same absolute value of the quadrupole moment. There may be different connections between these minima via the γ-degree of freedom. Either ^{32}S is γ-soft and the minima are connected by a valley in γ-direction or they may be separated by a potential well. Recently there have been reported triaxial calculations in the relativistic mean field model, but not yet for ^{32}S [28]. In any case the ground state will be a mixture of both configurations. The fact that the sign of the quadrupole moment of ^{32}S is not yet determined [27] supports this interpretation. A

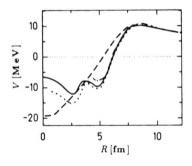

Figure 2: Fusion/Fission potential of ^{32}S as a function of the two-center distance $R \approx \sqrt{2Q/A}$. The potential is normalized such that $V(R \to \infty) = 0$. The full line corresponds to our relativistic mean field calculation with the parameter set NL075, the dotted curve to NL065. The results are compared to liquid drop calculation (dashed) and a non-relativistic Skyrme calculation (dash-dotted), see the case "new force" in fig. 4 of ref. [19].

similar situation seems to occur in ^{28}Si where the prolate and the oblate minima are almost degenerate.

For ^{32}S we have also explored a larger range of deformations. In fig. 2 the PES for fission/fusion is shown as a function of the distance R of the two centers. The R is calculated from the quadrupole moment via $R \approx \sqrt{2Q/A}$, see ref. [19]. The potential is corrected for the vibrational zero-point fluctuations where the correction has been taken over from ref. [19]. The potentials are normalized such that they approach zero for separated fragments. We show the relativistic results for the sets NL065 and NL075. Both display the outer Coulomb barrier and two prolate minima. The inner minimum is the ground state deformation and the outer minimum corresponds to an isometric nuclear molecular configuration (see fig. 2). The position of minima and barriers is independent of the force. But the depth of the minima is sensitive to the force, in particular for the ground state minimum.

The structure of the states is shown by equidensity plots in fig. 3 for two selected distances, $R = 4.6$ fm for the isomeric molecular configuration and $R = 10.0$ fm for an asymptotic configuration. The asymptotic case displays clearly two spherical ^{16}O nuclei. Other decay channels are energetically less favourable and they were excluded by reflection symmetry which has been used in our calculation. The configuration at $R = 4.6$ fm is already well merged and only a small remainder of a neck is seen. It is mainly the history of the entrance channel which suggests the name "quasi-molecular" configuration.

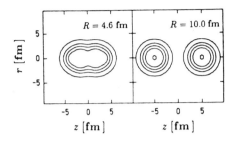

Figure 3: Equidensity plots for two selected configurations of ^{32}S, the nuclear molecular configuration at $R = 4.6$ fm and the two separated fragments at $R = 10.0$ fm. The contour lines are drawn at $\rho = 0.01$, 0.025, 0.05, 0.1, 0.15 fm^{-3}.

In fig. 2 we also show the result from a non-relativistic Skyrme force calculation and from the liquid drop model (LDM). The Skyrme curve is very close to the relativistic results although it has slightly less bound minima. The LDM curve agrees with all results in the Coulomb tail ($R > 8$ fm) but it misses the deformed minima which are obviously due to shell effects.

We have also explored large oblate deformations. In that direction of negative quadrupole moments the potential increases steadily. The corresponding shapes become more and more oblate until they reach a torodial configuration. With deformation increasing they get larger and thinner. For the investigated light systems no minimum in the potential for a toroidal structure could be found.

To summarize: The deformations of light nuclei are in fair agreement with the experimental values, independent of the (non-linear) parametrisation used. However, the relative energies within the PES's depend sensitively on the parametrisation. This means that the collective excitation properties of light nuclei will be a sensitive tool to discriminate various possible parametrisations. At larger prolate deformations we find isomeric minima and finally the fission barrier. At large oblate deformations toroidal configurations are found. For further investigations on collective dynamics and fission lifetimes one urgently needs to extend the calculations to compute also the corresponding cranking masses.

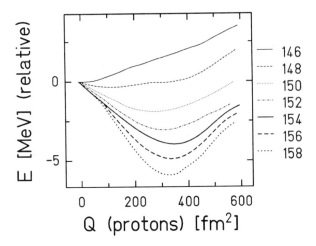

Figure 4: PES for the even Gd-isotopes 146 through 158 computed with the set NL075. The energy for each PES is rescaled such that all go through the same point at zero deformation.

3 The Gadolinium Isotopes

Fig. 4 shows the PES for different Gd-isotopes (^{146}Gd up to ^{158}Gd) calculated with the parameter set NL075. We plotted the cartesian quadrupole moment of the proton distribution ρ_{pr}, because the experimental data are evaluated from the charge distributions [22]. As expected ^{146}Gd comes out to be spherical, because it has a magic neutron number which drives the subshell closure for the protons [23]. Adding two neutrons deforms the nucleus only slightly, whereas the next two neutrons cause a jump in the deformation, which is due to a breakdown of the proton subshell closure. Adding further neutrons enlarges the deformation only little but enhances the barrier.

For Fig. 5 we performed the same calculations as for Fig. 4 but replaced the parameter set NL075 by the set NL06 with lower effective mass. In comparison with the calculation with the set NL075 the deformation minima are shifted to larger values whereas the energies are decreased. In particular the isotopes ^{150}Gd and ^{152}Gd develope second neighbouring minima with only slightly higher energy.

All calculations were made with a constant grid point distance of $\Delta r = \Delta z = 0.5\,\text{fm}$. To test the dependence of the results on this quantity (Δr, Δz) we performed for the isotope ^{152}Gd calculations on a finer grid ($\Delta r = \Delta z = 0.3\,\text{fm}$) with the set NL1. The deformations

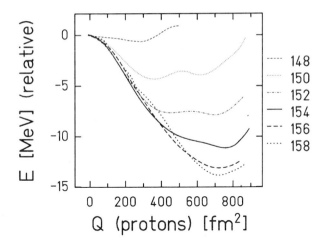

Figure 5: The results of a calculation corresponding to that of Fig. 4, where instead of the set NL075 the set NL06 is used.

come out to be unchanged, whereas the total energy increased by 40 MeV. But the relative changes within the PES are maximally 1 MeV.

Fig. 6 shows the deformation minima of different Gd-isotopes calculated with the forces NL1 and NL06 through NL075 as functions of the nucleon number in comparison with the experimental findings [22]. The curves for the sets NL075 and NL06 correspond to the deformation minima of Fig. 4 and Fig. 5. We see from Fig. 6 that indeed the deformation properties are sensitive to the m^*/m of the parametrisations. The stabilised deformation for the heavier isotopes differs dramatically. The best fit set NL1 and the sets NL06 and NL065 give slightly too large deformations whereas the somehow artifical sets NL07 and NL075 underestimate the deformations in this region by a factor of two. The results for NL1, NL06 and NL065 show some fluctuations. These may be smoothed if some dynamical deformations are added accounting for the vibrational zero-point motion and if a better treatment of pairing is used. All sets describe the transition from spherical to deformed shapes qualitatively right, but none of them finds the transition point at the proper place, which lies between ^{152}Gd and ^{154}Gd according to the experimental data. However, taking into account the dynamical deformation and a better pairing treatment the set NL065 may produce the right transition point.

We conclude that deformation properties of heavy nuclei add valuable selection criteria

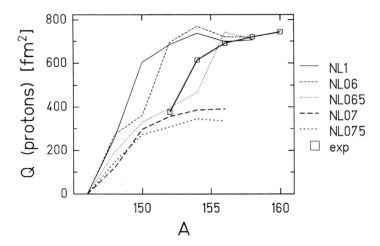

Figure 6: The quadrupole moment of the proton distribution at the prolate deformation minima of different Gd-isotopes versus their nucleon number. Shown are results from the set NL1 and the sets NL06 through NL075 in comparison with the experimental data.

for the relativistic mean-field model. The results for the Gd-isotopes presented here show that the deformation is very sensitive to the m^*/m of the parametrisation used. None of the parametrisations studied, NL1 and NL06 through NL075, is capable to place the transition to deformed nuclei properly. The heavier isotopes, however, are described quite well by the bestfit set NL1 and the sets NL06 and NL065, where the sets NL07 and NL075 fail to reproduce the data. Further investigations are needed to resolve the problem of the transitional Gd-isotopes.

4 Equation of State

The starting point again is a Lagrangian describing the interaction of nucleons and deltas via several meson fields. Since we are interested only in infinite symmetric nuclear matter, the ρ-meson and the photon need not be taken into account, but, on the other hand, we also allow for the presence of Δ-particles. The Lagrangian then takes the form

$$\mathcal{L} = \sum_{i=N,\Delta} \left[\bar{\psi}_i (i\gamma_\mu \partial^\mu - m_i)\psi_i - g_\sigma(i)\bar{\psi}_i \psi_i \Phi - g_\omega(i)\bar{\psi}_i \gamma_\mu \psi_i V^\mu \right]$$
$$- \frac{1}{2}\partial_\mu \Phi \partial^\mu \Phi - U(\Phi) - \frac{1}{4}G_{\mu\nu}G^{\mu\nu} - \frac{1}{2}m_v^2 V_\mu V^\mu \quad ,$$

with the nonlinear Φ-potential

$$U(\Phi) = \tfrac{1}{2}m_\sigma^2 \Phi^2 + \frac{b_2}{3}\Phi^3 + \frac{b_3}{4}\Phi^4.$$

In the following we will consider the thermodynamic quantities for infinite matter within the mean field approximation. Defining the effective masses of nucleons and deltas as

$$m^*(i) = m_i + g_\sigma(i)\Phi \quad , \quad i = N, \Delta \quad ,$$

the scalar densities read

$$\rho_s(i) = \frac{\gamma_i}{(2\pi)^3}\int_0^\infty d^3k \frac{m^*(i)}{\sqrt{k^2 + m^*(i)^2}}\left[n_i(T) + \bar{n}_i(T)\right] \quad ,$$

and the vector densitites

$$\rho_v(i) = \frac{\gamma_i}{(2\pi)^3}\int_0^\infty d^3k \left[n_i(T) - \bar{n}_i(T)\right] \quad ,$$

where the $n_i(T)$ are the Fermi-Dirac distributions of the particles and the $\bar{n}_i(T)$ the ones of the antiparticles. The energy density and pressure are then determined by

$$\begin{aligned}\epsilon &= \frac{g_\omega^2}{2m_v^2}\rho_B^2 + U(\Phi) \\ &+ \sum_{i=N,\Delta}\frac{\gamma_i}{(2\pi)^3}\int_0^\infty d^3k\sqrt{k^2 + m^*(i)^2}\left[n_i(T) + \bar{n}_i(T)\right] \\ p &= \frac{g_\omega^2}{2m_v^2}\rho_B^2 + U(\Phi) \\ &+ \frac{1}{3}\sum_{i=N,\Delta}\frac{\gamma_i}{(2\pi)^3}\int_0^\infty d^3k\frac{k^2}{\sqrt{k^2 + m^*(i)^2}}\left[n_i(T) + \bar{n}_i(T)\right] \quad ,\end{aligned}$$

where ρ_B is the total baryon number density.

The parameters of the model are the dimensionless coupling constants

$$C_s = \frac{g_\sigma(N)m_N}{m_s}, \quad C_v = \frac{g_\omega(N)m_N}{m_v}, \quad B = \frac{b_2}{g_\sigma^3 m_N}, \quad C = \frac{b_3}{g_\sigma^4}$$

and the relative strengths of the delta coupling constants defined by

$$\alpha = g_\omega(\Delta)/g_\omega(N) \quad , \quad \beta = g_\sigma(\Delta)/g_\sigma(N) \quad .$$

While the first four constants can be taken over from previous considerations, there is no information about the constants α and β determining the coupling of the deltas, and we will examine what happens if these values are varied. Quark counting arguments [24] suggest $\alpha = \beta = 1$, but in this case the effective masses can become negative, which can be overcome if at least β is made larger.

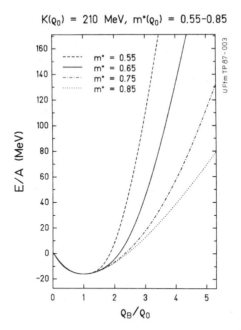

Figure 7: The equation of state computed in the mean-field theory for a fixed value of the incompressibility $K=210$ MeV and various values of the effective mass m^* at equilibrium density.

Fig. 7 shows equations of state without deltas for a fixed value of $K=210$ MeV, but varying m^*. Obviously the slope of the curves at higher densities is quite different, although they have, by construction, the same curvature in the ground state. Since a stiffer curve seems to be favored, we fix m^* at $0.55 m_N$ and vary K from 210 MeV to 400 MeV in a following examination. In this case the curves do almost not differ from each other. In this model, therefore, a measurement of $E_c(\rho)$ at higher densities measures the ground state value of m^* rather than the incompressibility, and the two properties of the equation of state are quite independent.

It should be mentioned that the model exhibits a very complicated phase structure, which is more extensively discussed in [25] and [26]. For example, taking K values below 200 MeV can produce unphysical density maxima.

The inclusion of deltas leads to the production of a delta matter phase. The location and depth of this additional minimum depends very strongly on the coupling parameters, and a reasonable variation of the coupling strength will allow anything from a minimum deeper than the ground state to a barely noticeable distortion of the slope at high densities.

Finally we come to the pion yields. Calculating them in the usual one-dimensional shock wave model [29] shows that small changes in the delta coupling strength β, implying no change at all in the equation of state, cause major changes in the pion yield. This is understandable, since the pions are directly influenced by the delta proportion in the compressed zone, but nevertheless it implies that the pion yields in this model are essentially useless for determining the equation of state; instead they seem to give information about the delta's coupling to the meson fields.

5 Acknowledgements

The authors would like to thank Dr. M.R. Strayer, Dr. A.S. Umar and Dr. Suk-Joon Lee for placing parts of the program underlying this work to our disposal and for many fruitful discussions.

This work was supported by the Bundesministerium für Forschung und Technologie through Contract Number 06OF772.

The Joint Institute for Heavy Ion Research has as member institutions the University of Tennessee, Vanderbilt University, and the Oak Ridge National Laboratory; it is supported by the members and by the Department of Energy trough Contract Number DE-AS05-76ERO-4936 with the University of Tennessee.

References

[1] H.P. Duerr, Phys. Rev. **103**, 469 (1956).

[2] L.D. Miller, Phys. Rev. Lett. **28**, 1281 (1972).

[3] J.D. Walecka, Ann. Phys. **83**, 491 (1974).

[4] J.Boguta and A.R. Bodmer, Nucl. Phys. **A229**, 414 (1977).

[5] P.-G. Reinhard, M. Rufa, J. Maruhn, W. Greiner and J. Friedrich, Z. Phys. **A323**, 13 (1986).

[6] M. Rufa, J. Maruhn, W. Greiner, P.-G. Reinhard, M.R. Strayer, Phys. Rev. **C5**, 390 (1988)

[7] S.J. Lee, J. Fink, A.B. Balantekin, M.R. Strayer, A.S. Umar, P.-G. Reinhard, J.A. Maruhn and W. Greiner, Phys. Rev. Lett. **57**, 2916 (1986), Phys. Rev. Lett. **59**, 1171 (1987).

[8] C.E. Price and G.E. Walker, Phys. Rev. **C37**, 354 (1987).

[9] W. Pannert, P. Ring and J. Boguta, Phys. Rev. Lett. **59**, 2420 (1987).

[10] J. Fink, J.J. Lee, A.S. Umar, M.R. Strayer, J. Maruhn, W. Greiner, P.-G. Reinhard, ORNL Preprint (1988)

[11] V. Blum, J. Fink, P. G. Reinhard, J. A. Maruhn, and W. Greiner, Phys. Lett. **223B**, 123 (1989).

[12] K.T.R. Davies and S.E. Koonin, Phys. Rev. **C23**, 2042 (1981).

[13] S.E. Koonin et. al., Phys. Rev. **C15**, 1359 (1977).

[14] C.M. Bender, K.A. Milton and D.H. Sharp, Phys. Rev. Lett. **51**, 1815 (1983).

[15] A.S. Umar, M.R. Strayer, R.Y. Cusson, P.-G. Reinhard and S.A. Bromley, Phys. Rev. **C32**, 172 (1985).

[16] P.-G. Reinhard and R.Y. Cusson, Nucl. Phys. **A378**, 418 (1982).

[17] Richard S. Varga, "Matrix Iterative Analysis", Prentice-Hall, Englewood Cliffs, New Jersey 1965

[18] R.Y. Cusson, P.-G. Reinhard, M.R. Strayer, J.Maruhn and W. Greiner, Z.Phys. **A320**, 475 (1985).

[19] P.-G. Reinhard, J. Friedrich, K.Goeke, F. Grümmer and D.H.E. Gross, Phys. Rev. **C30**, 878 (1984).

[20] J. Blocki and M. Flocard, Nucl. Phys. **A273**, 45 (1976).

[21] P.-G. Reinhard, Z. Phys. **A329**, 257 (1988)

[22] K.E.G. Löbner, M. Vetter and V. Hönig, Nucl. Data Tabl. **A7**, 495 (1970).

[23] R.F. Casten, D.D. Warner, D.S. Brenner and R.L. Gill, Phys. Rev. Lett. **47**, 1433 (1981).

[24] S. I. A. Garpman, N. K. Glendenning, and Y. J. Karant, Nucl. Phys. **A322**, 382 (1979).

[25] J. Theis, G. Graebner, G. Buchwald, J. A. Maruhn, W. Greiner, H. Stöcker, and J. Polonyi, Phys. Rev. **D28**, 2286 (1983).

[26] B. Waldhauser, J. Theis, J. A. Maruhn, H. Stöcker, and W. Greiner, Phys. Rev. **C36**, 1019 (1987),
B. Waldhauser, J. A. Maruhn, H. Stöcker, and W. Greiner, Phys. Rev. **C38**, 1003 (1988).

[27] G. Haouat, Ch. Lagrange, R. de Swiniarski, F. Dietrich, J.P. Delaroche, and Y. Patin, Phys. Rev. **C30**, 1795 (1984).

[28] W. Koepf, P. Ring, Phys. Lett. **212**, 397 (1988).

[29] J.A. Maruhn and H. Stöcker, Z. Physik **A327**, 75 (1987).

GRAND UNIFICATION, NUCLEAR STRUCTURE AND THE DOUBLE BETA-DECAY *

Amand Faessler
University of Tübingen
Institute of Theoretical Physics
D-7400 Tübingen
West-Germany

Abstract

Grand unified models of the electroweak and the strong interaction predict that the neutrino is a Majorana particle and therefore esssentially identical with its own antiparticle. In such models the neutrino has also a finite mass and a slight right-handed weak interaction, since the model is left-right symmetric. These models have also left handed and right-handed vector bosons to mediate the weak interactions. If these models are correct the neutrinoless double beta-decay is feasable. Thus if one finds the neutrinoless double beta-decay one knows that the standard model can not be correct in which the neutrino is a Dirac particle and therefore different from its antiparticle. Although the neutrinoless double beta-decay has not been seen it is possible to extract from the lower limits of the lifetime against the double neutrinoless beta-decay upper limits for the electron neutrino mass and for the mixing angle of the right-handed and the left-handed vector bosons mediating the weak interaction. One also can obtain an upper limit for the mass ratio of the light and the heavy vector bosons. The extraction of this physical quantities from the data is made difficult due to the fact that the weak interaction must not be diagonal in the representation of the mass matrix of the six neutrinos requested by such left-right symmetric models. The generalized " Kobayashi Maskawa matrix " must be known to extract these quantities. Thus the extraction is model dependent. If one assumes that this 6 x 6 Kobayashi Maskawa matrix is almost diagonal one can derive that the electron neutrino mass can not be larger than 2 eV.

---

* Lectures given at the Predeal Summerschool, September 1990 in Romania. Supported by the BMFT under contract no. 06 TÜ 716 and by the International Büro in Karlsruhe

1. Introduction

The neutrino has always been an interesting particle since it was predicted by Wolfgang Pauli on December 4th, 1930 in a letter to a conference in Tübingen

The neutrino is the only fermion from which we do not know if it is different from its antiparticle ($\nu \neq \bar{\nu}$) and therefore a Dirac particle or if it is identical with its antiparticle ($\nu = \bar{\nu}$) and therefore a Majorana particle.

In 1955 the physicists did believe that this question was solved in favour of a Dirac neutrino by an experiment of Davis[1]. He was using antineutrinos from the beta decay in a reactor to induce the inverse beta decay in ^{37}Cl for which one needs in the standard model a neutrino:

$$n \rightarrow p + e^- + \bar{\nu} \quad (1)$$

$$\nu +^{37} Cl \rightarrow\ ^{37}Ar + e^-$$

Since he did not observe the formation of ^{37}Ar by the inverse beta decay he assumed that the antineutrino $\bar{\nu}$ is different from the neutrino ν and thus the neutrino is a Dirac particle. But this conclusion was outdated already two years later. In 1956 Lee and Yang[2] proposed to test if in the weak interaction parity is conserved, and in 1957 Wu et al.[3] did find that indeed paritys is violated in the beta decay and soon it turned out that it is violated maximally and the weak interaction is purely left-handed. Therefore, the neutrino created in the beta decay of the neutron in eq. (1) has a positive helicity (is right-handed), while the neutrino needed for the inverse beta decay must have a negative helicity and therefore needs to be left-handed. Thus, even if the neutrino is a Majorana particle, reaction (1) would be forbidden due to helicity mismatch of the two reactions. Thus, we know already since about 30 years that the problem, if the neutrino is a Dirac or if it is a Majorana particle, is not solved. But why are we just today discussing this question so intensively? The reason is that the grand unified theories from which we think that they are most successful predict that the neutrino is a Majorana particle[4]. Measurements of the lower limit of the proton lifetime seem to exclude SU5 and thus one concentrated on SO10, which is also a subgroup of dynamical groups discussed in superstring theories. SO10 which has first been proposed by Fritzsch and Minkowski[5] predicts in improved versions[4] not only, that the neutrino is a Majorana particle, but automatically predicts also that the neutrino has a mass and a weak right-handed interaction. As we will see below these facts allow the double beta decay without neutrinos. Or inversely: The existence of the double neutrinoless beta decay would establish that the neutrino is a Majorana particle. Figure 1 shows the diagram for the double neutrinoless beta decay. This is naturally possible if the neutrino is identical with its antiparticle since at the first vertex, one would emit an antineutrino and at the second vertex one needs to absorb a neutrino in the standard model. Since we have only two particles in the continuum in the process in fig. 1, the phase space is bigger by a factor 10^6 compared to the double beta-decay with two neutrinos. Thus, even if the matrix element squared is reduced by a factor 10^{-6}, one has the same transition probability for the neutrinoless double beta decay as for the double beta decay with two neutrinos.

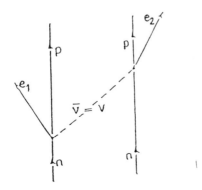

Figure 1
Diagrams for the double neutrinoless beta decay with a Majorana neutrino. By having only two particles in the final states in the continuum, the phase space is increased by a factor of about 10^6 compared to the $2\nu\beta\beta$ decay. Even with a Majorana neutrino this process is only possible if the neutrino has a finite mass and/or also a right-handed weak interaction.

But even if the neutrino is a Majorana particle, the process in fig. 1 can not happen since for a pure left-handed weak interaction theory, the emitted neutrino must be right-handed (positive helicity), while the absorbed neutrino must be left-handed (negative helicity). But grand unified theories predict also that the neutrino has a mass and a slight right-handed weak interaction. With a finite mass the neutrino has not any more a good helicity and the interference term between the leading helicity and the small admixtures allows a double neutrinoless beta decay. The same happens if we have also right-handed vector bosons which allow a right-handed weak interaction. If the first vertex is a left handed V-A interaction and the second vertex is a right-handed V+A interaction, the process in fig. 1 is also allowed. Thus, the double neutrinoless beta decay can either happen due to the finite mass of the neutrino or due to a small admixture of right-handed leptonic currents. (A close analysis of the problem shows that even for a right-handed interaction the finite mass of the neutrino is needed to make the double neutrinoless beta decay possible.) Thus the matrix element for the double neutrinoless beta-decay consists of two parts: One proportional to the neutrino mass and the other to the small right-handedness η of the weak interaction. η is equal to the mass ratio of the left and right vector bosons squared and to $sin^2 \vartheta$ of a mixing angle of the two vector bosons.

2. Description of the Two-Neutrino Double Beta Decay

Since there are measurements available for the two neutrino double beta decay with the geochemical method [6,7], and also one in the laboratory [8,9], one could try to calculate for a test of the theory the double beta decay with two neutrinos and compare them with the data. If one performs this calculation, one finds that the calculated transition probability for the double beta decay with two neutrinos ($2\nu\beta\beta$) is by a factor 10 to 100 too large[5,17]. This discrepancy which is typical for the $2\nu\beta\beta$ decay naturally casts doubt on the reliability of the calculations of the $0\nu\beta\beta$ transition probabilities.

The nuclear many-body methods for calculating the double beta decay are either shell model calculations[10] or extended shell model calculations based on the Hartree-Fock-Bogoliubov approach called MONSTER[11] or they are based on the Random Phase Approach (RPA). Let's try to analyse the $2\nu\beta\beta$ decay in the RPA approximation.

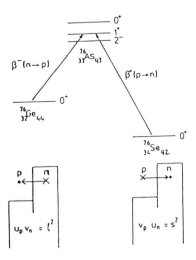

Figure 2
The upper part shows the way how in the Random Phase Approximation (RPA) the $2\nu\beta\beta$ decay is calculated. For the Fermi transitions the $\beta^-(n \to p)$ amplitude moves just a neutron into the same proton level and the $\beta^+(p \to n)$ amplitude moves a proton into the same neutron level. For the Gamow-Teller transitions it can also involve a spin flip, but the orbital part remains the same. One immediately realizes that the occupation and non-occupation amplitudes favour the β^- amplitude, but disfavour the β^+ amplitude. There one has a transition from an unoccupied to an occupied single particle state, which is two-fold small (s^2) first by the fact that the occupation amplitude for the proton v_p and secondly that the unoccupation amplitude for the neutron state u_n are both small. Therefore the $2\nu\beta\beta$ is drastically reduced.

Fig. 2 explains why the $2\nu\beta\beta$ decay amplitude is so drastically reduced. Therefore the small effects which normally do not play a major role can affect the $2\nu\beta\beta$ transition probability. If one looks to the $\beta^+(p \to n)$ amplitude one finds that the matrix elements involved in these diagrams are Pauli suppressed by a factor $(u_n v_p)^2 = (small)^4$. The second diagram shown in Fig. 3 shows the quasiparticle-quasiparticle excitations leading due to particle number non-conservation also to the intermediate nucleus ^{76}As. They are proportional to $(u_n u_p)^2 = (small)^2 (large)^2$. The neutron-particle proton-hole force in the isovector channel is repulsive while the particle-particle force is attractive. Therefore both excitations tend to cancel each other and therefore the amplitude β^+ is drastically reduced. In this way one obtains agreement with the experimental data.

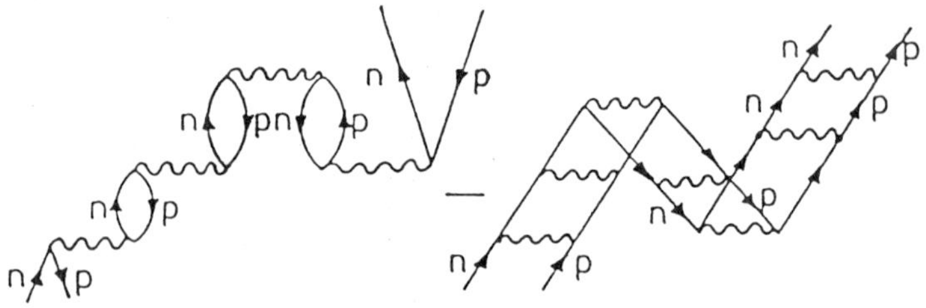

Figure 3
The left-hand side shows the isovector neutron-particle proton-hole excitations leading from ^{76}Se to ^{76}As. The particle-hole matrix elements are proportional to the unoccupation amplitude for the neutrons and the occupation amplitude for the proton squared $(u_n v_p)^2 = (small)^4$. The ground state correlations and the collectivity due to these excitations are therefore extremely weak. The right-hand side shows particle-particle excitations where the matrix elements have the Pauli factors $(u_n u_p)^2 = (small)^2 (large)^2$. Normally this particle-particle ground state correlation and excitation are neglected, but due to the weakness of the neutron-particle proton-hole excitations they play a major role in this case. Since the isovector particle-hole force is repulsive and the particle-particle (and the hole-hole) force is attractive the inclusion of the particle-particle (and hole-hole) correlations tend to quench the $2\nu\beta\beta$ transition probability.

We calculated the $2\nu\beta\beta$ decay half-lives for the following $0_i^+ \to 0_f^+$ transitions: $^{76}Ge \to {}^{76}Se$, $^{82}Se \to {}^{82}Kr$, $^{128}Te \to {}^{128}Xe$ and $^{130}Te \to {}^{130}Xe$.

We take the full $3\hbar\omega$ and $4\hbar\omega$ major oscillator shells for the description of the nuclei with A=76, 82 and a model space consisting of $1p_{3/2,1/2}$, the full $4\hbar\omega$ shell, $0h_{11/2,9/2}$, and $1f_{7/2,5/2}$ sub-shells for A= 128, 130. The single particle energies are calculated with the Coulomb-corrected Woods-Saxon potential. As a realistic two-body interaction we use the nuclear matter G-matrix calculated from the Bonn one boson-exchange potential. In order to take into account the renormalization for the finite basis, we multiply the pairing matrix elements of this force by factors g_{pair}^p and g_{pair}^n. These factors lie close to unity and deviate at most less than 20 % from the value obtained from the reaction matrix of the Bonn potential. These factors are adjusted to the odd-even mass differences. We also multiply the particle-hole matrix elements by a factor g_{ph} which we fix to the energy of the Gamow-Teller giant resonance (GTGR) in the intermediate odd-nucleus. This factor is also normally close to unity. The biggest deviation found is 30 %. To show the influence of the particle-particle correlations, we multiply also the particle-particle matrix elements by a factor g_{pp}. $g_{pp} = 0$ switches off the particle-particle correlations while $g_{pp} = 1$ gives the values predicted by the theory including these correlations.

3. The Neutrinoless Double β Decay

For the neutrinoless double β decay we have only lower limits for the lifetimes[12,13,14]. The theoretical transition probability for the zero neutrino double β decay can be written as

$$P_{0\nu\beta\beta} = const \cdot |m_\nu M_m + \eta M_\eta + \lambda M_\lambda|^2 \qquad (2)$$

with positive nuclear matrix elements M_m, M_η and M_λ. The experiment yields an upper limit for $P_{0\nu\beta\beta}$. Thus in the parameter space of the electron-neutrino mass m_ν and the right-left (lepton and baryon vertices) handedness $\eta \propto tg(\vartheta)$ and right-right (lepton and baryon vertices) handedness $\lambda \propto (M_{WL}/M_{WR})^2$ of the weak interaction relative to the left-handed one, one finds an ellipsoid, in which all allowed values for the neutrino mass $< m_{\nu e} >$ and the right-handedness $< \eta >$ and $< \lambda >$ must lie. Here M_{WL} and M_{WR} are the left and right-handed vector boson masses, respectively. ϑ is the mixing angle $W_1 = \cos \vartheta \cdot W_L + \sin \vartheta \cdot W_R$ of the left and right-handed vector bosons. The most stringent limits for $m_{\nu e}$ and η are obtained from the lower limit of the ^{128}Te lifetime $\tau_{1/2} > 5 \cdot 10^{24}$ years. The results are given in fig. 4.

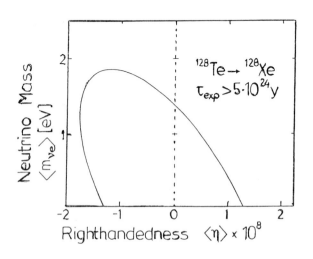

Figure 4
The allowed regions (inside the ellipses) deduced from the experimental bounds (Caldwell, Berkeley Conference [14] and refs. 6,8 and 15) for the neutrinoless double β-decay $^{128}Te \rightarrow {}^{128}Xe$ with $\tau_{1/2} > 5 \cdot 10^{24}$ years. The upper limit for the neutrino mass is $m_\nu \leq 1.9 eV$ and for the right-handedness $\eta \leq 1.8 \cdot 10^{-8}$ and $\lambda \leq 6 \cdot 10^{-6}$.

4. Particle-Particle Correlations and the Single β^+ Decay

The same reduction which one finds for the β^+-branch of the $2\nu\beta\beta$ decay one should also find for the β^+ decay from neutron deficient nuclei. Indeed it is well known[15,16] that the β^+ decay in neutron deficient nuclei are drastically quenched compared to the theoretical results.

Fig. 5 shows the reduced β^+ decay probability as a function of the particle-particle strength g_{pp} for the β^+ decay of $^{148}Dy \to {}^{148}Tb$. The same effect which yields agreement for the $2\nu\beta\beta$ decay can also explain the quenching in the β^+ decay of neutron deficient nuclei.

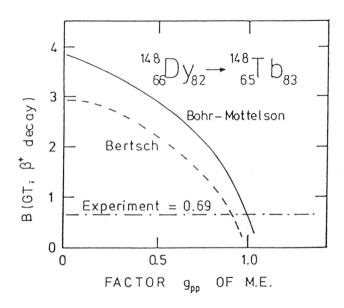

Figure 5
Reduced β^+ decay probability $B(GT)$ as a function of the particle-particle strength g_{pp} with which the particle-particle reaction matrix elements of the Bonn potential are multiplied. The solid line is calculated with single particle wave functions of a Saxon-Woods potential with parameters of Bohr and Mottelson[18] and the dashed line with Saxon-Woods parameters of Bertsch[19]. One finds a drastic reduction for the physical value of the particle-particle strength $g_{pp} = 1$. Around this value the theory agrees with the experimental data $B(GT) = 0.69$. The single-particle basis used in the RPA calculation is the full N=4 and 5 and the $i_{13/2}$ and $i_{11/2}$ shells.

5. Conclusions

The double neutrinoless β decay ($0\nu\beta\beta$) can distinguish if the neutrino is a Dirac particle, that means if the neutrino is different from the antineutrino, or if it is a Majorana particle and therefore identical with its antiparticle. Only in the case of a Majorana particle, the double neutrinoless β decay ($0\nu\beta\beta$) is possible. Grand unified theories predict that the neutrino is a Majorana particle, especially the ones which are built on the SO(10) group. But being a Majorana particle the neutrino has to have a finite mass and for SO(10) also a slight right-handed weak interaction. Since the $0\nu\beta\beta$ decay needs either a finite mass of the neutrino or a right-handedness of the weak interaction, one can derive from the lower limit of the half-lives of $0\nu\beta\beta$ decays upper limits for an averaged neutrino mass $< m_\nu > \leq 1.8 \cdot 10^{-8}$.

If one tries to test the calculation by the known $2\nu\beta\beta$ decays one finds that the theory is by a factor 10 to 100 larger than the experimental data. But if one includes the particle-particle correlations of protons with neutrons, which are attractive, they cancel by a large part the neutron-particle and proton-hole correlations. Including these additional ground state correlations one finds[20-23] a strong quenching of the $2\nu\beta\beta$ transitions in agreement with the experimental data.

For the $0\nu\beta\beta$ transition amplitude this quenching is for the leading recoil term only about a factor 0.7 or 30 %. This difference between the $2\nu\beta\beta$ and $0\nu\beta\beta$ decay stems from the fact that in the $0\nu\beta\beta$ one has for the transition operator a dependence on the distance between the two vertices and thus, by a multipole expansion one gets a transition operator which excites also higher multipoles in the intermediate nucleus[20]. These higher multipoles are not strongly quenched and thus the $0\nu\beta\beta$ transitions are not so drastically affected by the particle-particle correlations.

The quenching of the β^+ branch in the $2\nu\beta\beta$ decay can also be tested in the single β^+ decay of neutron deficient nuclei. Including the particle-particle correlations one finds also agreement for these transitions which could not be understood before.

References
1. Davis, R., Phys. Rev. **97** (1955) 766.
2. Lee, T.D. and Yang, C.N., Phys. Rev. **104** (1956) 254.
3. Wu, C.S., Ambler, E., Hayward, R.W., Hopper, D.D. and Hudson, R.P., Phys. Rev. **105** (1957) 1413.
4. Langacker, P., Phys. Rep. **72** (1981) 185.
5. Fritzsch, H. and Minkowski, R., Phys. Rep. **73** (1981) 67.
6. Kirsten, R., in Proceedings of the International Symposium on Nuclear Beta Decay and Neutrinos, Osaka, Japan 1986, Edited by T. Kotani, H. Ejiri, E. Takasugi (World Scientific, Singapore, 1986,p. 81)
7. Manuel, O.K., Proc. Int. Symp. on Nuclear Beta Decay and Neutrinos, Osaka, Japan, June 1986, eds. T. Kotani, H. Ejiri, E. Takasugi (World Scientific, Singapore, 1986, p. 71)
8. Elliott, S.R., Hahn, A.A. and Moe, M.K., Phys. Rev. Lett. **59** (1987) 2020.
9. Eliott, S.R. Hahn, A.A. and Moe, M.K., Proc. Int. Symp. on Weak and Electromagnetic Interactions in Nuclei, July 1986, ed. H.V. Klapdor (Springer, Berlin 1986) p. 692
10. Haxton, W.C. and Stephenson, G.J., Progr. Part. Nucl. Phys. **12** (1984) 409.
11. Tomoda, T., Faessler, A., Schmid, K.W. and Grümmer, F., Nucl. Phys. **A452** (1986) 591.
12. Avignone III, F.T., Brodzinski, R.L., Evans, J.C., Ir., Hensley, K., Miley, H.S. and Reeves, J.H., Phys. Rev. **C34** (1986) 666.
13. Caldwell, D.O., Eisberg, R.M., Grumm, D.M., Hale, D.L., Witherell, M.S., Goulding, F.S., Laudis, D.A., Madden, N.W., Malone, D.F., Pehl, R.H. and Smith, A.R., Phys. Rev. **D33** (1986) 2737.
14. Caldwell, D.O., Osaka Conference, June 1986 $(\tau_{1/2}^{0\nu}(^{76}Ge) > 3.9 \cdot 10^{23}$ years); Berkeley Conference, July 1986 $(\tau_{1/2}^{0\nu}(^{76}Ge) > 4.7 \cdot 10^{23}$ years).
15. Nolte, E. et al., Z. Physik **A306** (1985) 223.
16. Kleinheinz, P. et al., Phys. Rev. Lett. **55** (1985) 2664.
17. Civitarese, O., Faessler, A. and Tomoda, T., Phys. Lett. **B194** (1987) 11.
18. Bohr, A. and Mottelson, B.R., Nucl. Structure, Vol. 1 (Benjamin, New York,

1969).
19. Bertsch, G., The Practitioners Shell Model (American Elsevier, New York, 1972)
20. Tomoda, T. and Faessler, A., Phys. Lett. **B199** (1987) 2383.
21. Vogel, P., and Zirnbauer, M.R., Phys. Rev. Lett. **57** (1986) 3148.
22. Muto, K. and Klapdor, H.V., Phys. Lett.**201B** (1988) 420
23. Engel, J., Vogel, P., Civitarese, O. and Zirnbauer, M.R., Phys.Lett. **B208** (1988) 187.

359

PARITY NONCONSERVATION IN NUCLEAR REACTIONS AND ALPHA DECAY

Ovidiu Dumitrescu

Department of Fundamental Physics
Institute of Physics and Nuclear Engineering
Institute of Atomic Physics, POB MG-6, R-76900
Bucharest - Măgurele, ROMANIA

Abstract

Parity nonconservation (PNC) in some low energy nuclear processes, such as nuclear reactions and nuclear scattering induced by polarized protons as well as alpha decay is reviewed. Some comments on PNC-nucleon-nucleon (PNC - NN) potential are presented. Explicit expressions for the PNC-analyzing powers and α-widths are rederived. Applications for : $^{13}C(\vec{p},p)^{13}C$, $^{15}N(\vec{p},p)^{15}N$, $^{15}N(\vec{p},\alpha)^{12}C$ resonance reactions and $^{16}O^*(2-\alpha, E_x = 8.87$ MeV$) \to \alpha + ^{12}C(g \cdot s.)$ alpha decay are done. New experiments are proposed.

1. INTRODUCTION

Symmetries bring beauty and order into physical phenomena and their mathematical description.
Some of them can be and often are used as tools, helping us to hide our lack of deeper understanding. Yet, if absolute, they are somehow dead in their rigidity. However, if a symmetry is broken - as it is well known in art or archi - tecture - it springs to life and reveals a forceful personality . Through its breaking a symmetry can teach us more about the physical system than if it were absolute. Reflexion symmetry or parity conservation was for a long time just there to be used as a constraint. Since its breaking was found , however, it has become a great stimulus for experimental and theoretical investigations - exemplified e.g. by Carlo Rubia group long work and final establishment[1] of the existence of the heavy $W^{\pm}$ and Z^0 bosons mediators of the weak force that violates the parity conservation low.
In the nuclear physics parity violation is the only sign of weak interactions contribution to the nuclear structure.
In 1957, the same year that PNC was discovered in β and μ decay, it was reported[2] the first search for PNC in NN interaction. The following year Feynmann and Gell-Mann[3] pre - dicted , on the basis of their universal current-current theory of weak interactions, that in addition to the familiar weak processes of μ, β and hyperon decay, there should be a first-order weak PNC-NN interaction.
Preliminary evidence for this new form of weak interaction was presented by a russian group[4] six year later. However, the first persuasive result was obtained in 1967 by Lobashov and his group[5], who by pioneering the technique of integral detection (as opposed to pulse counting) obtained a result consistent with the qualitative predictions of the current-current theory[6]. In the years following these experiments,

great progress has been made in understanding the weak interactions : 1) the development of "Standard" SU(2)xU(1) electroweak theory; 2) the discovery of the neutral weak current and 3) the recent experimental detection of the $W^{\pm}$ and Z^{o} bosons, mediators of the weak force.

The strong quark interaction is believed to conserve parity strictly, but quarks also take part in the PNC-NN weak force, in which charged $W^{\pm}$ or neutral Z^{o} bosons are exchanged. The quark model predicts that the exchange of charged $W^{\pm}$ bosons will add to the NN force a small weak force component that does not conserve parity and that chiefly causes the isospin of a pair of interacting nucleons, either to remain the same or to change by two units. The neutral Z^{o} exchange gives rise to a weak force component that also does not conserve parity and that changes the isospin of a pair of nucleons by zero, one or two units. The weak PNC-NN interaction is only of the order of 10^{-7} compared to strong interactions. Therefore, only small effects are expected and measured in few nucleon systems. However, for complex nuclei substantial enhancements of PNC effects are predicted and observed. The enhancement of these effects originates from several reasons, the most important being the small level spacing between nuclear states of the same spin and opposite parity of the compound nucleus involved in the PNC low energy nuclear process (the so-called parity-mixed doublets (PMD)). The next important reason arises from the possible increase of the ratio between parity forbidden and parity allowed transition matrix elements caused by the nuclear structure of the states involved[7]. The third reason of enhancement of PNC effects sometimes could be the "coherence" contribution of different PNC terms (channels) [8-12] as e.g. in the case of low energy neutron scattering[8] from ^{139}La, were the p-wave resonance is embedded, in a dense "sea" of S-wave resonances, hence, there are many possibilities for parity admixing with very small energy denominators.

In this work, we shall focus ourself especially to some favoured PNC-cases in resonance scattering, nuclear reac - tions and alpha decay.

II. T- MATRIX AND DECAY WIDTHS

By applying the Fonda-Newton theory[13] generalized in Refs.[14-17] we separate the full spectrum of states of a parity conserving nonperturbed Hamiltonian ($H_{o(PC)}$) into four classes of states: 1º/ Bound states embedded in continuum (BSEC); 2º/ Active open channels; 3º/ Passive open channels and 4º/ Closed channels. Following the notation of Ref. 17:

$$P = |n><n| \qquad (1)$$

projects onto some doorway states (BSEC)[18] of $H_{o(PC)}$.
Q - onto the active open channels (the channels most important to the reaction or decay process),
q - onto the passive open channels (the channels which are inserted usually into the imaginary part of the optical potential and A - projects onto the closed channels (other bound states than BSEC-ones and continuum states whose thresholds are lying above the scattering or decay energy[15-17]).

The interaction governing the nuclear reaction or decay process we shall consider as a sume of strong parity con - serving (PC) and PNC-weak interaction potentials.

$$H = H_{o(PC)} + H' \quad ; \quad H' = H'_{PC} + H'_{PNC} \qquad (2)$$

The S-matrix on the energy shell is defined as

$$S(E) = 1 - 2\pi i T(E) \qquad (3)$$

where[14-17]

$$T(E) = H' + H'G(E)H' \qquad (4)$$

is the transition matrix operator defined in terms of the residual interaction H' that determines the nuclear reaction or decay process and the Green's operator defined by

$$(E-H)G(E) = 1; \quad G(E) = \lim_{\eta \to 0^+} (E + i\eta - H)^{-1} \qquad (5)$$

Using the separation of $H_o = H_{o(PC)}$-space states into four subspaces as above mentioned the transition operator (4) can be rewritten as follows:

$$T = R + RPD^{-1}PR \qquad (6)$$

where

$$D = E - H_o - R \qquad (7)$$

and

$$R = W + W G_Q W = L + (2i)^{-1}\Gamma \qquad (8)$$

with

$$W = V + V G_q V \quad ; \quad V = H' + H'G_A H' ; \qquad (9)$$

and

$$(E - H_o - AH'A) \; G_A = A = G_A(E - H_o - AH'A) \qquad (10)$$

$$(E - H_o - qVq) \; G_q = q = G_q(E - H_o - qVq) \qquad (11)$$

$$(E - H_o - QWQ)G_Q = Q = G_Q(E - H_o - QWQ) \qquad (12)$$

The S-matrix elements can be expressed in terms of the inverse of D operator (7) as follows:

$$S_{c_1 c_2} = (1 - 2\pi iR)_{c_1 c_2} -$$

$$- e^{i\xi_{c_2}} i [\sum_{nn} (D^{-1})_{nn}, (\Gamma_{nc_2} \Gamma_{n,c_1})^{1/2} \times e^{i\xi_{c_1}} \quad (13)$$

where

$$(1 - 2\pi iR)_{cc'} \cong \begin{cases} \delta_{cc'}, \quad e^{2i\xi_c} & \text{for elastic scattering} \\ \\ -2\pi i (T_{bg})_{cc'}, & \text{for nuclear reactions} \end{cases} \quad (14)$$

and [17]

$$\Gamma_{nc} = 2\pi | <c|R|n> |^2 \quad (15)$$

stands for the partial resonance width.

Here c stands for all channel quantum numbers[18,19] $\{\varepsilon_c, j_p \pi_p j_t \ell sJ\}$ (e.g. - ε_c - the channel energy, $j_p \pi_p (j_t \pi_t)$ the projectile (target) spin and parity, respectively, ℓ-the orbital angular momentum of the projectile - target relative motion, S- the channel spin, J - the total channel spin, which in the resonance reaction should be equal to the total compound nucleus state spin). |c> is the projectile-target (or emitted particle-residual nucleus) channel wave function:

$$\langle R, x_p x_t | c \rangle = \frac{1}{R} U_{c_1 \varepsilon_c}(R) [Y_{\ell_c} \otimes [\phi_{j_p \pi_p}(x_p) \otimes \phi_{j_t \pi_t}(x_t)]_S]_J$$

in which $\phi_{p(t)}(x)$ are the internal projectile (target) wave functions, $Y_{\ell m}(R)$ - is the ℓth spherical harmonic wave function and $U_{c, \varepsilon_c}(R)$ - is the radial part of the regular scattering solution of the channel c-problem[18]. The channel wave function should be normalized according to

$$\langle c| c'\rangle = \delta_{cc'} \delta(\varepsilon_c - \varepsilon_{c'}) \qquad (17)$$

The application of the renormalized R - operator onto the BSEC $|n\rangle$ - state, should be done according to the equation[17)

$$R|n\rangle = r|\lambda\rangle \qquad (18)$$

where $|\lambda\rangle$ should describe the compound nucleus state (by using some model Hamiltonian) and r stands for the transition operator describing the corresponding (to the channel c) decay process.

Expanding now the Green's operator to first order in H_{PNC}, assuming then that the projectile and the target states are parity conserving (PC) states of some nuclear structure Hamiltonians, ignoring any PNC effects arising from direct processes and keeping only those effects which are related to the closeness of the two resonances (PMD), we can write[20) the S-matrix elements as follows:

$$S_{c_1 c_2} = S_{c_1 c_2}(PC) + S_{c_1 c_2}(PNC) \qquad (19)$$

where

$$S_{c_1 c_2}(PC) = e^{i(\xi_{c_1}+\xi_{c_2})} [\delta_{\ell_1 \ell_2} \delta_{S_1 S_2} - 2\pi i (T_{bg})_{\ell_1 S_1, \ell_2 S_2} - i \frac{(\Gamma_{\ell_1 S_1}^{J^\pi} \Gamma_{\ell_2 S_2}^{J^\pi})^{1/2}}{E - E^{J^\pi} + \frac{1}{2}\Gamma^{J^\pi}}]$$

and $\qquad (20)$

$$S_{c_1 c_2}(PNC) = -i\, e^{i(\xi_{c_1}+\xi_{c_2})} \frac{(\Gamma_{\ell_1 S_1}^{J^\pi} \Gamma_{\ell_2 S_2}^{J^{-\pi}})^{1/2} \langle J^\pi |H_{PNC}| J^{-\pi}\rangle}{[E-E^{J^\pi} + \frac{i}{2}\Gamma^{J^\pi}][E-E^{J^{-\pi}} + \frac{i}{2}\Gamma^{J^{-\pi}}]}$$

$$(21)$$

there $\Gamma_{\ell S}^{J^\pi}$ stands for the partial channel width and
$\langle J^\pi | H_{PNC} | J^{-\pi} \rangle$ - for the PNC-matrix element.

The T-matrix elements are related to the S-matrix elements according to

$$T_{c_1 c_2} = e^{2i(\sigma_{\ell_1} - \sigma_o)} \delta_{c_1 c_2} - S_{c_1 c_2} \qquad (22)$$

where

$$\sigma_\ell - \sigma_o = \sum_{S=1}^{\ell} \tan^{-1}(\frac{\eta}{S}) \; ; \; \eta = Z_1 Z_2 m e^2 / \hbar^2 k \qquad (23)$$

determines the Coulomb phase.

III. CROSS-SECTIONS AND VECTOR ANALYZING POWERS

The cross section of the scattering or reaction process with polarized protons is defined by

$$\frac{d\sigma}{d\Omega} = \sigma_u + T_r(T 2\vec{S}_p T^+) \vec{\mathcal{P}}_p \tag{24}$$

in which $\vec{S}_p$ stands for the incident proton spin and $\vec{\mathcal{P}}_p = \langle 2\vec{S}_p \rangle$ - for its polarization vector. The T-matrix elements for any scattering or reaction process are:

$$T_{if} = (4k_i \hat{j}_i \hat{I}_i)^{-1} \otimes$$

$$\otimes \langle \tilde{\chi}^{(-)}(k_f) \phi_{j_f \mu_f} \phi_{I_f M_f} |T| \phi_{I_i M_i} \phi_{j_i \mu_i} \tilde{\chi}^{(+)}(\vec{k}_i) \rangle \tag{25}$$

where $I_i(I_f)$, $M_i(M_f)$ and $j_i(j_f)$, $\mu_i(\mu_f)$ are the spins of the target (residual) nucleus and projectile (outgoing particle) and their projections onto the laboratory Z-axis taken as usual along the $\vec{k}_i$ vector, ϕ wave functions describe the intrinsic motion of the corresponding atomic nuclei.

$$\tilde{\chi}^{(\pm)}(\vec{k}_{i(f)}) = \left[\frac{2m_{i(f)} k_{i(f)}}{\pi \hbar^2} \right]^{1/2} \chi^{(\pm)}(k_{i(f)}) \tag{26}$$

are the rescaled relative motion (e.g. DWBA or CCBA) wave functions for the initial (+) and the final (-) channels, $k_{i(f)}$ and $m_{i(f)}$ being the corresponding wave vectors and reduced masses, respectively.

The vector analyzing power is defined by

$$\vec{A} = Tr(T \, 2\vec{S}_p T^+)/\sigma_{un}$$

$$= A_L \vec{e}_L + A_n \vec{e}_n + A_b \vec{e}_b \tag{27}$$

where $\vec{e}_L = \vec{k}_i(k_i)^{-1}$; $\vec{e}_n = \dfrac{\vec{k}_i \times \vec{k}_f}{|\vec{k}_i \times \vec{k}_f|}$; $\vec{e}_b = \vec{e}_n \times \vec{e}_L$ (28)

are the three unit vectors which form the laboratory frame of reference given by the Madison convention[21]. A_n is the normal (PC) vector analysing power; A_L and A_b are the irregular PNC longitudinal and transverse vector analysing powers.

Defining by

$$\sigma^{(v)} = \text{Tr}(T\, 0^{(v)}\, T^+) \qquad (29)$$

with

$$0^{(o)} = 1 \qquad (30)$$

and $\vec{0}^{(1)} = \vec{S} = \sum_{\varkappa} \vec{e}^{\varkappa} S_{\varkappa}$; $0^{(1)}_{\varkappa} = S_{\varkappa}$ (31)

where $\vec{S}$ is the spin of the projectile
and

$$\vec{e}^1 = -2^{-1/2}(\vec{e}_b - i\vec{e}_n);\ \vec{e}^0 = \vec{e}_L;\ \vec{e}^{-1} = 2^{-1/2}(\vec{e}_b + i\vec{e}_n) \qquad (32)$$

are the cyclic unit vectors , we may write the expression:

$$\sigma_{un} = \sigma_o^{(o)} \qquad (33)$$

for the cross section of the reaction process induced by unpolarized protons and

$$\dfrac{d\sigma}{d\Omega} = \sigma_{un}(1+\vec{A}\vec{\mathcal{P}}_p) = \sigma_{un}(1 + 2\sum_{\varkappa}\sigma^{(1)}_{\varkappa}\,(\vec{e}^{\varkappa}\vec{\mathcal{P}}_p)(\sigma_o^{(o)})^{-1}) \qquad (34)$$

$$\vec{A}\vec{\mathcal{P}} = A_L\mathcal{P}_L + A_n\mathcal{P}_n + A_b\mathcal{P}_b$$

for the cross section of the reaction process induced by polarized protons, where

$$A_L = 2R_e \, \sigma_o^{(1)} \, (\sigma_o^{(0)})^{-1} \tag{35}$$

$$A_n = -2\sqrt{2} \, I_m \, \sigma_1^{(1)}(\sigma_o^{(0)})^{-1} \tag{36}$$

$$A_b = -2\sqrt{2} \, R_e \, \sigma_1^{(1)}(\sigma_o^{(0)})^{-1} \tag{37}$$

are the components of the vector analyzing power. Introducing the decomposition

$$\sigma_{\mathcal{X}}^{(v)} = \sum_{i=c,n} \sum_{j=c,n} (\sigma_{\mathcal{X}}^{(v)})_{ij} \tag{38}$$

where c(r) stands for the Coulomb (nuclear) part of the T-matrix, we find the following explicit formulas [20,22]

$$(\sigma^{(v)})_{cc} = \delta_{vo} \, \delta_{o\mathcal{X}} k_i^{-2} \, |C(\theta)|^2 \tag{39}$$

$$(\sigma_{\mathcal{X}}^{(v)})_{cn} + (\sigma_{\mathcal{X}}^{(v)})_{nc} =$$

$$= 2R_e \frac{i}{k_i^2} C(\theta) \sum_{\ell,\ell_1,j} \sum_{S,S_1 J} B^{(v)}(\ell,\ell_1,j,S,S_1,J)$$

$$* T_{j_i I_i \ell S J, \, j_i I_i \ell_1 S_1 J} \, Y^*_{L-\mathcal{X}}(k_f) \tag{40}$$

with

$$B^{(v)}(\ell,\ell_1,j,S,S_1,J) = \sqrt{\pi}\,\frac{(-1)^{\ell+\varkappa-S+S_1+\ell-\ell_1}}{\hat{I}_i^2\,\hat{\ell}^2\,\hat{\ell}_1}$$

$$<\tfrac{1}{2}\|O^{(v)}\|\tfrac{1}{2}>\hat{S}\,\hat{S}_1\,\hat{\ell}\,\hat{\ell}_1\,\hat{j}^2\,\hat{j}^2 \begin{Bmatrix} \tfrac{1}{2} & I_i & S_1 \\ \ell & S & J \\ j & j_i & \ell_1 \end{Bmatrix}$$

$$\sum_{\mu_i} C^{\tfrac{1}{2}\ \ v\ \ \tfrac{1}{2}}_{\mu_i\ \ \mu_i-\varkappa\ \ \mu_i}\ C^{j\ \ j\ \ \ell}_{-\mu_i\,\mu_i-\varkappa\,-\varkappa}\ C^{j\ \ \tfrac{1}{2}\ \ \ell_1}_{-\mu_i\ \mu_i\ 0} \qquad (41)$$

and

$$(\sigma^{(v)})_{nn} = \frac{1}{k_i^2}\sum_{L,j,\ell,\ell',\ell_1,\ell_2,S,S_1S_2,J,J'}\cdots\sum E(L,j,\ell,\ell_1',\ell_2,S,S_1S_2J,J')$$

$$T_{j_fI_f\ell SJ,j_iI_i\ell_1S_1J}\ T^*_{j_fI_f\ell'SJ',j_iI_i\ell_2S_2J'}\ Y^*_L(\hat{k}_f) \qquad (42)$$

with

$$E(L,j,\ell,\ell',\ell_1,\ell_2,S,S_1,S_2,J,J') =$$

$$= <\tfrac{1}{2}\|O^{(v)}\|\tfrac{1}{2}>\sqrt{\pi}\,\frac{(-1)^{I_i-\tfrac{1}{2}+S+S_2+\ell_1+J+L}}{4\,\hat{I}_i^2\,\hat{v}^2\,\hat{L}}$$

$$\hat{\ell}\,\hat{\ell}'\,\hat{\ell}_1\hat{\ell}_2\quad \hat{S}_1\,\hat{S}_2\,\hat{j}^2\hat{j}'^2\quad \hat{L}^2\,\hat{j}^2\,\hat{v}^2$$

$$\begin{pmatrix} j & v & L \\ 0 & \varkappa & -\varkappa \end{pmatrix} \begin{pmatrix} \ell_1 & \ell_2 & j \\ 0 & 0 & 0 \end{pmatrix} \begin{pmatrix} \ell & \ell' & L \\ 0 & 0 & 0 \end{pmatrix}$$

$$W(\tfrac{1}{2}\ \tfrac{1}{2}\ S_1\ S_2;\ v\ I_i) \qquad W(J\ \ell\ J'\ \ell';\ S\ L)$$

$$\begin{Bmatrix} \ell_1 & \ell_2 & j \\ S_1 & S_2 & v \\ J & J' & L \end{Bmatrix} \tag{43}$$

and e.g. $\hat{j} = \sqrt{2j+1}$; $j_i = \tfrac{1}{2}$

In the above expressions

$$< \tfrac{1}{2}\ \|S^{(1)}\|\ \tfrac{1}{2} > = \tfrac{\sqrt{3}}{2}\ ;\qquad < \tfrac{1}{2}\ \|1\|\ \tfrac{1}{2} > = 1 \tag{44}$$

$$C(\theta) = \frac{\eta}{2\sin^2 \tfrac{\theta}{2}}\ \exp\{-2i\eta\ \ln[\sin\tfrac{\theta}{2}]\};\ \eta = \frac{Z_1 Z_2\ me^2}{\hbar^2\ k_i}$$

$$\tag{45}$$

For the resonance reaction case the expressions for the cross section and vector analysing powers are obtained from eqs. (33-45) by putting [22] $C(\theta) = 0$ and

$$T_{C,C'} = (T_{bg})_{CC'} + (T_{res})_{CC'} \tag{46}$$

The resonance PC and PNC $(T_{res})_{CC'}$ - terms can be calculated according to the eqs. (20-22), while the background $(T_{bg})_{CC'}$ - term can be calculated by using some direct reaction code or by using the phase-shift analysis.

The background term (T_{bg}), the partial widths (Γ_c) and the phases (ξ_c) are supposed to vary slowly with the energy[23-27]. All these quantities must, however, have the proper zero energy threshold behaviour. When using the direct reaction code there have still to solve the problem of the phases of the resonance terms relative to the back - ground [23-27] ones. If denoting the background S-matrix by

$$B = 1 - 2\pi i\, T_{bg} \tag{47}$$

then[23])

$$S = B - i t \tilde{t}\, (E-E_r)^{-1} \tag{48}$$

with

$$E_r = E_o - \tfrac{i}{2}\, \Gamma_r \tag{49}$$

where E_r is the resonance energy, Γ_r is the total resonance width, t is a complex column vector and $\tilde{t}$ its transpose with their elements

$$t_c = e^{i\xi_c}\sqrt{\Gamma_c} \tag{50}$$

The unitarity condition for the B-matrix

$$B^+ B = 1 \tag{51}$$

together with the unitarity of the S-matrix

$$S^+ S = 1 \tag{52}$$

provide the conditions[23,24])

$$\sum_{c'} B^+_{cc'}\, t_{c'} = t^*_c \tag{53}$$

$$\sum_c |t_c|^2 = \Gamma_r \tag{54}$$

which should be fulfilled when a direct reaction code for the background is used. These conditions should be fulfilled also when within the phase-shift analysis ambiguous solutions are obtained[28].

IV. ALPHA DECAY WIDTH

The alpha decay width can be obtained as a particular case of the eqs. (15) and (18). Following the prescriptions of the Fermi liquid model of alpha decay[29] the transition operator r from the eq. (18) is choosen to be [30,31] :

$$r = T_{4 \to \alpha} = \mathcal{X} \delta(\xi_1) \delta(\xi_2) \delta(\xi_3) \tag{55}$$

where $\mathcal{X}$ is the coupling strenght of the vertex corresponding to the amplitude of the alpha particle formation in the four particle channel, while ξ_i ($i = 1, 2, 3$) are the usual Jacobi coordinates that describe the intrinsic motion of the nucleons inside the free alpha particle[17].

For the PNC alpha transition from the initial state $|J_i^{\pi_i} T_i\rangle$ to the final state $|J_f^{\pi_f} T_f\rangle$ (where J, π, T stand for the spin, parity and isospin of the state) the corresponding alpha decay width has the following expression[31] :

$$\Gamma_\ell(J_i^{\pi_i} T_i; J_f^{\pi_f} T_f) =$$

$$2\pi \mathcal{X}^2 \left| \int_0^\infty dR\ \tilde{U}_\ell(R)\ g_\ell^{J_i^{\pi_i} T_i;\ J_f^{\pi_f} T_f}(R) \right|^2 \tag{56}$$

where

$$g_\ell^{J_i^{\pi_i} T_i,\ J_f^{\pi_f} T_f}(R) = \sum_n \frac{\langle n, J^{-\pi_i} T_i | H_{PNC} | J^{\pi_i} T_i \rangle}{E_{J_i^{\pi_i} T_i} - E_{J_i^{-\pi_i} T_i}^{(n)}} *$$

$$* \langle \delta(R-R_\alpha) R^{-1} \varphi_\alpha [Y_\ell \otimes \phi_{J_f^{\pi_f} T_f}]_{J_i} | \delta(\vec{\xi}_1)\delta(\vec{\xi}_2)\delta(\vec{\xi}_3) | \phi_{J_i^{-\pi_i} T_i}^{(n)} \rangle$$

$$\tag{57}$$

with
$$\pi'_f(-)^\ell = -\pi'_i \qquad (58)$$

and
$$U_\ell(R) = U_\ell(R) - \int_0^\infty dR'\, K_\ell(R,R')\, U_\ell(R') \qquad (59)$$

where the Pauli-Kernel[31] K (R,R') comes from accurate normalization of the alpha particle - residual nucleus relative motion (U (R)) wave function and eliminates the spurious states due to the channel wave function antisymmetrization.

Its expression is :
$$K_\ell(R,R') = \int dx\, P_\ell(x)\, K(R,R',x)$$

where
$$\langle \vec{R} | 1-K | \vec{R}' \rangle = \langle \mathcal{A}\, \delta(\vec{R}-\vec{R}_\alpha)\varphi_\alpha \phi_A | \mathcal{A}\, \delta(\vec{R}'-\vec{R}_\alpha)\varphi_\alpha \phi_A \rangle$$

$P_\ell(x)$ is the ℓth Legendre polynomial and $x = (\vec{R}\vec{R}')/RR'$ The explicit expression of K (R,R') for ^{16}O case is given in Ref.[31] and for heavy nuclei in Ref.[30]

V. SHELL MODEL ESTIMATIONS OF THE ENERGIES, WIDTHS AND PNC MATRIX ELEMENTS

In this section we outline the basic recipe of the modern nuclear shell-model calculations incorporated in the OXBASH-code[32], we use for the calculations of the nuclear state energies, widths and PNC-matrix elements.

To obtain the effective two-body interaction we can e.g. use the g-matrix method[33,34] by solving the generalized Bethe-Goldstone[35] and Bethe-Faddeev equation[36] The method of solution is iterative[37]:

1) first a complete set of single-particle states is chosen;

2) the reaction g-matrix is then calculated and a first iterated effective two-body interaction (ETBI) is obtained

3) the Hartree-Fock equation with this ETBI is solved to yield a first iteration of the occupied single particle state energies and wave functions.;

4) the generalization Bethe-Goldstone and Bethe-Fadeev eqs. are calculated in order to establish the unoccupied state potential ;

5) the Schrödinger eq. for the unoccupied single-particle state energies and wave functions is solved;

6) the unoccupied single particle basis is orthogonalized to the occupied single particle states found in 3) to give the first iteration to the unoccupied single particle states.

Having in such a way a complete set of first iterated single-particle states we repeat a second cycle starting with pct. 2) and so on.

The ETBI -obtained in such a way is diagonalized in the last iterated single particle basis (m-scheme) or in a more sophisticated basis obtained by different coupling and projection procedures[38].

The ETBI can be extracted from experimental data also.

Our calculations use both procedures. Two different model spaces have been used: the Zuker, Buck and McGrory)(ZBM) model space which contains the $1p_{1/2}$, $1d_{5/2}$ and $2s_{1/2}$ orbits in the valence space and the PSD model space including in addition the $1p_{3/2}$ and $1d_{3/2}$ orbits. In order to maintain the matrix dimensions at a nonprohibited level, the nucleons have been considered to be frosen in the $1p_{3/2}$ orbit; thus a fixed $(1s_{1/2})^4 (1p_{3/2})^8$ configuration is assumed in all cases. It turns out that at least four particle-four hole calculations are needed in order to describe the 2^+ states in ^{16}O. Four different residual interactions have been used in ZBM model space: ZBM I and ZBM II are the models that use the interactions I^{39} (g-matrix[40])) and $II^{39)}$ of Zuker, Buck and McGrory; REWIL and ZWM are the F and Z interactions from Ref. 41. Two different combinations of interactions have been taken into account in the PSD model space: PSDMK is a Cohen-Kurath interaction[42]) for p-orbits, Preedom-Wildenthal[43]) for sd-orbits and Millener-Kurath[44]) for matrix elements between p and sd orbits. PSDMWK is similar to PSDMK except the W-interaction[45]) which is taken for the sd subspace. In the PSDMK+CM and PSDMWK+CM the contribution of spurious components were eliminated according to the procedure of Ref[46]), while in ZBM- according to Ref.[47]) one.

In the PNC-matrix elements, due to the short range contribution of the heavy meson exchange to H_{PNC}, short range correlations (SRC) of shell model wave functions must be implemented. SRC describes the effect of the distortion of the relative two-nucleon wave function at small distances due to the strong repulsive core in the nuclear interaction. The repulsion punches a hole in the wave functions at small relative distances and this has strong effects on two-body observables (when they are of short range). There remains also an effect on one body observables after averaging over the second particle. One finds that the charge form factor is enhanced by RSC[48,49]) for momentum transfers at $q > 3$ fm^{-1}.

The effects are negligible for the form factor at low q.

The techniques to include SRC are gradually developed in different many-body theories of nuclear structure, such as Brueckner-Hartree-Fock method[50], the method of Jastrow functions[51] or its extension to the hipernetted chain equations[33,52] and exp (S) formalism[35,36].

In our calculations SRC have been inserted by multiplying the radial two body wave function by a kind of Jastrow factor[53]:

$$1-\exp(-ar^2)(1-br^2); \quad a=1.1 \text{ fm}^{-2}; \quad b=0.68 \text{ fm}^{-2}$$

This procedure is in agreement with a much more elaborated treatment like the Bethe-Goldstone approach[6,54] and leads to a suppression of the pion exchange matrix element by 20%-30% and a decrease for ρ and ω exchange matrix elements by a factor of $3 \div 4$.

The PNC-observables that we calculated by using the OXBASH-code are mainly in the ^{16}O region. For illustration we show in Fig. 1 some $J^\pi T = 2^+0$ and 2^-1 experimentally observed and calculated[22] levels in ^{16}O. Fig. 2 shows the calculated[22] values of the PNC,

$$<2^-1, 12.9686 \text{ MeV} |H_{PNC}| 2^+0, 13.020 \text{ MeV}>$$

matrix element in ^{16}O, by using the above 8 mentioned OXBASH-code models for the strong interaction part and 4 weak PNC interaction potential models[22]. The general form of the weak PNC-NN interaction potential is a well established function[55] of NN radius and momentum as well as of strong meson-nucleon (g_M) and weak meson-nucleon (h_M) coupling constants.

While the strong couplings (g_M) are well established[55], the weak couplings (h_M) have been the subject of debate in recent years in particle physics. Investigating PNC-MNN vertices within the framework of nonlinear chiral effective lagrangean, Kaiser and Meissner[56] (KM) reported a conside - rably smaller value (1.9 x 10^{-8}) for the weak πN coupling constant (h_π) compared to the recent result (1.3 x 10^{-7}) obtained by Dubovik and Zenkin[57] (DZ) in the framework of the Weinberg-Salam theory plus quark model, both signifi - cantly lower than the often used Desplanques-Donoghue-Holstein (DDH) "best value"[58] (4.6 x 10^{-7}). Moreover, the weak couplings extracted from low energy experiments by Adelberger and Haxton[55] (AH) are close to the DZ values (see table 1). As it can be seen from Fig. 2, the results for different interactions agree within a factor of 2.5 and no large suppression appears when the model space is enlarged. The ρ and ω exchange contributions add coherently to the total matrix element in every case. The contribution from heavy bosons do not exceed 25% for DDH, AH and DZ cases, but increase up to 50% in the KM-model, reducing the contribution of the pion exchange. If this model is taken at face value, the chance to observe a trace of h_π is considerable decreased. Considering the present discrepancies between DDH-values and KM-values, the conservative choice of the PNC-matrix element above considered $\sim$ 0.1 eV is consistent with the DZ-model and is also supported by $\Delta T = 1$ PNC experiments[55].

VI. COMPARISON WITH EXPERIMENTAL DATA AND PREDICTION OF NEW EXPERIMENTS

An overview on experimental evidences concerning PNC observables is exhaustively done in Refs.[55] and [58]. We discuss here one of them, namely the PNC α-decay ^{16}O (2^-0, 8.87 MeV) → ^{12}C(g.s.) + α_0 experimentally reported with a very high experimental precision. This PNC α decay is sensitive only to the $\Delta T = 0$ part of the PNC interaction potential because both the initial and final states have T = 0. Adelberger and Haxton, in their last review paper[55] consider that nuclear structure theory is not adequate to take full advantage of the above mentioned experimental precision ($\Gamma_{PNC} = (1.03 \pm 0.28) \times 10^{-10}$ eV!)[60]. In our recent paper[31], however, we have shown that even in the case of not having parity mixed doublets (PMD), which simplify enormously [7,22,55] the nuclear structure calculations, by using nuclear structure models, as precise as OXBASH - code [32] is, we can describe experiments like above mentioned also.

In Table 2 by admixing to the 2^-0, 8.87 MeV level in ^{16}O the 2^+0 - levels shown in fig. 1 we could explain [31] rather well the experimental value, except for the KM-model which suppresses the PNC α-width by an order of magnitude. The calculations have been done within ZBM I - model and with the α-decay strength $\chi = 1.43 \times 10^6$ MeV -fm^9, which is taken to fit exactly the regular α-width for $^{16}O(2^+0$, 9.85 MeV) → ^{12}C(g.s.) + α_0 transition.

We have to mention that with such a choice of all the other regular α -transitions are described within less than 10% discrepancy as compared to the experimental data.

In our recent paper[22] we have chosen a $\Delta T = 1$ clean case for a favorable experiment, namely the resonance reaction $^{15}N(\vec{p},\alpha_0)^{12}C$ in which the 2^-1, 12.9686 MeV level in ^{16}O is populated. In this case the nuclear structure calculation is mainly reduced to the calculation of the PNC matrix element discussed in the preceeding section.

The PNC longitudinal (A_L) and transverse (A_b) analyzing powers (35) and (37) respectively, are functions of three resonance PC-T matrix ($T_{\alpha 101, p011}$; $T_{\alpha 101, p211}$; $T_{\alpha 202, p112}$); two resonance PNC-T-matrix ($T_{\alpha 202, p202}$; $T_{\alpha 202, p212}$) and one background PC-T-matrix ($T_{000, p110}$) elements. To calculate the unpolarized cross section (33) and normal analysing power (36) we need all the above mentioned T-matrix elements except the PNC-ones. All these T-matrix elements are expressed in terms of experimental observables[59] (proton energy, resonance energies, partial decay widths) and quantities (the PNC-matrix element, the background T-matrix element, the spectroscopic amplitudes and the proton-and α-phases ($\xi_{p\ell s}$, $\xi_{\alpha\ell s}$) calculated within nuclear structure (OXBASH) or reaction-codes or fitted to the experiment[61] (see for more details Ref. 22).

Fig. 3 (a,b,c,d) shows (on an expanded horizontal scale) the predicted size of the PNC longitudinal (A_L) and transverse (A_b) analyzing powers around the 2^-1, 12.9686 MeV resonance ($E_p \sim$ 898 KeV) level in ^{16}O, relevant for an experiment to determine the weak πN coupling constant (h_π).

These predictions are based on the size of 0.1 eV for the PNC matrix element, which is a conservative estimate, as can be verified e.g. from the Fig. 2. It should be mentioned that the most pesimistic KM-model predicts the energy anomaly in A_L for $150°$ equal to $\Delta A_L \sim 0.5 \times 10^{-5}$ within ZBM I and the most optimistic DDH-model predicts $\Delta A_L \sim 1.10^{-4}$ within ZWM, which seems to be in the possible range of the experiment[22].

Similar example is proposed in Refs.[7] and [20] by considering $^{13}C(\vec{p}, p)^{13}C$ resonance scattering of polarized protons in which the 0^+1, 8.618 MeV and 0^-1, 8.790 MeV levels in ^{14}N are populated. This is $\Delta T = 0$ case of PNC in which nuclear structure enhancements arise[7]. First the above mentioned levels constitute a 172 KeV PMD and secondly the ratio $\Gamma_p(0^-1)/\Gamma_p(0^+1) \sim 100$ acts as an enhancement factor in the PNC longitudinal analyzing power, which in this case is of the order$^{22)}$ of 2×10^{-5} within DDH - model (See Fig. 4) and $\sim 10^{-6}$ within KM-model. This case is a little less probable than the $^{15}N(\vec{p},\alpha_o)^{12}C$ case in which the $2^-1,12.9636$ MeV level in ^{16}O is populated.

An interesting case could be $^{15}N(\vec{p},p)^{15}N$ resonance scattering in which the $2^-1,12.9686$ MeV level in ^{16}O is populated.

Some preliminary estimations predict the PNC longitudinal analyzing power at least of the same order as compared to the $^{15}N(\vec{p},\alpha_o)^{12}C$ - case$^{22)}$ and $3 \div 4$ times greater than in the $^{13}C(\vec{p},p)^{13}C$ - case$^{7)}$. Such an enhancement factor could come from the ratio of the energy denominators ($\frac{\Delta E_C}{\Delta E_N} \sim \frac{172}{51}$), if bearing in mind that the ratio $\Gamma_{(2^+0)}/\Gamma_{(2^-1)} \sim 100$, is the same as in the $\vec{p} + ^{13}C$ case. The magnitude of the cross section is an important argument for performing the experiment. The elastic scattering cross section is always larger (~ 200 mb/Sr see Fig. 5) than the nuclear reaction cross section (~ 20 mb/Sr, see Refs.$^{22,61)}$. This could be an additional argument in favour of performing a measurement of the PNC analyzing powers for the resonance elastic scattering $\vec{p} + ^{15}N$.

VII. CONCLUSIONS

At the end it is worth summarizing some general features and properties which emerge from the already presented results.

Because of substantial experimental and theoretical difficulties in treating PNC effects for which $\Delta S = 0$, we have at this stage concentrated on the general broad outlines. All PNC-NN potentials can be divided in two groups, one with the isospin selection rule $\Delta T = 0,2$ (e.g. see Refs. 7, 20, 31, 55, 60) and the other allowing only $\Delta T = 1$ (e.g. see Ref. 22)). The most important contribution in the last group, is the pion exchange potential, whose strength coupling (h_π) is dominated by the neutral current contribution. It would be extremely useful to perform experiments which would measure the $\Delta T = 1$ alone (i.e. mainly h_π) as e.g. we proposed in Ref. 22). This type of experiments would definitely decide among quark picture[57,58] (and thus the contribution of the neutral currents) and solition picture[56].

ACKNOWLEDGMENTS

The author would like to thank Professor Luciano Fonda for many useful discussions and suggestions and also for the warm hospitality at Consorzium for Physics and ICTP, INFN and University of Trieste, where part of this work has been done.

He would also like to thank Professors Gunther Clausnitzer from Giessen University, Claudio Ciofi degli Atti from the INFN Roma for fruitful discussions concerning many aspects of this work. Many thanks to Professor Abdus Salam for the warm support at ICTP Trieste.

REFERENCES

1. G.Arnison et al., UA-1 Collaboration, Phys.Lett.B.122 (1983) 103.
   M.Banner et al., UA-2 Collaboration. Phys.Lett.B.122 (1983) 496.
2. N.Tanner, Phys. Rev. 107 (1957) 1203.
3. R.P.Feynman, M.Gell-Mann, Phys.Rev. 109 (1958) 193.
4. Yu.G. Abov, P.A.Krupchinsky, Yu.A.Oratovsky, Physics Letters 12 (1964) 25.
5. V.M.Lobashov, V.A.Nazarenko, L.F.Saenko, L.M. Smotriskii O.I.Kharkevitch, JETP Lett. 5 (1967) 59; Physics Letters B25 (1967) 104.
6. O.Dumitrescu, M.Gari, H.Kummel, J.G.Zabolitzky, Zeit. Naturforschungs 27A (1972) 733; Physics Letters 35B (1971) 19.
7. E.G.Adelberger, P.Hoodboy and B.A.Brown, Phys. Rev. C30 (1984) 456, C33 (1980) 1840.
8. Y.Masuda, KEK Preprint 88-123 (1989) H/M, Proc. 17th INS International Symposium on Nucl.Phys. at Intermediate Energy, Tokyo, Nov. 15-17 (1988).
9. A.K.Petuchov, G.A.Petrov, S.I.Stepanov, D.V.Nicolaev, T.D.Zvezdkina, V.I.Petrova, E.S.Markova, V.V.Ivanov and V.F.Morozov, JETP (Pisma) 32 (1980) 324; JETP Letters 32 (1980) 300.
10. V.A.Vesna, E.A.Kolomenskii, V.M.Lobashev, V.A.Nazarenko, A.N.Pirozhkov, L.M.Smotritzkii, Yu.Y.Sobolev and N.A.Titov, JETP (Pisma) 31 (1980) 704; JETP Letters 31 ((1980) 663.
11. V.A.Vesna, V.A.Knyazkov, E.A.Kolomenskii, V.M.Lobashev, A.N.Pirozhkov, L.A.Popeko, L.M.Smotritskii, S.M.Soloviev and N.A.Titov, JETP (Pisma), 36(1982) 169; 36(1982) 209; 35 (1982) 351; 35 (1982) 433.

12. M.Avenier, G.Bagien, H.Benkoula, J.V.Cavaignac, A. Idrissi, D.H.Koang and B.Vignon, Nucl. Phys. A 436 (1985) 83.
13. L.Fonda and R.Newton, Ann. Phys. (N.Y.) 10(1960) 490.
14. O.Dumitrescu and H.Kummel, Ann. Phys. (N.Y.) 71 (1972) 556.
15. O.Dumitrescu in "Interaction Studies in Nuclei" Eds. H.Jochim and B.Ziegler, North Holland, 1975, p.125.
16. O.Dumitrescu, Rev. Roum. Phys. 18 (1973) 277.
17. O.Dumitrescu, Fiz. Elem. Chastits At. Yadra 10(1979) 377; Sov. J.Part.Nucl. 10 (1979) 147.
18. C.Mahaux, H.A.Weidenmuller, "Shell Model Approach to Nuclear Reactions" (North Holland 1969).
19. A.M.Lane, R.G.Thomas, Rev. MOd. Phys. 30 (1958) 257.
20. O.Dumitrescu, M.Horoi, F.Carstoiu, G.Stratan, Phys.Rev. C41 (1990) 1562.
21. H.H.Barshall and H.Haerbelli "Polarization Phenomena in Nuclear Reactions" (University of Wisconsin Press, Madison, 1971).
22. N.Kniest, M.Horoi, O.Dumitrescu, G.Clausnitzer, "Isovector Parity Mixing in ^{16}O Investigated via the $^{15}N(\vec{p}, \alpha_o)^{12}C$ Resonance Reaction" , Preprint Giessen University (1990); Proceedings of the 1990 (July) Paris Conference on Polarization Phenomena in Nuclear Physics, to be published in Phys. Rev. C.
23. R.H.Dalitz and R.G.Moorhouse , Proc. Royal Soc. London Ser. A 318 (1970) 279.
24. C.J.Goebel and K.W.McVoy, Phys. Rev. 164 (1967) 1932.
25. L.Fonda "Fundamentals in Nuclear Theory", IAEA, Vienna 1967, p.793.
26. L.Fonda, "Many-Channel Collision Theory", in "Scattering Theory" p.129 (editor A.O.Barut, Gordon and Breach, N.Y. 1969).

27. L.Fonda, "Basic Concepts on Nuclear Reactions" in "Winter College on Fundamental Nuclear Physics" Eds. K.Dietrich, M. di Toro and H.J.Mang - World Scientific (1985) Vol. I, p.129.
28. I.Brandus, F.Carstoiu, O.Dumitrescu, M.Horoi and F.Nichitiu, Rev. Roum. Phys. 35 (1990).
29. A.Bulgac and O.Dumitrescu, Progrese in Fizica (ICEFIZ, Bucharest, 1979) I p.30.
30. A.Bulgac, F.Carstoiu, O.Dumitrescu, S.Holan, Proc.1981 Trieste Nucl. Phys. Workshop Eds. C.H.Dasso, R.A.Broglia, A.Winther (North Holland, Amsterdam, 1982) p. 295; Nuovo Cimento A 70 (1982) 142.
31. F.Carstoiu, O.Dumitrescu, G.Stratan, M.Braic, Nucl.Phys. A 441 (1985) 221.
32. B.A.Brown, A.Etchegoyen and W.D.M.Rol, MSU-NSLL, Report 524 (1985).
33. B.D.Day, Rev. Mod. Phys. 39 (1967) 719; 50(1978) 495.
34. R.Rajaraman and H.A.Bethe, Rev. Mod. Phys. 39 (1967) 745.
35. H.Kummel, Nucl. Phys. A 146 (1971) 205.
36. H.Kummel, K.H.Luhrman, J.G.Zabolitzky, Physics Reports, 36 (1978) 1.
37. J.M.Irvine, Rep. Progr. Phys. 51 (1988) 1181.
38. B.A.Brown and B.H.Wildenthal, Ann. Rev. Nucl. Part. Sci. 38(1988) 29.
39. A.P.Zuker, B.Buck and J.B.McGrory, Phys.Rev.Lett. 21(1968) 39.
40. T.T.S. Kuo and G.E.Brown, Nucl. Phys. 85 (1966) 40.
41. J.B.McGray and B.H.Wildenthal, Phys. Rev. $\underline{C7}$ (1973) 954; Ann. Rev. Nucl. Sci. 30 (1980) 383.
42. S.Cohen and D.Kurath, Nucl. Phys. A73(1965) 1.
43. O.Preedom and B.H.Wildenthal, Phys. Rev. C6(1972) 1633.
44. D.J.Millener and D.Kurath, Nucl. Phys. A255 (1975) 315.
45. B.H.Wildenthal, Progr. Part. Nucl. Phys. 11(1984) 5; Ed. D.H.Wilkinson, Pergamon - Oxford.

46. D.H.Gloekner and D.R.Lawson, Phys. Lett. 53B (1974) 313.
47. J.B.McGrory and B.H.Wildenthal, Phys. Lett. 60B (1975) 5.
48. C.Ciofi degli Atti, Nucl. Phys. A129 (1969) 350.
49. M.Gari, H.Hyuga, J.G.Zabolitzky, Nucl. Phys. A271(1976) 365.
50. H.A.Bethe Ann. Rev. Nucl. Sci. 21 (1971) 93.
51. E.Feenberg, "Theory of Quantum Fluids" (1969, New York, Academic) 8 and 9.
52. J.G.Zabolitzky, Phys. Rev. A16 (1977) 1258.
53. G.A.Miller and J.E.Spencer, Ann. Phys. (N.Y.) 100 (1976) 562.
54. M.Gari, Physics Reports C6 (1973) 317.
55. E.G.Adelberger and W.C.Haxton, Ann. Rev. Nucl. Part. Sci. 35 (1985) 501.
56. N.Kaiser and U.G.Meissner, Nucl. Phys. A499 (1989) 699.
57. V.M.Dubovik and S.V.Zenkin, Ann. Phys. (N.Y.) 172 (1986) 100.
58. B.Desplanques, J.F.Donoghue and B.R.Holstein, Ann. Phys. (N.Y.) 124 (1980) 449.
59. F.Ajzenberg - Selove, Nucl. Phys. A460 (1986) 1.
60. K.Neubeck, H.Schober and H.Wafler, Phys. Rev. C 10 (1974) 320.
61. G.H.Pepper and L.Brown, Nucl. Phys. A260 (1976) 163.

TABLE CAPTIONS

Table 1 : Weak meson-nucleon coupling constants within different weak interaction models (in units of 10^{-7}). The abbreviations are given in the text.

Table 2 : PNC α-width for the ^{16}O (2^-0, 8.87 MeV) → ^{12}C(g.s.) + α_o transition in the framework of different weak interaction models (in units of 10^{-10} eV). The abbreviations are given in the text. The experimental value is $\Gamma_{PNC} = (1.03 \pm 0.28) \times 10^{-10}$ eV.

TABLE 1

ΔT h_{meson}	AH[55]	KM[56]	DDH[58]	DZ[57]
h_π^1	2.09	0.19	4.54	1.3
h_ρ^0	-5.77	-3.70	-11.40	-8.3
h_ρ^1	-0.22	-0.10	-0.19	0.39
h_ρ^2	-7.06	-3.30	-9.50	-6.7
$h_\rho^{1\prime}$	0	-2.20	0	0
h_ω^0	-4.97	-1.40	-1.90	-3.9
h_ω^1	-2.39	-1.00	-1.10	-2.2

TABLE 2

Model	PNC α width
AH	0.526
KM	0.093
DDH	0.523
DZ	0.566

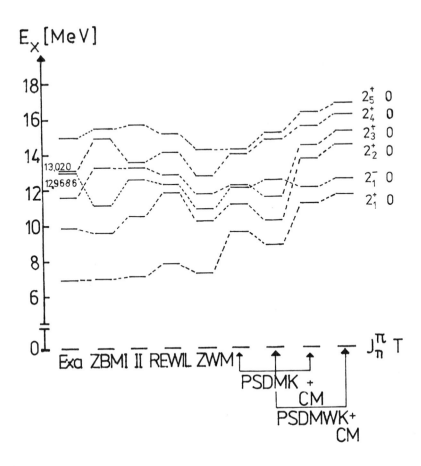

Fig. 1. The first five 2^+0 and the first 2^-1 excited levels in ^{16}O, taken from experiment [59] and calculated within different models of the OXBASH-code as explained in the text.

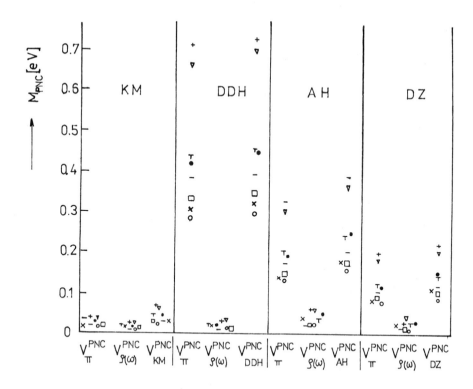

Fig. 2. The PNC matrix element between 2^-1 12.9686 MeV and 2^+0, 13.02 MeV states of ^{16}O calculated within different strong and weak interaction models.

The abbreviations are given in the text. The symbols refer to the following models:

o ZBM; I	ZBM II;	REWIL
+ ZWM	— PSDMK;	x PSDMK + CM;
T PSDMWK;	• PSDMWK+CM	

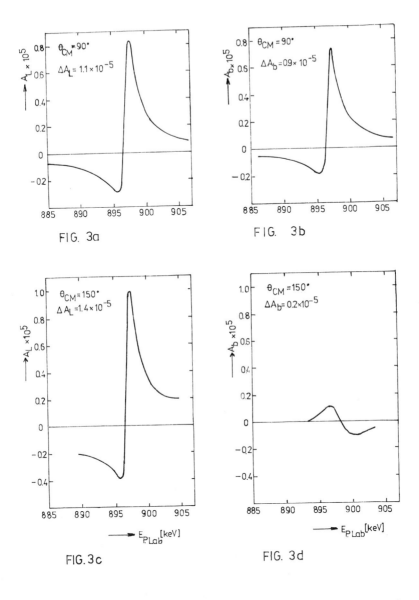

Fig. 3. PNC analyzing powers versus proton energy around 2^-_1, 12.9686 MeV ($E_p \simeq$ 898 keV) resonance level in ^{16}O;

a) A_L at $\theta = 90°$; b) A_b at $\theta = 90°$
c) A_L at $\theta = 150°$; d) A_b at $\theta = 150°$

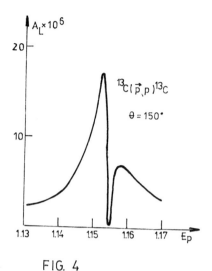

Fig.4. The irregular PNC analyzing power for elastic resonance scattering of polarized protons by ^{13}C.

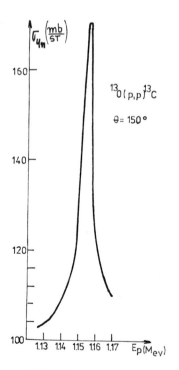

Fig.5. The cross section of the elastic scattering of unpolarized protons by ^{13}C.

CLUSTER RADIOACTIVITIES IN VARIOUS REGIONS OF PARENT NUCLEI

Dorin N. POENARU
Institute of Atomic Physics, P. O. Box MG-6, RO-76900 Bucharest, Romania
and
Institut für Theoretische Physik der Universität, Frankfurt am Main, Germany

Abstract

The main results obtained within analytical superasymmetric fission model for the region of parent nuclides with measured masses are briefly reviewed. Theoretical predictions are compared with experimental confirmations. New regions of cluster radioactivities (CR) far from the line of β-stability are found by using the mass tables published in 1988 in order to calculate the released energy. The cluster preformation probability is interpreted as a penetrability of the inner (prescission) part of the barrier. Consequently a preformed cluster model is equivalent with a fission model. A universal curve for each kind of cluster radioactivities of even-even nuclei is derived. The logarithm of symmetric spontaneous fission halflife calculated in the framework of the liquid drop model is a linearly decreasing function of fissility, but another dependence is obtained for asymmetric fission. Fissility is not a suitable parameter to describe any asymmetric fission.

1. INTRODUCTION

Nuclear disintegrations by spontaneous or beta-delayed emission of charged or neutral particles from nuclei have been studied intensively [1, 2] during the last decade. In 1984 Rose and Jones measured the ^{14}C radioactivity of ^{223}Ra, confirming our predictions of 1980 that ^{14}C should be the most probable emitted cluster from $^{222,224}Ra$. The unified approach of cluster radioactivities, cold fission and α decay had been reviewed recently [3, 4].

Spontaneous fission had been discovered in 1940 by Petrjak and Flerov, shortly after induced fission. Nevertheless, some properties including fragment mass asymmetry, have been explained only by taking into account both collective and single-particle nucleon motion [5]. Fragmentation theory and the asymmetric two center shell model [6] have been particularly succesful in this respect.

From the mass asymmetry parameter point of view, cluster radioactivities (CR) are intermediate phenomena between fission and α-decay. Consequently one natural way to develop the theory was to extend toward extremely large mass asymmetry what was known for almost symmetric fission. Three of the four models used in 1980 to predict CR, are based on this philosophy [2 - 4]. In the same time, we had also been using the other alternative - to apply to heavier clusters the traditional theory of a preformed α - particle.

There are many historical accounts on the development of CR and of the other kinds of nuclear disintegrations (for example Refs. [2, 7 - 9]). We made the first predictions of the nuclear lifetimes against spontaneous cluster emission by using our analytical superasymmetric fission model (ASAFM), developed since 1980. In 1984, before any other model was published, we gave the estimations of the half-lives and branching ratios relative to α decay for more than 150 decay modes, including all cases experimentally confirmed until now on ^{14}C emission from $^{222,224,226}Ra$; $^{24-26}Ne$ emission from ^{230}Th , ^{231}Pa and $^{232-234}U$; $^{28,30}Mg$ emissions from ^{234}U and $^{236,238}Pu$; and ^{32}Si radioactivity of ^{238}Pu . A comprehensive table [10] was produced by performing calculations within this model. ASAFM has been improved by taking properly into account the even-odd effect [11] and the variation of the correction energy [12] with the mass number, A_e, of the emitted cluster for $A_e > 24$. Cold fission fragments have also been considered in a new edition of the tables [13].

Recently, the half-life estimations within ASAFM, have been updated [14] according to the last improvements of the model and the calculations have been extended in the regions of nuclei far from stability and superheavies, by using the 1988 mass tables as input data.[1] In the present paper we shall present some of the results obtained in this way.

The close connection of CR with cold fission phenomenon [3] and its inverse process - the cold fusion [15] allowing to produce the heaviest elements [15 - 17], was realized very soon [4, 18 - 20]. A unified description of α-decay, CR and cold fission [3] is best illustrated on ^{234}U nucleus for which all of these decay modes have been measured.

Some authors claimed that the observed events could be produced by ternary fission. By examining a typical energy spectrum measured in such a process it is evident that any misinterpretation of CR as a particle accompanied fission should be ruled out due to experimental evidence of cluster monoenergeticity, demonstrating the two-body character of the output channel.

Several fission or preformation cluster models have been developed [21 - 36] since 1985, explaining what was already measured and (some of them) making predictions. P. B. Price did recently [37] a quantitative comparison between various theories and experiments. The overall uncertainty of the ASAFM was found to be the smallest among the unified (fission) models.

During the last two years, new experiments have been reported [38 - 43]. An elaborated study of the fission dynamics in a wide range of mass asymmetry [44 - 46]

---

[1]The quantity $\log T_t$ in the table of the Ref. [14] (pp. 12 - 153) should be divided by 10.

led us to the conclusion that some confusions and mistakes have been made in the literature. One should care about center of mass motion when the inertia tensor is calculated. Also, the cluster-like shapes (intersected spheres with a constant radius R_2 of the light fragment) are more suitable (the action integral takes lower values) than the more compact ones (intersected spheres with constant volumes of both fragments) for $A_e < 34$, and the smoothed neck influence is stronger for lower mass asymmetry.

We have compared [47] fission and preformation cluster models and our predictions with experimental data [37, 48 - 50], pointing out that at present there is no method available allowing to measure preformation probabilites or quantum penetrabilities, in oder to check the validity of the two kinds of theories. As an argument for the equivalence of these models, in the following we present a new interpretation of cluster preformation probability : it is the penetrability of the inner (prescission) part of the barrier, within a fission theory. Also we shall discuss whether the parameter Z^2/A proportional to the fissility is suitable or not to be used as a variable in a plot of nuclear lifetimes against cluster radioactivities.

2. CONFIRMED HALFLIVES

The nuclear half-life, T, for cluster radioactivities is estimated in the framework of the ASAFM with even-odd effects taken into account. The branching ratios $B = T_\alpha/T$, relative to α decay are calculated by using the measured α decay half-lives, T_α, selected in Ref. [51], or our semiempirical formula based on the fission theory of α decay, developed in 1980 (see Refs. [2] or [10]).

From the energetic point of view the spontaneous cluster emission is allowed if the released energy

$$Q = [M - (M_e + M_d)]c^2 \qquad (1)$$

is a positive quantity. In this equation M, M_e, M_d are the atomic masses of the parent, emitted and daughter nuclei and c is the light velocity.

Classically the process is forbidden due to the potential barrier. It is essentially a quantum-mechanical phenomenon taking place by tunnelling through the barrier. Within one-dimensional semiclassical WKB theory, the barrier penetrability, P, is well approximated by $\exp(-K)$, where K is the action integral along the fission path.

The half-life of a parent nucleus AZ against the split into a cluster $A_e Z_e$ and a daughter $A_d Z_d$, is calculated with analytical relationships derived from

$$T = [(h \ln 2)/(2E_v)] exp(K) \qquad (2)$$

where the action integral

$$K = \frac{2}{\hbar} \int_{R_a}^{R_b} \{2\mathcal{B}(R)[(E(R) - E_{cor}) - Q]\}^{(1/2)} dR \qquad (3)$$

in which h is the Plank constant, $\mathcal{B}(R)$ is the nuclear inertia (equal to the reduced mass $\mu = mA_eA_d/A$ for separated fragments), m is the nucleon mass, $E(R)$ is the interaction energy of the two fragments separated by the distance R between centers, and R_a, R_b are the turning points of the WKB integral.

The two parts of the action integral, corresponding to the overlapping and separated fragments, respectively, are given explicitly in the Ref. [10], or (in an equivalent improved version) in Ref. [2].

Table 1: Comparison Between Measured and Calculated Half-lives

Parent		Emitted		Daughter		$\log T(s)$		
						ASAFM		Experiment
Z	A	Z_e	A_e	Z_d	N_d	Early	Ref. 13	
88	222	6	14	82	126	12.6	11.2	11.02 ± 0.06
88	223	6	14	82	127	14.8	15.2	15.20 ± 0.05
88	224	6	14	82	128	17.7	15.9	15.90 ± 0.12
88	226	6	14	82	130	22.4	21.0	21.33 ± 0.20
90	230	10	24	80	126	24.9	25.3	24.64 ± 0.07
91	231	10	24	81	126	22.0	23.4	23.38 ± 0.08
92	232	10	24	82	126	20.4	20.8	21.06 ± 0.10
92	233	10	24	82	127	23.1	24.8	24.82 ± 0.15
92	233	10	25	82	126	23.3	25.0	
92	234	10	24	82	128	25.7	26.1	25.25 ± 0.05
92	234	12	28	80	126	24.6	25.8	25.75 ± 0.06
94	236	12	28	82	126	19.8	21.0	21.68 ± 0.15
94	238	12	28	82	126	24.8	26.0	25.70 ± 0.25
94	238	12	30	82	128	24.4	25.7	
94	238	14	32	80	126	23.7	25.1	25.30 ± 0.16

It is known that the fission barrier heights are too large within liquid drop model. E_{cor} is a correction energy including contributions coming from both the statics and the dynamics of the process. $E_v = h\nu/2$ is the zero point vibration energy. For practical reasons (to reduce the number of fitting parameters) we took $E_v = E_{cor}$ though it is evident that, owing to the exponential law, any small variation of E_{cor} induces a large change of T, playing a more important role compared to the preexponential factor variation due to E_v.

The energy conservation is expressed by

$$E_k = QA_d/A \qquad (4)$$

where E_k is the kinetic energy of the light fragment.

Both shell and pairing effects have been included in E_{cor} in order to obtain the best agreement with experimental results :

$$E_{cor} = a_i(A_e)Q \quad (i = 1, 2, 3, 4) \quad (5)$$

From a fit with about 380 α-emitters selected in four groups according to the even-odd character of the proton and neutron numbers, we have obtained four values of the coefficient a_i - the largest for even-even nuclei and the smallest for odd-odd ones. The coefficients decrease smoothly for heavier emitted ions up to $A_e = 50$ and then increase slightly toward $A_e = 100$. The present correction energies for $A_e > 24$, are different from those adopted in Ref. [13], by the smooth character of the extrapolation [12].

Several successful experiments have been performed since 1984. The results compiled by Price [37] are presented in Table 1, where the recently performed experiments [42] on ^{28}Mg emission from ^{236}Pu is also included. They are compared with our early predictions and with those [13] of 1986, after taking into account the even-odd effect. One can see the good agreement between the calculated and experimentally determined half-lives in a range of 15 orders of magnitude, though at the beginning we gave optimistic numbers for $A_e \geq 24$.

The strong shell effect predicted by the theory has been confirmed. It is evident from the variation of the half-life for ^{14}C radioactivity of Ra isotopes with the neutron number, N_d of the daughter and from the variation with both Z_d and N_d for ^{24}Ne radioactivity. The even-odd effect is also very clearly seen for ^{14}C radioactivity of Ra isotopes.

3. PARENT NUCLEI FAR FROM STABILITY

The released energy, Q, given by equation (1) has to be known with a very good accuracy. The action integral decreases when Q-value increases, producing a large variation of the half-life due to the exponential dependence on K. Owing to this large sensitivity we need the massess of the three partners as close as possible to the true values. For this reason we adopted the following strategy. The measured masses [53] of 1659 nuclides (with a code number C = 0) and those of 679 nuclides determined from the systematics (C = 1), are used with the first priority for any of the three partners. When a nuclide is not present on the table of Wapstra et al, we try to use other table from Ref. [52], adopting conventionally the following mass codes : C = 3 for the 3571 nuclides of the Ref. [54] (beyond the Wapstra et al region); C = 2 for the 2158 nuclides of the Ref. [55]; C = 4 for the 2759 nuclides of the Ref. [56]; C = 5 for the 4868 nuclides of the Ref. [57]; C = 6 for the 1224 nuclides of the Ref. [58]; C = 7 for the 4227 nuclides of the Ref. [59]; C = 8 for the 2270 nuclides of the Ref. [60], and C = 9 for the 2270 nuclides of the Ref. [61]. According to the figure 9 of the Ref. [62], one can expect for C = 3, mass values very close to the real ones.

The number of input masses, obtained by adding the above mentioned contributions of various mass tables is quite large : 25685 ! By multiplying with about 200 possible candidates to be emitted we get a total number of combinations parent - cluster of the order of 5×10^6. Consequently we need some selection criteria.

By taking into account the presently available techniques allowing to measure long half-lives and small branching ratios, we have selected only those parent nuclei leading to $\log T(s) < 35$, and maximum branching ratios $B = T_\alpha/T > 10^{-18}$.

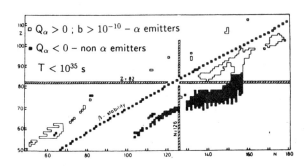

Figure 1: Regions of CR with estimated branching ratios relative to α - decay larger than 10^{-10} at least for one value of the mass code C.

We get the results plotted in Figure 1 if we select further only the parent nuclides for which the branching ratio $B \geq 10^{-10}$, at least for one C-value. As can be seen, except for some isolated cases, there are mainly five important islands of CR with high branching ratios relative to α-decay : three of α-emitters (one on the neutron deficient side, and two on the neutron rich side) and two of nuclides stable against α-decay on the neutron rich side.

3.1. Cluster radioactivities of α-emitters with $Z < 104$

The most probable clusters emitted from nuclei are grouped in a way which assures a number of protons and/or neutrons of the daughter equal or very close to a magic number. For example such a shell effect is present at N = 88 and 90, where for ^{12}C and ^{16}O emissions respectively, one has $N_d = 82$, and at Z = 56 and 58, where $Z_d = 50$ for the same clusters.

In a similar way, the effect of $N_d = 126$ can be seen at N = 130 for ^8Be, at N = 134 for ^{14}C, at N = 140 for ^{24}Ne, at N = 142 for ^{28}Mg, at N = 146 for ^{34}Si, etc. The proton magic number $Z_d = 82$ appear somewhat less pronounced at Z = 88 for C, at Z = 90 for O, at Z = 92 for Ne, at Z = 94 for Mg and at Z = 96 for Si emission.

A pairing effect is also present - the number of even Z and even N parent nuclei is larger than that of the corresponding ones with odd N and Z.

The decimal logarithm of the halflife expressed in seconds and of the branching ratio of some CR are given in Table 2. From the parent nuclei of Fig. 1, we have selected those for which there is no discrepancy from mass table to mass table concerning the most probable emitted cluster. For example in all cases, from $^{120}_{58}Ce$ the most probable emitted cluster is ^{16}O. Nevertheless, Q and Q_α are different, hence $logT(s)$ takes a whole range of about 8 units from 13.1 to 20.8 . The corresponding $logB$ lies in a much larger interval (of about 24 units) from -5.5 to 18.3.

Table 2: Cluster radioactivities of some α - emitters

				a		b		c		d		e	
Z	A	Z_e	A_e	lgT	-lgB	lgT	-lgB	lgT	-lgB	lgT	-lgB	lgT	-lgB
56	114	6	12	10.6	7.2	9.7	8.8	9.5	9.1	12.1	4.0	7.9	11.0
	115			14.7	6.4	15.2	5.0	13.6	8.9	15.8	3.2	12.3	11.0
	116			15.5	8.3	16.6	5.9	14.3	10.5	16.0	7.2	14.3	10.5
58	118	8	16	11.9	5.2	14.5	-16.7	10.6	5.6	12.3	8.5	11.0	-2.7
	119			15.9	7.5	20.0	-16.5	15.4	5.4	16.5	9.1	15.3	-0.2
	120			16.6	5.5	20.8	-18.3	15.9	3.6	17.3	4.7	16.6	0.2
59	124	8	16	27.2	-7.3	33.2	-6.7	27.7	-3.9	28.2	0.6	26.0	5.7
	125			28.4	-3.4	34.7	-3.8	28.3	-1.3	29.4	5.4	27.1	2.8
62	127	14	28	20.9	6.1	24.9	7.4	20.8	3.3	21.2	9.5		
	128			21.7	8.8	25.4	9.6	21.3	5.0	22.0	12.7		
93	225	6	12	10.2	13.1	13.5	9.9	11.0	12.5	10.7	12.8	10.1	13.2
94	226	6	12	8.8	13.0	12.7	14.2	10.4	14.6	9.4	13.4	9.7	10.0

a- C=0, 1; b- C=3;c- C=4; d- C=6; e- C=7.

A positive value for $logB$ means CR more probable than α-decay. Very likely Q_α-values obtained for Ce isotopes by using the atomic masses from Spanier and Johansson are too low, leading to extremely large T_α.

When the neutron number of the daughter increases over the closed shell value (50 or 126), the halflives became longer and longer.

Following emitted clusters have been noticed as the most probable: ^5He ; $^{8,10}Be$; $^{12,16}C$; $^{15,16,20-22}O$; ^{23}F ; $^{24-26}Ne$; $^{24,28-30}Mg$; $^{31,32}Al$; $^{28,33,34,36}Si$; ^{37}P ; $^{40,43-46}S$; ^{45}Cl ; $^{46-48}Ar$; $^{49,50}K$; $^{50-53}Ca$; ^{53}Sc ; $^{55,70}Fe$; ^{56}Co and $^{58,74,76-79}Ni$. Almost all are neutron rich nuclei.

For a given combination (parent - emitted cluster), we get various halflives by using different mass tables to calculate the released energy Q. A small difference in the mass value of one, two or three partners (parent and two fragments) obtained with different mass formulas, produces corresponding shifts in the Q_α and Q-values and induces a large variation of the lifetimes T_α and T. The dispersion of the branching ratios, B, is of course much higher. In some cases even the most probable emitted cluster may differ from table to table. When the mass of both daughter and α - daughter nuclei are available on the Wapstra et al. table (see for example Figure 2a), only the parent mass is changed. Another typical example is shown in Figure 2b, corresponding to a change of both parent and α - daughter masses.

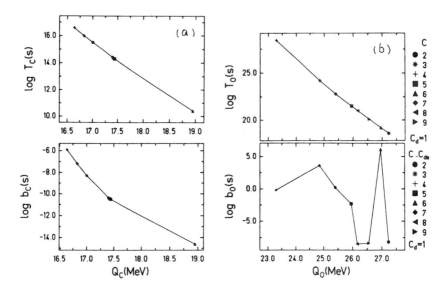

Figure 2: Variation of the halflife and branching ratio for ^{12}C decay of ^{116}Ba ($C_d = C_{d\alpha} = 1$) (a), and ^{16}O radioactivity of ^{124}Nd ($C_d = 1$) (b), with the Q-value for cluster emission calculated by using different mass tables.

In spite of the large discrepancies from one C-value to another one, it seems that ^{12}C and ^{16}O radioactivities have a good chance to be detected in some very neutron deficient Ba, Ce or Pr isotopes. Other possible candidates can be found in Table 2 and in Ref. [14].

3.2. Superheavy nuclei

Two mass tables [60, 61] are extended in the region of superheavy nuclei up to Z = 122 and N = 196. They have been used to calculate Q-values.

For nuclides with $Z \geq 104$ and $Q_\alpha > 0$, there are two groups of emitted clusters :

1) ^8Be and $^{12,14}C$;

2) ^{52}Ca ; ^{54}Ti ; ^{55}V ; $^{56-58,60}Cr$; ^{59}Mn ; $^{58,60,62,64,68,76}Fe$; ^{77}Co and $^{66-80}Ni$.

The first group of lighter clusters is frequently met in the neutron deficient region of nuclides and the second group is located mostly around the β-stability line and on the neutron rich side. ^8Be and the neutron rich Ni isotopes, respectively, are the main representatives of these two groups.

Table 3: Cluster emission from some superheavy nuclei

Z	A	Z_e	A_e	E_α	lgT_α	lgT	$-lgB$	E_α	lgT_α	lgT	$-lgB$
				\multicolumn{4}{c	}{C = 8}	\multicolumn{4}{c	}{C = 9}
112	304	28	78	6.07	13.7	24.2	10.5	6.70	10.3	23.6	13.3
	306			6.54	11.1	23.7	12.5	6.96	9.0	23.1	14.1
	308			6.23	12.8	23.8	11.0	6.61	10.7	23.4	12.6
119	299	4	8	11.55	-2.8	10.1	12.9	12.05	-3.9	8.1	12.1
120	300			12.43	-4.4	7.1	11.6	12.95	-5.5	5.2	10.8
	304			12.48	-4.5	8.6	13.2	13.04	-5.7	6.6	12.3
	305			12.08	-3.0	10.7	13.8	12.65	-4.3	8.7	13.0
121	305			12.32	-3.9	8.7	12.6	12.87	-5.0	6.7	11.8
	306			11.91	-1.5	10.8	12.4	12.46	-2.9	8.8	11.7
	307			11.26	-2.0	10.7	12.7	11.82	-3.3	8.6	11.9
122	305			12.32	-2.9	9.2	12.1	12.86	-4.1	7.2	11.4
	306			12.41	-3.8	8.7	12.5	12.93	-4.9	6.7	11.6

Table 4: Cluster emission from some nuclides stable against α-decay

Z	A	Z_e	A_e	Q_α	Q	lgT	C	Q_α	Q	lgT	C
56	160	6	16	-5.47	13.5	31.9	2				
58	162	8	22	-5.74	23.6	30.0	2				
60	168	10	26	-6.81	30.8	34.8	2				
66	186	16	46	-9.44	72.8	17.8	2				
67	181			-4.53	65.4	34.2	6				
70	196	20	56	-10.50	87.0	29.7	2				
72	204	28	78	-10.44	118.7	20.5	2	-3.09	115.5	25.2	3
73	205			-8.77	117.6	27.1	2	-3.01	116.5	28.7	3
75	223	26	76	-10.79	119.4	23.3	3	-4.98	115.1	29.9	7
76	212	28	78	-7.34	123.5	29.6	2	-4.49	122.9	30.4	3
	222	26	76	-8.98	119.7	26.3	3	-3.51	115.3	33.1	7
77	225	27	77	-8.88	122.5	30.3	3	-3.78	122.5	30.4	7
78	216	28	78	-6.07	126.9	32.4	2	-3.81	125.6	34.3	3

Some of the CR of superheavy nuclei are given in Table 3. Maximum emission rates are expected in this region from α-decay and spontaneous fission. If the masses used to compute Q-values are reliable enough, one can conclude that CR are not responsible for the fact that superheavies are still not found.

3.3. Cluster radioactivities of nonα-emitters

The neutron rich nuclides which are stable relative to α-decay (see some examples in Table 4) are not as good cluster emitters as the neutron deficient α-emitting nuclei.

The most probable emitted clusters : $^{16,22}O$; ^{26}Ne ; ^{28}Si ; $^{44,46,48}S$; $^{50,52}Ar$; ^{53}K ; ^{56}Ca ; ^{62}Ti ; ^{74}Cr ; ^{75}Mn ; $^{74,76}Fe$; $^{75-78}Co$ and $^{57.58,74,76-82,84,86}Ni$ are further away from the line of β-stability, compared to the previously discussed two cases.

There is a very steep decrease of the lifetime when the neutron number increases (a trend which is reversed in comparison with that from the Table 2), if the emitted cluster is lighter than Fe, and a much smaller variation for heavier clusters.

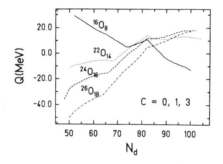

Figure 3: Variation of the Q-value for emission of various Oxygen isotopes from the corresponding Cerium parents, with the neutron number of the daughter

This behaviour may be explained easily by taking into account that the Q-value is a decreasing function of N_d when the emitted cluster is not very neutron-rich, and it is an increasing one when the emitted cluster has a large number of neutrons (see Figure 3).

In conclusion, in spite of the high uncertainties originating from the lack of a very precise mass formula, one can draw some reliable conclusions from the study of CR emissions in the regions of parent nuclides far off the line of β-stability.

Cluster emitters with measurable branching ratios relative to α-decay could be found not only around the line of β-stability, but also far removed from this line, on the neutron deficient side, e.g. ^{12}C, ^{16}O and ^{28}Si radioactivities of some Ba (A = 114 - 116), Ce (A = 118 - 125) and Sm(A = 127, 128) isotopes, respectively. In this region lifetimes decrease when the neutron number of the parent decreases.

Very likely CR are not responsible for the negative results in the search for superheavy nuclei. Nevertheless some ^8Be and $^{12,14}C$ radioactivities are expected for neutron deficient nuclei with $Z > 106$. Also around $Z = 110$, $N = 176$ there should be an island of $^{78}_{28}Ni_{50}$ emission because the corresponding daughter is the doubly magic ^{208}Pb and one can believe that the neutron rich fragment ^{78}Ni is doubly magic too.

Neutron rich parents far removed from the line of β-stability, with $Z = 58 - 78$ are estimated to be stable with respect to α-decay, but they could exhibit CR with emitted particles having very high neutron excess. Unlike in the neutron deficient region, the lifetimes are decreasing functions of the parent neutron number.

Some of the decay modes by cluster emission mentioned above have not been predicted previously, when the region of interest had been confined to nuclides with measured masses. The following clusters : $^{44-50}S$; $^{47-51}Cl$; $^{48-54}Ar$; $^{52-55}K$; $^{56,58}Ca$; $^{62,64}Ti$; $^{57-60}Cr$; ^{61}Mn ; $^{64-78}Fe$; $^{69-81}Co$ and $^{70-86}Ni$ with extremely high neutron excess, are among such new radioactivities.

4. FISSION and PREFORMATION MODELS

In our papers[63, 64] of 1980 we presented three fission models and (a part of) one preformed cluster model. The other approaches[21 - 36] developed since 1985 are either of one of these two categories, or they are hybrides with elements from both.

4.1. Decay constant and penetrability

The essential difference[47] between a fission model (see for example Refs. [3, 21, 22, 26]) and a "pure" preformed[2] cluster model (Ref. [27, 28]) appears in the expression of the decay constant :

$$\lambda_f = \nu_f P_f \; ; \; \lambda_p = \nu_p S P_p \qquad (6)$$

where the subscripts f and p denote fission and preformation, respectively, ν is the frequency of assaults on the barrier (the zero-point vibration energy $E_v = h\nu/2$ with h the Planck constant), P is a potential barrier penetrability, and S is the preformation probability (to get a preformed cluster at the nuclear surface).

Alternatively, in a hybrid model,[24] the Fermi's Golden Rule for the cluster transition rates is used to compute the decay constant

$$\lambda = \frac{2\pi}{\hbar} |<\psi_E | \psi_0(\xi_n) >|^2 v^2 | \psi_0(\xi_{n-1})|^2 \frac{dN}{dE} \qquad (7)$$

where the initial state is described by the eigenfunction $\psi_0(\xi_n)$ of the Hamiltonian at the touching configuration and ψ_E is a continuum wavefunction of the fragments

---
[2]We have to make this distinction, because some of the models using a potential barrier shape typical for preformed clusters, assume $S = 1$. It is very difficult to understand this hypothesis.

treated as particles. The matrix element connecting these two states is $v = -\Delta^2/(4G)$ with Δ the pairing gap and G the pairing force coupling constant. $|\psi_0(\xi_{n-1})|^2$ is the probability that the system deforms into a configuration just one step in ξ previous to ξ_n. The variable ξ is zero at the parent ground state and 1 at the touching point. $dN/(dE)$ is the density of final states.

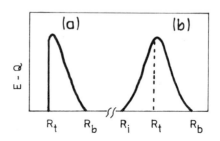

Figure 4: Typical potential barrier shape within a preformed cluster (a) and a fission (b) models.

In a preformed cluster model (Fig.4a) the potential barrier $E(R)$ is thinner than in a fission model; it is mainly of Coulomb nature and extends from the "channel radius" $R_t = R_1 + R_2$ up to the outer turning point $R_b = Z_1 Z_2 e^2/Q$ (where e is the electron charge) if spherical shapes of nuclei are assumed. Usually the finite cluster extension is not taken into account. We are using the subscript 2 or e for the emitted cluster and 1 or d for the daughter. The hadron numbers are conserved : $A = A_1 + A_2$ and $Z = Z_1 + Z_2$.

By considering the finite radius R_2 of the light fragment, the potential barrier in a fission model (Fig. 4b) has an important inner part, coming from the early stage of the deformation process, when the fragments are not separated (overlapping or prescission region), from the initial $R_i = R_0 - R_2$ to the touching point (or scission) configuration R_t. For heavier clusters the relative width of this inner part is broader.

Except some few models using the wavefunction ψ in order to find the penetrability as $|\psi|^2$, both P_f and P_p are usually calculated within semiclassical WKB approximation :

$$P_f = \exp(-\frac{2}{\hbar} \int_{R_i}^{R_b} \{2B(R)E(R)\}^{1/2} dR) \tag{8}$$

$$P_p = \exp(-\frac{2}{\hbar} \int_{R_t}^{R_b} \{2\mu E(R)\}^{1/2} dR) \tag{9}$$

where $\mathcal{B}$ is the nuclear inertia (the effective mass), equal to the reduced mass $\mu = mA_e A_d/A$ for separated fragments, m is the nucleon mass, E(R) is the interaction energy of the two fragments separated by the distance R between centers, from which the Q-value has been subtracted out.

4.2. Equivalence of the two kinds of models

In our fission models we have always made a distinction between the penetrabilities P_{ov} and P_s, by expressing P_f as a product of two probabilities : $P_f = P_{ov}P_s = \exp[-(K_{ov} + K_s)]$. P_{ov} corresponds to the inner part of the barrier (overlapping or prescission stage) and is calculated by changing R_b with R_t in the eq. (8). By substituting R_i with R_t and $\mathcal{B}(R)$ with μ in the same equation, we get the expression for P_s - the penetrability in the outer part of the barrier (region of separated fragments or postcission stage). The equations (6) look now

$$\lambda_f = \nu_f P_{ov} P_s \; ; \; \lambda_p = \nu_p S P_p \qquad (10)$$

Both ν_f and P_s from a fission model, have their counterparts ν_p and P_p in a preformation cluster model. Consequently one can say that *in a fission model the preformation probability is the penetrability of the inner part of the barrier* :

$$P_{ov} = \exp(-K_{ov}) = S_f \qquad (11)$$

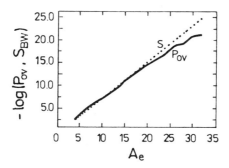

Figure 5: Comparison of the ASAFM $-\log P_{ov}$ with Blendowske and Walliser's $-\log S$ versus A_e for even-even parent nuclei and ^{208}Pb daughter.

Ideally both kinds of models should reproduce the true (experimental) value $\lambda_f = \lambda_p = \lambda_{exp}$. In a particular case, if $\nu_f = \nu_p$ and $P_s = P_p$, one is left with $P_{ov} = S$. Alternatively, if $\nu_f P_s = \nu_p P_p$, we get also $P_{ov} = S$. By chance, this condition is almost fulfilled when we compare ASAFM with Blendowske and Walliser's preformed cluster model for even-even parent nuclei.

A linear variation of $\log S$ with A_e has been found in the Ref. [28]; for even-even parent nuclei

$$-\log S = -\frac{(A_e - 1)}{3}\log(0.0063) = 0.73355(A_e - 1) \qquad (12)$$

We have performed within ASAFM, a calculation of the corresponding quantity

$$-\log P_{ov} = 0.43429 K_{ov} \qquad (13)$$

where the action integral in the overlapping region is given by :

$$K_{ov} = 0.2196(\mu_A E_b^0)^{1/2}(R_t - R_i)\{(1 - b^2)^{1/2} - b^2 \ln[b^{-1} + (b^{-2} - 1)^{1/2}]\} \qquad (14)$$

with $b^2 = E_{cor}/E_b^0$, $E_b^0 = e^2 Z_1 Z_2/R_t - Q$, $\mu_A = A_1 A_2/A$, and E_{cor} - the correction energy (see Refs. [12, 14]). Calculations have been done for emission of ^{4}He, ^{8}Be, ^{14}C, ^{16,18,20}O, ^{24,26}Ne, ^{28,30}Mg, and ^{32,34}Si, from various parent nuclei, leading to the ^{208}Pb daughter. The results, plotted in Fig. 5, show that we also got a linear function $\log P_{ov} = f(A_e)$ with a slightly smaller slope for $A_e > 20$.

4.3. Universal curves of cluster radioactivities

The nuclear halflife is directly related to the decay constant $T = \ln 2/\lambda$, hence

$$\log T = 0.43429 K_s - (\log \nu_f + 0.159) + 0.43429 K_{ov} \qquad (15)$$

within ASAFM. In the region of separated fragments, for a vanishing contribution of the angular momentum, one has :

$$K_s = 0.527(\mu_A Z_1 Z_2 R_b)^{1/2}[\arccos\sqrt{r} - \sqrt{r(1-r)}] \qquad (16)$$

where $r = R_t/R_b$. This action integral is called sometimes Gamow-factor[3] and is denoted by G.

In the first approximation, one may assume that for a given cluster radioactivity (given $A_2 Z_2$), both ν_f and K_{ov} in the eq. (10) are constant quantities, independent on $A_1 Z_1$. In fact we are mostly interested in decay modes with $Z_1 \simeq 82$, $A_1 \simeq 208$. From eqs. (12) and (13) one has $K_{ov} \simeq 1.68908(A_e - 1)$ for even-even parent nuclei. Consequently, for any cluster radioactivity there is one and only one curve - a linear variation of $\log T$ with K_s. In the preformed cluster model of Ref. [28]

$$\log T = 0.43429 K_p - (\log \nu_p + 0.159) + 0.7335(A_e - 1) \qquad (17)$$

---

[3]The corresponding equation of Ref. [42] should be corrected by substituting the numerical constant 0.629 with 0.63244. Also the branching ratios given in the table 2 of this Ref. for ^{26}Ne and ^{30}Mg radioactivities of ^{236}U should be recalculated by introducing the right value of λ_α which differs by almost two orders of magnitude.

The slope of this universal curve is $\log e = 0.43429$. The vertical distance between two universal curves corresponding to A_{e1} and A_{e2} is about $0.7335(A_{e2} - A_{e1})$ if ν is constant.

The equation of universal curves based on ASAFM, obtained by taking ν constant and $\log P_{ov}$ proportional to A_e is

$$\log T = -\log P_s - 22.169 + 0.598(A_e - 1) \qquad (18)$$

For separated fragments

$$-\log P_s = 0.22873(\mu_A Z_1 Z_2 R_b)^{1/2}[\arccos\sqrt{r} - \sqrt{r(1-r)}] \qquad (19)$$

where $r = R_t/R_b$, $R_t = 1.2249(A_1^{1/3} + A_2^{1/3})$, $R_b = 1.43998 Z_1 Z_2 / Q$.

A closer estimation of the deviations from experimental results in case of α - decay of 124 emitters shows that the shell effects not taken into account in ν, are very clear present, leading to a minimum around the magic number of neutrons of the daughter nucleus. Consequently, it is important to consider the variation of ν with A_1, Z_1, A_2, Z_2, as we have shown previously.

5. FISSILITY and SUPERASYMMETRIC FISSION

In 1952, when some spontaneous fission halflives have been measured, it was observed[65, 66] that a straight line with a negative slope was obtained by plotting $\log T$ of even-even parent nuclei versus Z^2/A, which is proportional to the fissility parameter X. Substantial deviations from this linear decrease have been noted very soon[67 - 69].

Recently some authors (see for example a paper cited in Ref. [42]) have used the same parameter Z^2/A to plot the halflives of cluster radioactivities. An increasing trend has been observed, which has been interpreted as an argument against fission theory of α decay and cluster radioactivities.

It is evident that in the ratio Z^2/A there is no information about an asymmetric split; it is a suitable parameter only for symmetric fission. The corresponding quantity for asymmetric fission can be borrowed from the theory of heavy ion fusion reactions[70] :

$$X_{as} \sim Z_1 Z_2 / [A_1^{1/3} A_2^{1/3} (A_1^{1/3} + A_2^{1/3})] \qquad (20)$$

Intuitively, one can guess that a linear decrease of $\log T$ with X comes from the liquid drop model. By taking into account only the deformation energy obtained within this model, we have obtained the results plotted in Fig. 6a for symmetric fission and in Fig. 6b for a superasymmetric fission (in which the heavy fragment is ^{208}Pb and for the light fragments we considered $A_e = 4 - 61$). This figure clearly shows that a linear decrease of $\log T$ is a characteristic feature of a symmetric division of the charged drop, but for a superasymmetric fission with $A_2 = 22 - 61$ and $A_1 = 208$, one has an

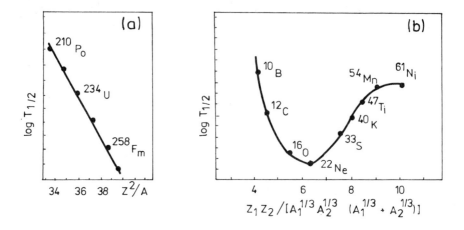

Figure 6: Halflives for symmetric fission (a) and superasymmetric fission with ^{208}Pb the heavy fragment (b), calculated within liquid drop model.

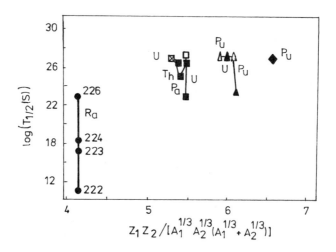

Figure 7: Systematics of experimental halflives for cluster radioactivities (● ^{14}C, ■ ^{24}Ne, □ ^{25}Ne, ⊠ ^{26}Ne, ▲ ^{28}Mg, △ ^{30}Mg, ◆ ^{32}Si).

increase of $\log T$ with X_{as}. By introducing the shell effects one can reproduce very well (see Table 1) the experimental trend shown in Fig. 7.

The Green approximation for the line of beta-stability :

$$A = (Z - 100)/0.6 + \sqrt{(Z - 100)^2/0.36 + 200Z/0.3} \qquad (21)$$

had been used to select the nuclei considered in Fig. 6a.

Acknowledgments

This paper is based on research work performed in cooperation with prof. dr. W. Greiner, I. Căta, D. Mazilu, and D. Schnabel. It was partly supported by the Institute of Atomic Physics, Bucharest and partly by Bundesministerium für Forschung und Technologie, Bonn. One of us (DNP) is grateful to the German - Romanian programme for scientific research and technological development directed by the Stabsabteilung Internationale Beziehungen of the KfK Karlsruhe for their support of the mutual collaboration.

REFERENCES

1. *Treatise on Heavy Ion Science*, Vol. 8, edited by D. A. Bromley (Plenum Press, New York, 1989).
2. *Particle Emission from Nuclei*, Vol. I - III, edited by D. N. Poenaru and M. Ivaşcu (CRC Press, Boca Raton, Florida, 1989).
3. D. N. Poenaru, M. Ivaşcu, and W. Greiner, in : Ref. [2], Vol. III, p. 203.
4. W. Greiner, M. Ivaşcu, D. N. Poenaru, and A. Săndulescu, in : Ref. 1, p. 461.
5. V. M. Strutinsky, Nucl. Phys. **A 95**, 420 (1967).
6. J. A. Maruhn, W. Greiner and W. Scheid, in *Heavy Ion Collisions* edited by R. Bock (North Holland, Amsterdam, 1980), Vol. 2, p. 399.
7. D. N. Poenaru and M.Ivaşcu, in *Atomic and Nuclear Heavy Ion Interactions, 2nd Part*, edited by G. Semenescu et al. (Central Inst. of Phys., Bucharest, 1984), p. 277.
8. E. Ye. Berlovich and Yu.N. Novikov, in *Modern Methods of Nuclear Spectroscopy* (in Russian) (Nauka, Leningrad, 1988), p. 107.
9. G. Herrmann, Nucl. Phys., A 502 141c (1989).
10. D. N. Poenaru, W. Greiner, K. Depta, M. Ivaşcu, D. Mazilu, and A. Săndulescu, Atomic Data and Nuclear Data Tables, **34** 423 (1986).
11. D. N. Poenaru, W. Greiner, M. Ivaşcu, D. Mazilu, and I. H. Plonski, Z. Phys. **A 325**, 435 (1986).
12. D. N. Poenaru, M. Ivaşcu, D. Mazilu, I. Ivaşcu, E. Hourani, and W. Greiner, in *Developments in Nuclear Cluster Dynamics* (Proceedings International Symposium, Sapporo, Japan, 1988), edited by K. Akaishi, K. Kato, H. Noto, and S. Okabe (World Scientific, Singapore, 1989), p.76.

13. D. N. Poenaru, M. Ivaşcu, D. Mazilu, R. Gherghescu, K. Depta, and W. Greiner, Central Institute of Physics, Bucharest, Report NP–54–86, 1986.
14. D. N. Poenaru, D. Schnabel, W. Greiner, D. Mazilu, I. Căta , Preprint, GSI 90-28, Darmstadt, 1990.
15. Yu. Ts. Oganessyan, in *Lecture Notes in Physics* (Springer, Heidelberg, 1974) **33**, 221.
16. G. Münzenberg, Rep. Prog. Phys. **51**, 57 (1988).
17. P. Armbruster, Annu. Rev. Nucl. Part. Sci. **35**, 135 (1985).
18. D. N. Poenaru, M. Ivascu and W. Greiner, Proc. Int. Conf. on SSNTD, Rome, 1985, in Nucl. Tracks, **12**, 313 (1986).
19. P. Armbruster, invited talk International Conf. "Cluster 88", Kyoto, Japan, J. Phys. Soc. Jpn. **58**, Suppl. 232 (1989).
20. F. Gönnenwein, B. Börsig and H. Löffler, in Proc. Internat. Symp. on Collective Dynamics, Bad Honnef, edited by P. David (World Sci., Singapore, 1986), p. 29.
21. Y. J. Shi and W. J. Swiatecki, Nucl. Phys. A **464**, 205 (1987).
22. G. A. Pik-Pichak, Yad. Fiz. **44**, 1421 (1986).
23. M. Greiner, W. Scheid, and V. Oberacker, J. Phys. G. **14**, 589 (1988).
24. F. Barranco, R. A. Broglia, and G. F. Bertsch, Phys. Rev. Lett. **60**, 507 (1988).
25. S. S. Malik and R. K. Gupta, Phys. Rev. C **39**, 1992 (1989).
26. G. Shanmugam and B. Kamalaharan, Phys. Rev. C **41**, 1184 (1990).
27. M. Iriondo, D. Jerrestam, and R. J. Liotta, Nucl. Phys. A **454**, 252 (1986).
28. R. Blendowske and H. Walliser, Phys. Rev. Lett. **61**, 1930 (1988).
29. V. A. Rubchenya, V. P. Eysmont, and S. G. Yavshits, Izv. A. N. SSSR, ser. Fiz. **50**, 1017 (1986).
30. S. G. Kadmensky, S. D. Kurgalin, V. I. Furman, and Yu. M. Tchuvilsky, Yad. Fiz. **51**, 50 (1990).
31. I. Rotter, J. Phys. G. **15**, 251 (1989).
32. M. Ivaşcu and I. Silişteanu, Nucl. Phys. A **485**,93 (1988).
33. H. G. de Carvalho, J. B. Martins, and O. A. P. Tavares, Phys. Rev. C **34**, 2261 (1986).
34. N. Cindro and M. Bozin, Phys. Rev. C **39**, 1665 (1989).
35. B. Buck and A. C. Merchant, Phys. Rev. C **39**, 2097 (1989).
36. R. Herrman, K. Depta, D. Schnabel, H. Klein, W. Renner, D. N. Poenaru, A. Săndulescu, J. A. Maruhn, and W. Greiner, in Proceedings International Symposium on Physics and Chemistry of Fission, Gaussig, G. D. R, 1988, (Teubner Texte, Leipzig, 1990), to be published.
37. P. B. Price, Annu. Rev. Nucl. Part. Sci. **39**, 19 (1989) ; P. B. Price and S.W.Barwick, in : Ref. [2], Vol. II, p. 206.
38. S. D. Wang, D. Snowden-Ifft, P.B. Price, K.J. Moody, and E.K. Hulet, Phys. Rev. C **39**, 1647 (1989).

39. L. Brillard, A. G. Elayi, E. Hourani, M. Hussonnois, J. F. Le Du, L. H. Rosier, and L. Stab, C. R. Acad. Sci. **309**, 1105 (1989).
40. S. P. Tretyakova, Yu. S. Zamyatnin, V. N. Kovantsev, Yu. S. Korotkin, V. L. Mikheev, and G. A. Timofeev, Z. Phys. **A 333**, 349 (1989).
41. R. Bonetti, E. Fioretto, C. Migliorino, A. Pasinetti, F. Barranco, E. Vigezzi, and R. A. Broglia, Phys. Lett. **B 241**, 179 (1990).
42. A. A. Ogloblin et al, Phys. Lett. **B 235**, 35 (1990).
43. D. Weselka, P. Hille, and A. Chalupka, Phys. Rev. **C 41**, 778 (1990).
44. D. N. Poenaru and M. Ivaşcu, J. Phys. Soc. Jpn. **58** Suppl., 249 (1989) ; in Ref. [2], Vol. II, p. 73 ; D. N. Poenaru,J. A. Maruhn, W. Greiner, M. Ivaşcu, D. Mazilu, and I. Ivaşcu, Z. Phys. **A 333**, 291 (1989).
45. D. N. Poenaru, M. Ivaşcu, I. Căta, M. Mirea, W. Greiner, K. Depta, and W. Renner, in *50 Years with Nuclear Fission* (Proceedings International Conference, Gaithersburg, Maryland, 1989), American Nuclear Society, La Grange Park, 1989, p. 617.
46. D. N. Poenaru, M. Mirea, W. Greiner, I. Căta, and D. Mazilu, Modern Physics Letters A, **5**, 2101 (1990).
47. D. N. Poenaru, W. Greiner, and M. Ivaşcu, Nucl. Phys., **A 502**, 59c (1989).
48. E. Hourani and M. Hussonnois, in : Ref. [2], Vol. II, p. 171 ; E. Hourani, M. Hussonnois, and D. N. Poenaru, Ann. Phys. (Paris) **14**, 311 (1989).
49. W. Henning and W. Kutschera, in : Ref. [2], Vol. II, p. 190.
50. D. Haşegan and S. P. Tretyakova, in Ref. [2], Vol. II, p. 234.
51. D. N. Poenaru and M. Ivaşcu, Rev. Roum. Phys. **28**, 309 (1983); **29**, 587 (1984).
52. Atomic Data and Nuclear Data Tables, special editor P. E. Haustein, **39**, 185 - 393 (1988).
53. A. H. Wapstra, G. Audi, and R. Hoekstra, in Ref. [52], p. 281.
54. J. Jänecke and P. J. Masson, in Ref. [52], p. 265.
55. P. J. Masson and J. Jänecke, in Ref. [52], p. 273.
56. L. Spanier and S. A. E. Johansson, in Ref. [52], p. 259.
57. T. Tachibana, M. Uno, M. Yamada, and S. Yamada, in Ref. [52], p. 251.
58. L. Satpathy and R. K. Nayak, in Ref. [52], p. 241.
59. E. Comay, I. Kelson, and A. Zidon, in Ref. [52], p. 235.
60. P. Möller, W. D. Myers, W. J. Swiatecki, and J. Treiner, in Ref. [52] p. 225.
61. P. Möller and J. R. Nix, in Ref. [52], p. 213.
62. P. E. Haustein, in Ref. [52], p. 185; in Ref. [2], Vol. I, p. 233.
63. A. Săndulescu, D. N. Poenaru, and W. Greiner, Sov. J. Part. Nucl. **11**, 528 (1980).
64. D. N. Poenaru and M. Ivaşcu, in Proc. International Summer School, Poiana Braşov, Central Institute of Physics, Bucharest, 1980, p. 743.
65. G. T. Seaborg, Phys. Rev. **85**, 157 (1952).
66. W. J. Whitehouse and W. Galbraith, Nature **169**, 494 (1952).
67. A. Ghiorso et al., Phys. Rev. **87**, 163 (1952).
68. J. R. Huizenga, Phys. Rev. **94**, 158 (1954).
69. W. J. Swiatecki, Phys. Rev. **100**, 937 (1955).
70. R. Bass, Phys. Lett. **47 B**, 139 (1973).

EXOTIC RADIOACTIVITY :

CLUSTERS AS SOLITONS ON THE NUCLEAR SURFACE

A.Ludu[*] and A.Sandulescu[**,+]

[*] University of Bucharest, Faculty of Physics, Bucharest, Romania

[**] Institut fur Theoretische Physik der J.W.Goethe Universitat, Frankfurt am Main, Germany

Abstract : The exotic decays, i.e. spontaneous emission of light nuclei like carbon, neon, magnezium and silicon from heavy nuclei, indicates a large enhancement of such clusters on the nuclear surface of the most rigid double magic spherical nucleus ^{208}Pb. By introducing nonlinear terms in the liquid drop model, i.e. in the hydrodynamical equations, we could explain this enhancement as stable solitons on the surface of a sphere. This lead to a new minimum in the total energy which, due to the fact that these decays are spontaneous decays, must be degenerate in energy with the usual ground state minimum. We conclude that the present description of the exotic decays is very similar with the description of shape isomers in fission.

[+] Permanent address: Institute of Atomic Physics, Department of Theoretical Physics, P.O.Box MG-6, Bucharest, Romania

1. INTRODUCTION

It is well known that only the nuclei close to the magic numbers are spherical all others situated between two magic numbers being deformed. This is theoretically explained by minimizing the potential energy of a nucleus, computed by the liquid drop model with shell and pairing corrections, as function of the nuclear shape. The main hypothesis is the fact that the external nucleons modify the spherical self consistent field to a deformed one. By using a multipole expansion of the nuclear shape results that the main deformation is the quadrupole deformation. For some nuclei the octupole and the hexadecapole deformations could become important. The stability to different deformations is given by the behaviour of the potential energy as function of these deformations. For a step dependence we have a "rigid" nucleus and for a shallow dependence a "soft" nucleus. The most rigid nucleus is the double magic nucleus ^{208}Pb. These results describe excelently not only the ground state properties but also the first excited states.

On the other hand the experimental discovery of the exotic decays [1], i.e. spontaneous emission of carbon, neon, magnezium and silicon from heavy nuclei, indicates a large enhancement of such clusters on the nuclear surface of the double magic nucleus ^{208}Pb. Alpha decay was the first example of such an enhancement the nuclear shell model underestimating the preformation factor in a R-matrix description of this process by at least 2-3 orders of magnitude.

This suggest that the external nucleons joint together to form the emitted cluster leaving the residual nucleus unpolarized. Such a picture, which is contrary to the usual description of heavy nuclei by multipole expansion of the shape, is supported by two facts: first the nucleus ^{208}Pb could be hardly polarized being the most rigid nucleus and second the many body correlations are not included in the actual calculations which may have excluded up to now such configurations. In a phenomenological description we have to find some arguments of collective type which may justify the existence of such new shapes.

In the present paper, by including nonlinear terms in the hydrodynamical equations, we have shown that a stable soliton could exist on a surface of a sphere. This is based on the hypothesis that the outside nucleons do not polarize the double magic core ^{208}Pb. Evidently this new shape conduct in addition to the usual ground state minimum, to a new minimum of the potential energy. Due to the fact that the exotic decays are spontaneous decays we impose the condition that both minima are degenerate in energy. Evidently we have to consider these minima in a many dimensional space.

Due to the existence of two completely different configurations the ground state of a heavy nucleus could be approximately described by the superposition of these two states. The first configuration correspond to an usual spherical or deformed nucleus and the second configuration to a cluster-like state described by a soliton moving on a sphere. This conclusion is similar with the basic hypothesis of the coexistence model which assumes spherical and deformed states in the same nucleus. Due to the large barrier between the two minima the amplitude of the cluster-like state is much smaller than of the usual ground state. The ratio of the square of the two amplitudes could be interpreted as the preformation probability of the corresponding cluster on the nuclear surface.

We conclude that this picture is completely different that the present descriptions of the exotic decays either as a fission process where first the nucleus deforms and later on neck-in in two fragments in their ground states or similar with alpha decay where first the cluster is preformed from individual nucleons and later on penetrates through the barrier [2]). It is much more similar with the description of the shape isomers in fission the only difference being the degeneracy in energy of the two wells.

Evidently in order to show that the new description is correct we have to find all the consequences like the corrections to the binding energies around the magic numbers or the quantum spectrum of the nonlinear equations corresponding to the second well.

2. HYDRODYNAMIC EQUATIONS

We consider a small propagating perturbation on the free surface of a liquid drop. By assuming an axial nucleus with the symmetry axis in the direction of the perturbation we reduce our problem to an one dimensional model: a small perturbation on a circle of radius R. The surface Σ could be written in the form:

$$r = R + \eta(\theta,t) \qquad (2.1)$$

where by r and θ we denoted the corresponding cylindrical coordinates.

We make the assumptions:

1) The amplitude of the perturbation $a = \max |\eta(\theta,t)|$ is small compared to the radius, so that we can introduce a small parameter $\varepsilon = \frac{a}{R} \ll 1$.
2) The core of the liquid is unperturbed up to a radius $r = R-h$, with $h \ll R$.

We treat the case of an ideal, incompressible (the mass density ρ_m = const) and irrotational fluid (the curl of the field of velocities rot v = 0). This is the case of a potential flow so that the field of velocities results from a scalar potential ϕ, v = $\Delta\phi$. This velocity potential satisfies the Laplace equation, $\Delta\phi$ = 0.

Now it is natural to ask for solutions of Laplace equation written as power series into the small parameter $\xi = \frac{r-R}{R}$. The potential becomes :

$$\phi(r,\theta,t) = \sum_{n \geq 0} \xi^n f_n(\theta,t) \qquad (2.2)$$

In this way the r dependence is written explicitly.

On the surface Σ the parameter ξ becomes $\frac{\eta}{R} = \epsilon\eta'$ written with a dimensionless function η' defined as $\eta' = \eta/a$. The Laplace equation in cylindrical coordinates is :

$$\phi_{rr} + \frac{1}{r}\phi_r + \frac{1}{r^2}\phi_{\theta\theta} = 0 \qquad (2.3)$$

Using the relation (2.2) and expanding also the coefficients in ξ

$$\frac{1}{r} = \frac{1}{R}(1 - \xi + \xi^2 - \xi^3 + \ldots)$$
$$\frac{1}{r^2} = \frac{1}{R^2}(1 - 2\xi + 3\xi^2 - 4\xi^3 + \ldots) \qquad (2.4)$$

we obtain, by choosing the coefficients of different powers of ξ zero, an infinite system of recurrence relations for the functions f_n :

$$f_2 = \frac{1}{2}(-f_1 - f_{0,\theta\theta}),$$
$$f_3 = \frac{1}{3}f_1 - \frac{1}{6}f_{1,\theta\theta} + \frac{1}{2}f_{0,\theta\theta}, \qquad (2.5)$$
$$f_4 = -\frac{1}{4}f_1 + \frac{1}{4}f_{1,\theta\theta} - \frac{11}{24}f_{0,\theta\theta} + \frac{1}{24}f_{0,\theta\theta\theta\theta}, \text{ etc}$$

The two components of the velocity of the fluid (the radial $v = \phi_r$ and the tangential one $u = \phi_\theta/r$) are depending only on two independent functions : $f_{0,\theta}$ and f_1 denoted in the following by g and j respectively. Thus, using equation(2.5) we can write in the second order in ξ the components of the velocity in the form :

$$u = \frac{1}{R}[g + \xi(-g + j_\theta) + \frac{\xi^2}{2}(2g - g_{\theta\theta} - 3j_\theta) + \ldots]$$
$$v = \frac{1}{R}[j + \xi(-j - g_\theta) + \xi^2(j - \frac{1}{2}j_{\theta\theta} + \frac{3}{2}g_\theta) + \ldots] \quad (2.6)$$

Now, we set the equations ruling the dynamics of the perturbation. The first equation occurs from the geometric condition expressed by the Lagrangean derivative of the radial coordinate [3] :

$$\dot{r} = v|_\Sigma = \frac{dr}{dt}|_\Sigma = \eta_t + \frac{1}{r}u\eta_\theta \quad (2.7)$$

The second equation is given by Euler equation of the momentum balance :

$$\rho_m(\frac{\partial V}{\partial t} + (V \cdot \nabla)V) = -\nabla p + f \quad (2.8)$$

where p is the pressure and **f** is the volume force density. From the continuity equation :

$$\frac{\partial \rho_m}{\partial t} + \text{div}(\rho_m v) = \rho_m \text{ div } v = 0$$

and the condition of irrotationality (rot **V** = 0) we can put the equation (2.8) in a gradient form :

$$\nabla(\phi_t + \frac{1}{2}|\nabla\phi|^2 + \frac{p}{\rho_m}) = \frac{1}{\rho_m}f$$

If the force density is derived from a scalar potential field ϕ_F then we have :

$$\phi_t + \frac{1}{2}(\nabla\phi)^2 + \frac{p}{\rho_m} + \frac{F^\phi}{\rho_m} = \bar{f}(t) \qquad (2.10)$$

where $\bar{f}(t)$ is a constant of integration depending only of time. On the free surface the pressure is given by the following expression in cylindrical coordinates [4]) :

$$p_{/\Sigma} = \sigma \frac{r^2 + 2r_\theta^2 - rr_{\theta\theta}}{(r^2 + r_\theta^2)^{3/2}} \qquad (2.11)$$

where σ is the surface pressure coefficient. Using the surface equation (2.1) and expanding the above expression into powers of ε we obtain the form :

$$p_{/\Sigma} = \frac{\sigma}{R}[1 - \varepsilon(\eta' + \eta'_{\theta\theta}) + \ldots] \qquad (2.12)$$

The force density is given by the Coulomb interaction. To calculate the electrostatic force density $f = \rho_{el} \times E = -\rho_{el}\nabla\phi_{el}$ we need the charge density ρ_{el} and the expression for the electrostatic potential ϕ_{el}, generated by this charge. Based on the experimental fact that charge equilibration in deep inelastic reactions is a very rapid process [5]), we make the assumptions that the sphere of radius R is uniformly charged with the density ρ_{sph} corresponding to the residual nucleus and the disturbance is, also, uniformly charged with another electric charge density ρ_{sol}, corresponding to the emitted cluster.
The charge distribution is [6,11])

$$\rho_{el}(r,\theta,t) = (\rho_{sph}-\rho_{sol}) H(1-\frac{r}{R}) + \rho_{sol} H(1-\frac{r}{R(1+\varepsilon\eta')}) \qquad (2.13)$$

where $H(x)$ are the Heaviside distributions. Keeping only the terms up to second order in ε we have

$$\rho_{el}(r,\theta,t) = \rho_{sph} H(1-\frac{r}{R}) - \rho_{sol}(\frac{r_\eta'}{R}\varepsilon + \frac{r_\eta'^2}{R}\varepsilon^2)\delta(1-\frac{r}{R})$$
$$+ \rho_{sol} \frac{r_\eta^2{}'^2}{2R^2}\varepsilon^2 \delta'(1-\frac{r}{R}) \qquad (2.14)$$

The electrostatic potential is given by the Poisson equation with the charge distribution (2.13). We look, like in the previous case, for a solution in a power series in the parameter $\xi = \frac{r-R}{R}$

$$\phi_{el} = \sum_{n \geq -2} \xi^n f_n(\theta) \qquad (2.15)$$

The action of the Laplace operator in spherical coordinates on (2.15) results in two singular terms which are in connection with the second and the third term from (2.14),

$$f_{-1}(\theta) = \frac{\rho_{sol}}{4\pi\varepsilon_o R} \varepsilon\eta'(\theta) \; ; \; f_{-2}(\theta) = -\frac{\rho_{sol}}{8\pi\varepsilon_o R^2} \varepsilon^2 \eta'^2(\theta) \qquad (2.16)$$

Thus we obtain, by imposing that the coefficients of different powers of ξ are zero, a system of five recurrence relations for f_o, f_1, f_2, f_3 and f_{-3} which lead to the following final form of the potential

$$\phi_{el}\Big|_{\Sigma} = \frac{R^2 \rho_{sph}}{3\varepsilon_o}(1 - \varepsilon\eta') - \varepsilon^2 \frac{\rho_{sol} R^2}{2\varepsilon_o} \eta'^2 \qquad (2.17)$$

Further we differentiate the eq.(2.10) with respect to the angle θ and we introduce the eqs.(2.12) and (2.17) obtaining a final dynamic equation :

$$[(ru)_t + uu_\theta + vv_\theta - \frac{\sigma\varepsilon}{\rho_m R}(\eta'_\theta + \eta'_{\theta\theta\theta}) + \varepsilon^2 \frac{\rho_{sol} R^2}{\rho_m \varepsilon_o}(\rho_{sol} - \frac{\rho_{sph}}{3})\times$$
$$\eta'\eta'_\theta - \varepsilon \frac{\rho_{sol}\rho_{sph} R^2}{3\varepsilon_o \rho_m} \eta'_\theta]_\Sigma = 0 \qquad (2.18)$$

Thus, we can write eqs. (2.7) and (2.18), in the second order in ε, only in terms of three functions η', j and g :

$$j - aR\eta'_t - \varepsilon(\eta'g)_\theta - \varepsilon \eta'j = 0 \qquad (2.19)$$

$$g_t + \varepsilon\eta'_t j_\theta + \varepsilon \eta' j_{\theta t} + \frac{1}{R^2} gg_\theta + \frac{1}{R^2} jj_\theta -$$
$$- \frac{\sigma\varepsilon}{\rho_m R}(\eta'_\theta + \eta'_{\theta\theta\theta}) - \varepsilon \frac{\rho_{sol}\rho_{sph} R^2}{3\varepsilon_o \rho_m} \eta'_\theta + \qquad (2.20)$$
$$+ \varepsilon^2 \frac{\rho_{sol} R^2}{\rho_m \varepsilon_o}(\rho_{sol} - \frac{\rho_{sph}}{3}) \eta'\eta'_\theta = 0$$

In order to have an additional condition for the above three functions we apply the second assumption, asking that the radial velocity of the fluid to be zero for $r = R-h$:

$$v\Big|_{r=R-h} = f_1 + \frac{h}{R}(f_1 + f_{0,\theta\theta}) + \ldots \qquad (2.21)$$

This condition, which is equivalent with the condition of shallow water for usual solitons, becomed in the first order

$$j(1 + \frac{h}{R}) = -\frac{h}{R} g_\theta + \ldots \qquad (2.22)$$

condition which requires that j must be of order ε. Thus, keeping only the terms up to the second order in ε, the above eqs.(2.19) and (2.20) become :

$$j - aR\eta'_t - \varepsilon(\eta'g)_\theta = 0 \qquad (2.23)$$

$$g_t + \frac{1}{R^2} g g_\theta - \frac{\sigma\varepsilon}{\rho_m R}(n'_\theta + n'_{\theta\theta\theta}) - \varepsilon \frac{\rho_{sol}\rho_{sph} R^2}{3\varepsilon_o \rho_m} n'_\theta +$$

$$+ \varepsilon^2 \frac{\rho_{sol} R^2}{\rho_m \varepsilon_o}(\rho_{sol} - \frac{\rho_{sph}}{3}) n'n'_\theta = 0 \qquad (2.24)$$

From eq.(2.22) we observe that the function j can be expressed in the form :

$$j = -\frac{h}{R} g_\theta + O_2(\varepsilon) \qquad (2.25)$$

where $O_2(\varepsilon)$ denotes the terms quadratic in ε and higher.

We should like to stress that this condition is equivalent with the fact that the radial component of the velocity v to be smaller then the tangential component u, up to ε as order of magnitude. This is a typical behaviour of solitons[7]).

In order to solve the system of eqs.(2.23), (2.24) and (2.25) we make the transformation :

$$g = \chi n' + \xi(\theta, t) \qquad (2.26)$$

where χ is an arbitrar real parameter and $\xi(\theta,t)$ is an arbitrary function. Substitution of eqs.(2.26) and (2.25) into eqs.(2.23) and (2.24) results in a system of two nonlinear differential equations in the unknown functions n' and ξ.

$$k\chi n'_\theta - aRn'_t - 2\varepsilon\chi n'n'_\theta + k\xi_\theta - \varepsilon(n'\xi)_\theta = 0 \qquad (2.27)$$

$$\chi n'_t + \xi_t + \frac{\chi^2}{R^2} n'n'_\theta + \frac{\chi}{R^2}(n'\xi)_\theta + \frac{\xi\xi_\theta}{R^2} + \alpha\, n'n'_\theta$$

$$+ \beta\, n'_\theta - \gamma(n'_\theta + n'_{\theta\theta\theta}) = 0 \qquad (2.28)$$

with the constants $\alpha = \varepsilon^2 \frac{2R^2 \rho_{sol}}{\rho_m \varepsilon_o}(\frac{\rho_{sph}}{3} - \rho_{sol})$, $k = \frac{h}{R}$,

$$\beta = -\varepsilon(\frac{\sigma}{R\rho_m} + \frac{R^2 \rho_{sph} \rho_{sol}}{3 \varepsilon_o \rho_m}) \text{ and } \gamma = \frac{\sigma \varepsilon}{\rho_m R} \text{ . We make a}$$

functional transformation :

$$\xi(\theta,t) = \frac{k\chi\eta' - \varepsilon\chi\ \eta'^2 - aR \int \eta'_t d\theta + \xi^{(1)}}{k - \varepsilon\eta'} \quad (2.29)$$

which satisfies the eq.(2.27). Here $\xi^{(1)}$ is an arbitrary function. Choosing $\xi^{(1)}$ so that :

$$\xi_t - \frac{\beta}{\chi}\xi_\theta + \frac{1}{R^2}\xi\xi_\theta + (\frac{\chi}{R^2} + \frac{\varepsilon\beta}{k\chi})(\eta'\xi)_\theta = 0 \quad (2.30)$$

we obtain, after introducing (2.29) and (2.30) in (2.28), a Korteweg-de Vries equation [8] in the function η'

$$(\chi - \frac{aR^2\beta}{h\chi})\ \eta'_t + (\frac{\chi^2}{R^2} + \alpha + \frac{2a\beta}{h})\eta'\eta'_\theta - \gamma\eta'_{\theta\theta\theta} = 0 \quad (2.31)$$

This equation has a soliton solution [3,9,10]

$$\eta' = \eta_o \text{ sech}^2 \frac{\theta + Vt}{L} \quad (2.32)$$

with the following expressions for the halfwidth L and velocity V as functions of the four independent parameters χ, a (a << R), h (h < R) and $\eta_o \simeq 1$:

$$L = \left[\frac{\eta_o R \rho_m}{12\ a\sigma}(\frac{2\ a^2}{R^2 h\ \rho_m}(\frac{\sigma}{R} + \frac{\rho_{sol}\rho_{sph}R^2}{3\ \varepsilon_o} + \right.$$
$$\left. + \frac{a^2 \rho_{sol}}{\rho_m \varepsilon_o}(\rho_{sol} - \frac{\rho_{sph}}{3}) - \frac{\chi^2}{R^2})\right]^{-1/2} \quad (2.33)$$

and

$$V = \frac{4a\sigma}{R^2 \rho_m L^2}\left[\chi - \frac{a^2 R^2}{h\chi R}(\frac{\rho_{sph}\rho_{sol}R^2}{3\ \varepsilon_o\ \rho_m} - \frac{\sigma}{R\rho_m})\right]^{-1} \quad (2.34)$$

So, we have obtained soliton type propagating solutions on the surface of the nucleus in one dimensional case. General properties of the soliton, like the typical dependence between the amplitude a η_o, velocity V and width L are satisfied. The periodic condition of the solution is almost fulfilled by the condition of rapid decreasing of the solution.

We see that for some values of the physical constants and free parameters involve in exp.(2.33) the shape of the propagating deformation is a pure soliton. For example, if the parameter χ^2 is less than some critical values the square root exists for $\eta_o > 0$. For large values of χ^2 we must have $\eta_o < 0$ in order to have real solutions for L which lead to antisoliton solutions. We note that a whole class of soliton solutions of different halfwidths L and velocities V can be obtained from exps. (2.33) and (2.34) depending on the parameters χ, a, h and η_o.

We would like to close this paragraph with the remark that the final system of dynamic eqs. (2.27), (2.28) is only a second order approximation in terms of ε. The addition of higher terms will probably lead to more sophisticated solutions like "boomerons" [11]) and other. The reason for keeping only the second order terms in the dynamic equations is the requirement for the existence of stable soliton solutions on the surface.

3. THE POTENTIAL ENERGY

Let us consider, like before, an arbitrary shape of an axial deformed nucleus with the symmetry axis along the z-axis (2.1). The potential energy E_p consists of two terms: the surface energy E_s and the repulsive Coulomb energy E_c.

The surface energy can be written in the form [4]):

$$E_s = \sigma \oint_\Sigma ds \qquad (3.1)$$

where Σ is the surface of the deformed nucleus. In spherical coordinates we have:

$$E_s = 2\pi\sigma R^2 \int_0^\pi r\sqrt{r^2+r_\theta^2}\, \sin\theta d\theta \qquad (3.2)$$

and taking into account (2.1) we can write in the second order in ε

$$E_s = 2\pi r^2 \sigma \int_0^\pi (1+2\varepsilon\eta' + \varepsilon^2(\eta'^2 + \frac{\eta_\theta'^2}{2}))\sin\theta d\theta \ . \qquad (3.3)$$

The Coulomb energy consists of three terms: the proper energies of the sphere E_{sph} and of the soliton E_{sol} and the interaction energy between them E_{int}. Correspondingly we have[11]):

$$E_{sph} = \frac{4\pi \rho_{sph} R^5}{15\varepsilon_0} \qquad (3.4)$$

and respectively [6]):

$$E_{int} = \frac{1}{2}\rho_{sol} \int_{V_{sol}} \phi_{sph}\, dV \qquad (3.5)$$

where ρ_{sph} and ρ_{sol} are like before the charge densities of the sphere respectively of the soliton and ϕ_{sph} is the electric potential produced by the sphere in exterior. From equation (3.5) we have, in the second order:

$$E_{int} = \frac{\rho_{sol}\rho_{sph} R^3}{6\varepsilon_0} \int_0^\pi \int_0^{2\pi} \int_R^{R(1+\varepsilon\eta')} r\, \sin\theta dr\, d\theta d\varphi =$$

$$= \frac{\rho_{sol}\rho_{sph} R^5 \pi}{6\varepsilon_0} \int_0^\pi (2\varepsilon\eta' + \varepsilon^2\eta'^2)\sin\theta d\theta$$

$$E_{sol} = \frac{\rho_{sol}^2}{2\varepsilon_o} \int_o^\pi \sin\theta' d\theta' \int_R^{R(1+\varepsilon\eta')} r'^2 dr' \int_R^{r'} \frac{r^2 dr \sin\theta d\theta}{\sqrt{r^2+r'^2-2rr'\cos\theta}} \qquad (3.6)$$

which can be approximated up to order ε^5 with :

$$E_{sol} \simeq \frac{\pi \rho_{sol}^2 a^5}{120 \varepsilon_o} \qquad (3.7)$$

On the other hand the conservation of the volume gives :

$$V_o = \frac{4 R_o^3}{3} = \frac{4\pi R^3}{3} + V_{sol}$$

which can be written in the form ^{3}) :

$$R^3 = R_o^3(1 - 3 \ln 2 \; \varepsilon L^2 R_o^2) \qquad (3.8)$$

Consequently, from the equations (3.3), (3.4) and (3.6) with the additional condition (3.8) we obtain for the potential energy the expression :

$$E_p = 2\pi\sigma R_o^2(1-2\ln 2 \; \varepsilon L^2 R_o^2 - \ln^2 2 \; \varepsilon^2 L^4 R_o^4) \int_o^\pi (1 + 2\varepsilon\eta' +$$

$$+ \varepsilon^2(\eta'^2 + \frac{\eta_\theta'^2}{2}))\sin\theta d\theta + \frac{4\pi\rho_{sph}^2 R_o^5}{15\varepsilon_o} [1 - 5\ln 2 \; \varepsilon L^2 R_o^2 +$$

$$+ 5\ln^2 2 \; \varepsilon^2 L^4 R_o^4] + \frac{\rho_{sol}\rho_{sph} R_o^5}{6 \; \varepsilon_o} [1-5\ln 2 \; \varepsilon L^2 R_o^2 + 5\ln^2 2\varepsilon^2 L^4 R_o^4] \times$$

$$(3.9)$$

$$\times \int_o^\pi (2\varepsilon\eta' + \varepsilon^2\eta'^2)\sin\theta d\theta + \frac{\pi \rho_{sol}^2 R_o^5}{120\varepsilon_o} \varepsilon^5$$

The above integrals could be evaluated approximatively by taking into account the rapid decreasing shape of the soliton. Indeed, let us denote the first integral with :

$$C_1 = \frac{1}{\eta_o} \int_o^\pi \eta' \sin\theta d\theta \sim \frac{1}{\eta_o} \int_o^{L/2} \eta' \sin\theta d\theta \qquad (3.10)$$

Changing the variable θ with $x = \frac{\theta}{L}$ and expanding the sin function, the principal terms are the following :

$$C_1 = 4L \int_o^{1/2} \frac{\sin(Lx)}{(e^x+e^{-x})^2} dx = 4L^2 \int_o^{1/2} \frac{xdx}{(e^x+e^{-x})^2} -$$

$$- \frac{4L^4}{3!} \int_o^{1/2} \frac{x^3 dx}{(e^x+e^{-x})^2} + \ldots$$

$$= 4 A_1 L^2 - 2/3 \, A_3 L^4 + \ldots \qquad (3.11)$$

where A_1 and A_3 are the corresponding integrals. Proceeding in the same way for the other integrals we obtain :

$$C_2 = \frac{1}{\eta_o^2} \int_o^\pi \eta'^2 \sin\theta d\theta = 16L^2 \int_o^{1/2} \frac{xdx}{(e^x+e^{-x})^4} -$$

$$- \frac{16L^4}{3!} \int_o^{1/2} \frac{x^3 dx}{(e^x+e^{-x})^4} + \ldots$$

$$= 16 B_1 L^2 - \frac{8}{3} B_3 L^4 + \ldots \qquad (3.12)$$

and

$$C_3 = \frac{1}{\eta_o^2} \int_o^\pi \eta_\theta'^2 \sin\theta d\theta = 64 \int_o^{1/2} \frac{x(e^x-e^{-x})^2}{(e^x+e^{-x})^6} dx -$$

$$- \frac{64}{3!} L^2 \int_o^{1/2} \frac{x^3(e^x-e^{-x})^2}{(e^x+e^{-x})^6} dx + \ldots = 64 D_1 - \frac{32}{3} D_3 L^2 + \ldots \qquad (3.13)$$

where B_1, B_3, D_1 and D_3 are the corresponding integrals. Introducing the above integrals C_1, C_2 and C_3 in equation (3.9) we obtain, in the second order in ε the following expression for the potential energy :

$$\Delta E_p = E_p(\varepsilon, L^2) - E_p(0, L^2) = \\ = \varepsilon(AL^2 + BL^4) + \varepsilon^2(C + DL^2 + EL^4 + FL^6) \qquad (3.14)$$

where :

$$A = 8\pi\sigma R_o^2 (2n_o A_1 - \ln 2) + \frac{4\rho_{sph} R_o^5}{3\varepsilon_o} (\rho_{sol} n_o A_1 - \pi \ln 2) < 0$$

$$B = -\frac{2}{3}(4\pi\sigma + \frac{\rho_{sol}\rho_{sph} R_o^3}{3\varepsilon_o}) n_o A_3 R_o^2 < 0$$

$$C = 64 \pi\sigma R_o^2 n_o^2 D_1 > 0$$

$$D = 4\pi\sigma R_o^2 (8B_1 n_o^2 - \frac{8}{3} A_1 n_o - \frac{8}{3} D_3 n_o^2 - 32 D_1 \ln 2 n_o^2 - \ln^2 2) + \\ \frac{8}{3} \rho_{sol}\rho_{sph} R_o^5 n_o^2 B_1 \qquad (3.15)$$

$$E = 4\pi\sigma R_o^2 (\frac{16}{3} \ln 2 \, D_3 n_o^2 - \ln^2 2 + \frac{4}{9} A_3 n_o - 16 B_1 \ln 2 n_o^2 - \frac{4}{3} B_3 n_o^2) \\ + \frac{20 \ln 2}{3} \cdot \frac{sol \, sph R_o^5}{\varepsilon_o} n_o A_1 - \frac{4\pi}{3} \cdot \frac{\rho_{sph}^2 R_o^5}{\varepsilon_o} \ln^2 2$$

$$F = \frac{32 \ln 2\pi}{3} \sigma R_o^2 B_3 n_o^2 + \frac{10 \ln 2}{9} \cdot \frac{\rho_{sol}\rho_{sph} R_o^5}{\varepsilon_o} A_3 n_o > 0$$

with the corresponding signs.

We should like to stress that the above expression of the potential energy (3.14) depends only of two variables ε and L^2. Evidently as function of ε has a minimum at $\varepsilon = 0$.

This minimum corresponds to the spherical ground state. A second minimum corresponding to the soliton solution with $\varepsilon > 0$ could exists if $\Delta E_p(\varepsilon, L^2)$, as a third order polynomial in L^2, has another minimum as function of L^2. Correspondingly the maxima and minima of ΔE_p are given by the roots of a polynomial of second order. Due to the fact that A is negative and F is positive one root is positive and one negative. Consequently only one minimum of the potential energy ΔE_p as function of $L^2 > 0$ is existing. This could also be seen by the fact that the potential energy for small and large values of L^2 has positive values $\Delta E_p(\varepsilon, L^2 \to 0) \to \varepsilon^2 C > 0$ and $\Delta E_p(\varepsilon, L^2 \to \infty) \to \varepsilon^2 FL^6 > 0$ which gives a minimum at a fixed value of L^2 at L_{min} zero. The corresponding value of the potential energy (3.14) at the second minimum is given by $\Delta E_p(\varepsilon, L_{min}^2)$. Due to the volume conservation of the soliton (3.8) which gives a relation between ε and L^2 the value of the second minimum is fixed.

4. THE KINETIC ENERGY

The kinetic energy is given by the expression :

$$E_k = \frac{\rho_m}{2} \int_{V_{sol}} |v|^2 \, dv \qquad (4.1)$$

where the integration is performed over all the volume of the soliton V_{sol}. In order to calculate the above expression we can use the Gauss' formula with the values of the velocity potential ϕ on the surface Σ

$$E_k = \frac{\rho_m}{2} \int_{V_{sol}} |\nabla \phi|^2 \, dv = \frac{\rho_m}{2} \int_{\Sigma} \phi \, \Delta\phi \cdot d\mathbf{s} \qquad (4.2)$$

The vectorial element of the aria can be written in the form[12]):

$$d\vec{S} = \vec{n}dS = [1-\epsilon\eta' +\epsilon^2(\eta'^2 - \frac{\eta'_\theta}{2})]\cdot(1+\epsilon\eta', -\epsilon\eta'_\theta, 0)dS \quad (4.3)$$

with dS having the same expression as that used in the eq. (3.2). From the expansion of the velocity potential ϕ (eq. (2.2)) :

$$\phi|_\Sigma = f_0(\theta) + \epsilon\eta'j(\theta) + \epsilon^2 \frac{\eta'^2}{2}(-j(\theta) - g_\theta(\theta)) + \ldots \quad (4.4)$$

we obtain the integral (4.2) in the form :

$$E_k = 2\pi R \int_0^\pi [jf_0 + (\eta'j^2 + \eta'jf_0 - (\eta'g)_\theta f_0)] \sin\theta d\theta \quad (4.5)$$

Now, we use again the fact that the core of the nucleus is not affected by the dynamics of the soliton up to a depth of order h << R. So, the tangential component of the velocity u must vanish too on the surface r = R-h :

$$u|_{r=R-h} = f_{0,\theta} - \frac{h}{R} f_{1,\theta} + \ldots = 0 \quad (4.6)$$

Choosing the arbitrary constant f_0 equal to zero, the kinetic energy E_k becomes :

$$E_k = 2\pi R \int_0^\pi \eta'j^2 \sin\theta d\theta \quad (4.7)$$

From the eqs.(2.25) and (2.26) we obtain the following form for the j function :

$$j = k(\chi \eta'_\theta + \xi_\theta) \quad (4.8)$$

with k = h/R. Using the transformation (2.29) we obtain :

$$\xi_\theta = \frac{k(k\chi + aRV)\eta'_\theta - 2\epsilon\chi \, k\eta'\eta'_\theta + \epsilon^2\chi\eta'^2\eta'_\theta}{(k - \epsilon\eta')^2} \quad (4.9)$$

where we have used the fact that η' is in the form $(2.33)_{(1)}$. In addition we choosed the initial condition $\xi_{(t=0)} = 0$, for the Riccati equation from Chapter 2. After introducing eq. (4.9) into eq.(4.8) and eq.(4.8) into eq.(4.7) we have, in the order εk^2 the following form for the kinetic energy :

$$E_k = 2\pi R_o \varepsilon k^2 \int_o^\pi \frac{\eta'}{(k-\varepsilon\eta')^4} \{ \chi\eta'_\theta(k-\varepsilon\eta')^2 + [(k^2\chi+kaRV)\eta'_\theta - 2\varepsilon\chi kn'\eta'_\theta]^2 \} \sin\theta d\theta \quad (4.10)$$

A rough estimation of the kinetic energy gives :

$$E_k \lesssim \frac{m_{sol}V^2}{2} \sim \frac{\rho_m \cdot 2\ln 2 R_o^2 L^2}{2} V^2$$

$$\frac{E_k}{E_p} \sim \frac{L^2}{64\pi D_1} << 1$$

This shows that the addition of the kinetic energy E_k to the potential energy ΔE_p does not affect the existence of the minimum in the potential energy but modifies only the value of the minimum $\Delta E_{p min}$. We mention that the derivative of E_k with respect to k is positive :

$$\frac{dE_k}{dk} = 4\pi R_o \varepsilon \int_o^\pi \eta' j \frac{\partial j}{\partial k} \sin\theta d\theta$$

$$\frac{\partial j}{\partial k} = \frac{j}{k} + k \frac{\partial \xi_\theta}{\partial k}$$

so that :

$$\frac{dE_k}{dk} = 4\pi R_o \varepsilon \int_o^\pi \eta' (\frac{j^2}{k} + kj \frac{\partial \xi_\theta}{\partial k}) \sin\theta d\theta > 0$$

because $j < 0$ (for $\theta>o, \eta_\theta<0$ but $v>0$ and $v \sim j$ in the first order) and $\frac{\partial \xi_\theta}{\partial k} < 0$ as can be seen by direct differentiation. As a consequence if we write the sum between potential and kinetic energy it is possible to find such a value for k so that, at the minimum point, the total energy to be zero :

$$E_k(k, L^2_{min}, \varepsilon_{vol}) + \Delta E_p(k, L^2_{min}, \varepsilon_{vol}) = 0$$

where ε_{vol} is obtained from the volume conservation.
In this way, the second minimum is degenerate in energy with the ground state minimum, condition which is asked by the fact that the cluster decays are spontaneous decays.

5. CONCLUSIONS

We have shown that the hydrodynamical equations with nonlinear terms have soliton-type solutions. These solutions may explain the large enhancement of clusters, observed experimentally, on the surface of the spherical double magic nucleus ^{208}Pb. We concluded that the existence of natural radioactivities with emission of carbon, neon, magnezium and silicon is a proof of the existence of solitons on the nuclear surface. Further consequences must be deduced and compared with the experimental data in order to reach such an exotic conclusion.

REFERENCES

1. P.B.Price, Annu.Rev.Nucl.Part.Sci. $\underline{39}$, 19 (1989)
2. A.Sandulescu, J.Phys.G : Nucl.Part.Phys. $\underline{15}$, 529 (1989)
3. E.Lamb, "Theory of solitons", Academic Press, New-York, 1985
4. L.Landau, E.Lifschitz, "Mecanique des fluides", MIR, Moscow, 1971
5. A.Pop, A.Sandulescu, H.Scutaru, W.Greiner, Z.Phys. $\underline{329}$, 357 (1988)
6. J.D.Jackson, "Classical Electrodynamics", John Willey, New-York, 1962
7. Plasma Physics, ed. B.Kadomtsev, MIR Publ.,Moscow,1981
8. S.Novikov et al, "Theory of Solitons", Consultants Bureau, New-York, 1986
   M.Laksnmanan, "Solitons-Introduction and Applications", Proc.Winter School Bharathidasan Univ., 1987, Springer-Verlag
9. L.D.Fadeev, L.A.Takhtajan, "Hamiltonian Methods in the Theory of Solitons", Springer-Verlag, 1987
10. V.I.Arnold, "Mathematical Models of Classical Mechanics", Nauka, Moskwa, 1974
11. G.Alaga, "Selected Topics on Vibrational Nuclei", Cargese School, 1966
12. B.Budak, S.Fomine, "Multiple Integrals, Field Theory and Series", Mir, Moskow, 1973.

HEAVY FRAGMENT RADIOACTIVITY

I.Silisteanu, M.Ivascu

Institute of Physics and Nuclear Engineering

Bucharest, P.O.Box MG-6, ROMANIA

and

I.Rotter

Zentralinstitut für Kernforschung, Rossendorf
Postfach

ABSTRACT : The effect of collective mode excitation in heavy fragment radioactivity (HFR) is explored and discussed in the light of current experimental data. It is found that the coupling and resonance effects in fragment interaction and also the proper angular momentum effects may lead to an important enhancing of the emission process. New useful procedures are proposed for the study of nuclear decay properties. The relations between different decay processes are investigated in detail. We are also trying to understand and explain in a unified way the reaction mechanisms in decay phenomena.

INTRODUCTION

The renewed interest in heavy fragment radioactivity(HFR) processes sprang up because some experimental data could not be understood within the framework of the solutions proposed earlier. These data include the "fine structure", similar to the alpha decay, and the evidence of the important role of excited states in HFR.

In spite of remarkable progress in untangling and understanding of these new phenomena $^{1-10}$), some interesting and incomprehensible questions still remain. The most intricating question concerns the significant interplay between the reaction mechanism and the nuclear structure. This question will be investigated in detail in the present paper. For this purpose we would like to extend the lifetime calculations in the spirit of references ^{7,8}) to all HFR cases so far observed. In our investigations we try to search for answers to the following urgent questions : what are the possibilities for enhancing HFR and how much is the probability of induced HFR (similar to the induced fission).

1. An outline of the theory

We shall adopt for a description of HFR processes the multichannel approach ^{8}) which joins together the advantages of the microscopic decay theory and resonance theory of nuclear reactions with the optical model for the emitted fragment. If we consider the decay of the metastable state k of a nucleus into the set c of two body channels, the total width is given by

$$\Gamma^k = 2\pi \sum_c \left| \frac{<u_c^o(r)|I_c^k(r)>}{<u_c^k(r)|I_c^k(r)>} \right|^2 \qquad (1)$$

where u_c are solutions of systems of differential equations

$$[\frac{\hbar^2 d^2}{2M dr^2} - V_c(r) + \varepsilon_c] u_c^{o(k)} - \sum_{c'} V_{cc'}(r) u_{c'}^{o(k)}(r) = \begin{Bmatrix} 0 \\ I_c^k(r) \end{Bmatrix}$$

where :

$M = A_1 \cdot A_2/(A_1+A_2)$ is the reduced mass of the system ;

V_c and $V_{cc'}$ are the diagonal and coupling parts of the interaction potential ;

I_c^k is the cluster formation amplitude (CFA) ;

$\varepsilon_c = E - E_1 - E_2$ is the released energy.

The boundary conditions for the homogeneous system (2) are those imposed for the scattering states. The systems (2,3) are solved numerically in the adiabatic approximation.

Thus, the decay width naturally results from the evolution of formation and penetration probabilities. The usual approach to these probabilities includes the average effects on single particle degrees of freedom in order to obtain CFA and the collective degree of freedom (relative distance between the fragments) for obtaining the interaction potential and finally the decay width.

We construct CFA in the simple shell model approximation [7]) and also in the resonance approximation when the metastable state is identified with the lowest resonance state of the compound system at a given decay energy ε_c. The first approximation corresponds to the fragments having the internal degrees of freedom associated to nucleons, while in the second these degrees of freedom are suppressed.

We shall use for the fragment interaction the nuclear potential in conjunction with the Coulomb one. The input parameters used are those which reproduce the measured fusion touching cross sections or reaction cross section data [11-13]) at energies near the Coulomb barrier. Because the released energies ε_c in HFR are somewhat different from experimental energies we shall modify only the depth of the nuclear potential in order to recover a resonance state in the potential at a given energy. In other words the decaying state is identified with the lowest resonance state of the compound system. In this case the CFA is given by resonance wave function [7]). This in fact is just the asymptotic Coulomb CFA. The parameters used for some HFR cases are given in Fig.1. It should be mentioned that the potentials resulting from the optical analysis of the fusion and scattering experiments are expected to be very relevant for the calculations of HFR rates.

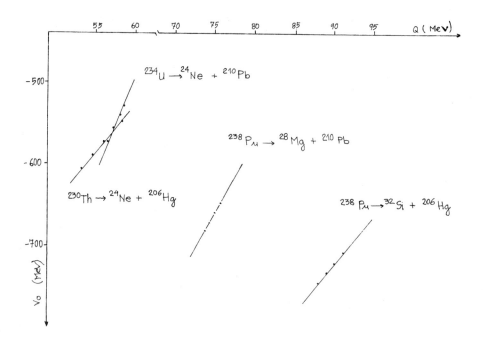

Fig.1. Nuclear resonance depths used in lifetime calculations for ^{24}Ne, ^{28}Mg and ^{32}Si emissions. The values r_o and a of the real optical potential are taken from the ref. [11-13]).

2. Results

The lifetimes calculated with the prescriptions of sections 2 and 3 for some HFR cases suggested by Price[14]) are presented in Table 1. Column 3 gives the results of the single channel analysis [7]) (one decaying state one open channel) "without" the coupling induced by nuclear deformation. The results of the multichannel analysis (one decaying state - many open channels) which includes the coupling induced by nuclear deformation are given in Column 4.

TABLE 1. *Halftime calculations for HFR cases ^{14}) within two different procedures ^{7,8}) without and with nuclear deformation. The resonance parameters of the potential are used in both types of calculations (see text).*

Decay			Log T^{calc} (s)		Log T^{exp} (s)
			Ref.^{7})	Ref.^{8})	Ref.^{10})
^{221}Fr	^{14}C + ^{107}Tl	31.285	20.18	18.74	>15.77
^{221}Ra	^{14}C + ^{207}Pb	32.400	19.17	17.62	>14.35
^{225}Ac	^{14}C + ^{211}Si	30.470	23.51	22.13	>18.34
^{231}Pa	^{23}F + ^{208}Pb	51.831	25.63	24.42	>24.61
^{230}Th	^{24}Ne + ^{206}Hg	57.766	26.42	25.16	14.64±0.07
^{232}Th	^{26}Ne + ^{206}Hg	56.581	28.89	26.71	>27.94
^{233}U	^{25}Ne + ^{208}Pb	60.832	26.15	24.24	24.83±0.15
^{234}U	^{24}Ne + ^{210}Pb	58.830	27.99	25.78	25.25±0.05
^{234}U	^{26}Ne + ^{208}Pb	60.088	28.84	26.07	25.25±0.05
^{237}Np	^{30}Mg + ^{207}Tl	75.706	28.56	27.69	>27.27
^{238}Pu	^{28}Mg + ^{210}Pb	75.915	25.87	24.93	25.70±0.25
^{238}Pu	^{30}Mg + ^{208}Pb	76.716	28.47	26.15	25.70±0.15
^{238}Pu	^{32}Si + ^{206}Hg	91.208	27.74	26.33	25.30±0.16

Comparing the results of Columns (4,5) and one may observe that the corrections arising essentially from low multipole distorssions of the potential at the scission configuration are quite important. The magnitude of these corrections clearly depends on the magnitude of fragment deformations, the maximal corrections corresponding to the large values of nuclear deformations. As a rule, these corrections produce a lifetime decrease by 1-2 orders of magnitude.

Note that the results of Table 1 column 4, together with those of ref. [8]), table 1 column 5, represent a complete lifetime multichannel account which covers all current experimental information on HCR.

The present study also confirms the conclusion of ref. [8]) - the HFR may be enhanced if the rotational degrees of freedom of the fragments are excited.

3. Comparison of cluster models

Table 2 shows in the columns 3-5 the emission rates predicted by the simple cluster models [5-7]) and in the column 6 the experimental rates. The common approach to these models includes the average over single particle degrees of freedom in order to obtain a reasonable estimation for CFA. In spite of rather different approaches for CFA the agreement of the calculated and experimental rates is remarkable.

The CFA results in the semi-empirical approach [6]) from fitting the α-formation date, while in our approach CFA is obtained numerically by averaging over the shell model α-formation amplitudes. In the former case the preformation factors used $S_\alpha = 6.3 \times 10^{-3}$ (even parent), $S_\alpha = 3.2 \times 10^{-3}$ (odd parent).

In the second case S_α depends on the structure of the single particle configuration used, finally the CFA being a Gaussian function of the relative distance between fragments. For completness we also calculate the so called experimental spectroscopic factor (ESF)

TABLE 2. Nuclei with measured partial halflives for HFR and the results predicted by the cluster models.

Decay		E (MeV)	log T (s) 5	6	7	log T^{exp} (s)
^{221}Fr	^{14}C	29.28	14.00	15.5	20.18	15.77
^{221}Ra	^{14}C	30.34	12.40	14.2	19.17	14.35
^{222}Ra	^{14}C	30.87	11.2	11.8	10.5	11.02 ± 0.06
^{223}Ra	^{14}C	29.85	15.3	15.1	14.9	15.2 ± 0.05
^{224}Ra	^{14}C	28.63	16.1	16.2	15.3	15.9 ± 0.12
^{225}Ac	^{14}C	28.57	18.8	18.6	23.51	18.34
^{226}Ra	^{14}C	26.46	21.2	21.1	21.7	21.33 ± 0.2
^{231}Pa	^{23}F	46.68	-	-	25.63	24.61
^{230}Th	^{24}Ne	51.75	24.4	24.8	26.42	24.64 ± 0.07
^{232}Th	^{26}Ne	49.70	28.7	27.9	28.89	27.94
^{231}Pa	^{24}Ne	54.14	21.6	23.4	23.9	23.38 ± 0.08
^{232}U	^{24}Ne	55.86	20.2	20.8	19.8	21.06 ± 0.1
^{233}U	^{24}Ne	54.27	23.7	25.4	24.4	24.83 ± 0.15
^{233}U	^{25}Ne	54.32	-	-	26.15	24.83 ± 0.15
^{234}U	^{24}Ne	52.81	25.5	25.6	27.99	25.25 ± 0.05
^{234}U	^{26}Ne	52.87	25.9	26.4	28.84	25.25 ± 0.05
^{234}U	^{28}Mg	65.26	25.7	25.4	25.7	25.75 ± 0.06
^{237}Np	^{30}Mg	65.52	27.3	29.9	26.47	27.27
^{238}Pu	^{30}Mg	67.00	25.6	25.8	28.47	25.7 ± 0.15
^{238}Pu	^{28}Mg	67.32	25.7	26.9	25.87	25.7 ± 0.25
^{238}Pu	^{32}Si	78.95	26.0	25.7	27.74	25.3 ± 0.16
^{241}Am	^{34}Si	80.60	25.3	28.8	28.8	25.3

$$S_c = \Gamma^{exp} / \Gamma^{res} \qquad (4)$$

where Γ^{exp} is the experimental cluster decay width and Γ^{res} is the resonance decay width arising only from the proper transmission probability (the correct asymptotic CPA being the resonance wave function of the compound system (see ref.[7])).

The ratio (7) may be looked at as a global measure of the contribution of all structure effects to the decay width. The calculated ESF are presented in the Table 3 (column 4). Here one may observe :

    i) The ESF is practically constant for a given type of HFR,
    ii) The ESF decreases with the increasing mass of the emitted fragment,
    iii) Even-odd effects in HFR are clearly transposed in ESF.

TABLE 3. *The experimental and resonance decay widths for HFR and their corresponding spectroscopic factors (see text)*

	Decay		Γ^{exp} (MeV)	Γ^{res} (MeV)	S
1	^{221}Fr	^{14}C	0.775 (-37)	0.913 (-26)	0.848 (-11)
2	^{221}Ra	^{14}C	0.204 (-35)	0.671 (-24)	0.304 (-11)
3	^{222}Ra	^{14}C	0.440 (-32)	0.202 (-21)	0.217 (-10)
4	^{223}Ra	^{14}C	0.303 (-36)	0.228 (-26)	0.132 (- 9)
5	^{224}Ra	^{14}C	0.629 (-37)	0.858 (-27)	0.773 (-10)
6	^{225}Ac	^{14}C	0.209 (-39)	0.769 (-28)	0.271 (-11)
7	^{226}Ra	^{14}C	0.285 (-42)	0.105 (-31)	0.276 (-10)
8	^{231}Pa	^{23}F	0.112 (-45)	0.713 (-31)	0.157 (-14)
9	^{230}Th	^{24}Ne	0.104 (-45)	0.257 (-31)	0.404 (-14)
10	^{232}Th	^{26}Ne	0.523 (-49)	0.670 (-35)	0.780 (-14)
11	^{231}Pa	^{24}Ne	0.176 (-44)	0.781 (-29)	0.225 (-15)
12	^{232}U	^{24}Ne	0.460 (-42)	0.901 (-27)	0.510 (-15)
13	^{235}U	^{24}Ne	0.699 (-46)	0.649 (-30)	0.107 (-15)
14	^{233}U	^{25}Ne	0.691 (-46)	0.504 (-29)	0.137 (-16)
15	^{234}U	^{24}Ne	0.257 (-46)	0.280 (-33)	0.917 (-13)
16	^{234}U	^{26}Ne	0.257 (-46)	0.139 (-31)	0.184 (-15)
17	^{234}U	^{28}Mg	0.606 (-47)	0.517 (-31)	0.117 (-15)
18	^{237}Np	^{30}Mg	0.245 (-48)	0.994 (-32)	0.246 (-16)
19	^{238}Pu	^{30}Mg	0.910 (-47)	0.397 (-31)	0.229 (-15)
20	^{238}Pu	^{28}Mg	0.910 (-47)	0.522 (-31)	0.174 (-15)
21	^{238}Pu	^{32}Si	0.292 (-46)	0.766 (-31)	0.381 (-15)
22	^{241}Am	^{34}Si	0.113 (-46)	0.407 (-30)	0.277 (-16)

If we introduce the following parametrisation in complete analogy with the one for α-decay

$$S_c = S_\alpha^{n_\alpha + (A_2 - 4n_\alpha)/4} \qquad (5)$$

where n_α is the number of α-particles, then from S_c values (Table 3) one may deduce the mean value of the alpha spectroscopic factor S_α. For all the observed HFR S_α varies between $0.17 \cdot 10^{-3}$ (^{34}Si from ^{241}Am) and $6.8 \cdot 10^{-3}$ (^{24}Ne from ^{236}U). These values are in good agreement with those calculated by Blendowske and Wallisser [6]).

4. Short comment on HFR data and theoretical models

A detailed comparison of the data with the results predicted by theoretical models is made in ref.[10]). In addition, we want to point out that :

1. The Geiger-Muttal law, $\text{Log } T = AQ^{-1/2} + B$, is strictly verified for measured HFR cases, as well as in α-decay (see fig.2).

2. The dependence of $\text{Log } T_c$ on the fissionability parameter $X = Z_1 Z_2 / A^{1/3}$ is very different from the one observed for experimental fission lifetimes (see fig.3). Note that the released energy Q increases with the barrier high (fig.4) as well as the mean kinetic energy in fission.

3. With the increasing mass of the emitted cluster lifetimes of HFR vary between α and fission lifetimes (fig.5).

From a theoretical point of view all the decay rates (alpha, cluster and fission) may understood as a barrier penetration phenomenon showed down by the formation probability factor.

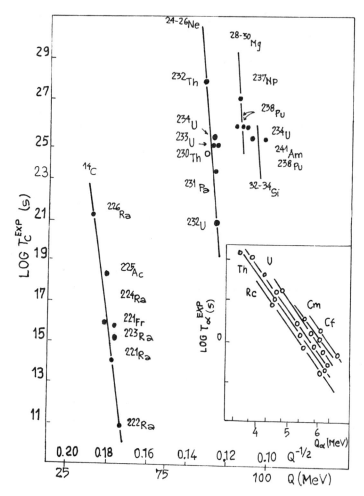

Fig.2. Geiger-Nuttal plott for nuclei with measured partial halflives for HFR and alpha decay (right lower part).

4. The inclusion of degrees of freedom such as deformation [2], pairing [15], collective rotation [8] leads to important corrections to the decay rates that may improve the accord to the data.

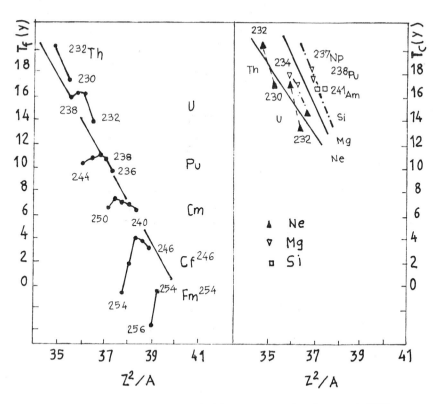

Fig.3. Experimental halflives for fission and HFR plotted as a function of fissionability parameter Z^2/A.

Usually in refs. [1,4] the exact boundary conditions at small and large distances are not imposed in the dynamical description of the process. Therefore, the resonant aspects of the phenomenon and asymptotic normalizations of formation and penetration probabilities are also neglected. Furthermore, the excited states [18] are not explicitely treated.

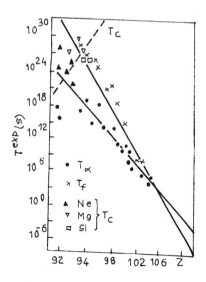

Fig.4. Experimental halflives for α-decay, HFR and fission plotted as a function of the proton number Z. The halflives correspond to the most long living isotopes.

In our method, these difficulties in the dynamical treatment of the decay are eliminated from the beginning. But, the shell model approach for CFA usually needs [7] tedious calculations and finally, a very long computing time.

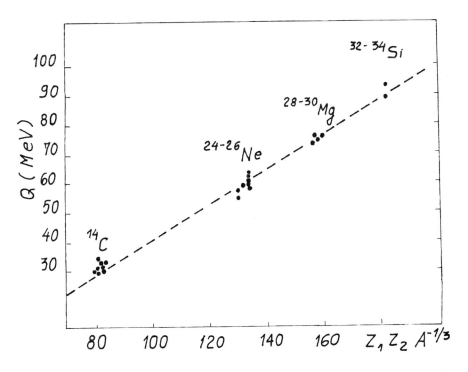

Fig.5. The released energy in HFR as a function of the barrier high.

The major part of lifetime calculation in both the macroscopic and microscopic treatments is contained in penetrability integral associated with the post-scission barrier, determined by the interaction energy of final fragments in the regime between last contact and infinite separation. This major part can be estimated accurately provided realistic nuclear radii and nucleus α-nucleus potentials are used.

The more difficult part of lifetime calculation concerns the region between the ground state of the parent and the last contact (scission) configuration.

In the macroscopic treatment this region is treated as a simple extension of the barrier (one dimensional) penetrability problem by using empirically information on the barrier at both ends of the regime. It should to be stressed here that the informations on the left end (inner) of the barrier remains uncertain and quite conventional, while that the informations on the right end are more complete being supported by scattering data.

In the microscopic treatment this part of calculation is estimated in general case by considering the preformations as a part of the process of barrier penetration itself. Thus, the CFA controlates the penetrability of this regime.

5. *Statistical coupling between orbital and intrinsic angular momenta*

In the spirit of simplicity let us assume that the exit channel configuration consists of two touching inequal rigid spheres with all the associated rotational degrees of freedom. This simple model leads to analytical predictions for relevant statistical distributions in HFR.

First, we consider the equilibrium between the intrinsic rotation of the fragments and their orbital rotation, assuming that : the relevant angular momenta are all parallel to the other and there are no sliding and rolling forces. As we shall see, such an equilibrium partition involves the excitation of the collective modes known as wriggling [16]).

If the total angular momentum is I and the fragment spins s_1 and s_2, the energy for an arbitrary partition between intrinsic and orbital angular momentum is

$$E(s_1, s_2) = \frac{s_1^2}{2J_1} + \frac{s_2^2}{2J_2} + \frac{(I - s_1 - s_2)^2}{2J_3} \qquad (6)$$

where J_1 and J_2 are the momenta of inertia of the fragments and J_3 is the relative moment of inertia. The partition function is expressed as

$$F \propto \iint ds_1\, ds_2\, \exp[-E(s_1 s_2)/T] = \\ = 2T \left| \frac{J_1 J_2 J_3}{J_1 + J_2 + J_3} \right|^{1/2} \exp\left| -\frac{I^2}{2T} \frac{1}{J_1 + J_2 + J_3} \right| \qquad (7)$$

where T is the nuclear temperature.

The average orbital momentum is given by

$$\bar{L} = \overline{(I - s_1 - s_2)} = F^{-1} \iint ds_1\, ds_2 (I - s_1 - s_2) \exp{-E(s_1, s_2)/T} \qquad (8)$$

After simple calculations one obtains

$$\bar{L} = \frac{J_3}{J_1 + J_2 + J_3} I \qquad (9)$$

Similarly, the average fragment spins arising from rigid rotation are

$$\bar{s}_1 = \frac{J_1}{J_1 + J_2 + J_3} I \quad ; \quad \bar{s}_2 = \frac{J_2}{J_1 + J_2 + J_3} \qquad (10)$$

while the individual orbital momenta are

$$I_1 = \frac{R_2}{R_1+R_2} \frac{J_3}{J_1+J_2+J_3} \quad ; \quad I_2 = \frac{R_1}{R_1+R_2} \frac{J_3}{J_1+J_2+J_3} \tag{11}$$

where R_1 and R_2 are the fragment radii.

The distribution of the intrinsic and orbital angular momenta between the two fragments with the mass A_1 and A_2 is

$$\frac{\bar{s}_1}{\bar{s}_2} = \left|\frac{A_1}{A_2}\right|^{5/3} \quad \text{and} \quad \frac{\bar{\ell}_1}{\bar{\ell}_2} = \left|\frac{A_2}{A_1}\right|^{1/3} \tag{12}$$

In other words the energy is mainly converted in the rotation energy of the lead fragment and the orbital energy of the light fragment.

The average module of the orbital momentum relevant for decay processes [16]) is obtained from

$$\overline{|L|} = E^{-1} \int\int ds_1\, ds_2\, |I-s_1-s_2|\, \exp - E(s_1,s_2)/T \tag{13}$$

If one approximates the module $|I-s_1-s_2|$ for small and large I by values $(s_1+s_2)1 - \frac{I}{s_1+s_2}$ for I s_1+s_2 and $I(1-\frac{s_1+s_2}{E})$ for $I = s_1+s_2$ and use the continuity of the integrand at $I = s_1+s_2$ one obtains the dimensionless form

$$\frac{\overline{|L|}}{(JT)^{1/2}} = \left[\frac{8}{\pi}\right]^{1/2} \left[\frac{J_2+J_3}{J_2}\right]^{1/2} \exp\left[-\frac{x^2}{2} \frac{J_1^2\, J_2}{J_3^2(J_2+J_3)}\right] +$$

$$+ \frac{J_1}{J_3} X \{ \text{erf} \left[X \left(\frac{J_2}{2(J_2+J_3)} \right)^{1/2} \right] \frac{J_1}{J_3} - 1 \}$$

$$+ \left[\frac{8}{\pi} \right]^{1/2} \left(\frac{J_2+J_3}{J_1} \right)^{1/2} \exp \left[-\frac{X^2}{2} \frac{J_1 J_2^2}{J_3^2(J_1+J_3)} \right]$$

$$+ \frac{J_2}{J_3} X \{ \text{erf} \left[X \left(\frac{J_1}{2(J_1+J_3)} \right)^{1/2} \frac{J_2}{J_3} \right] - 1 \} \tag{14}$$

for small I

$$\frac{\overline{|L|}}{(JT)^{1/2}} = X \tag{15}$$

for large I

where

$$J = \frac{J_1 J_2}{J_1+J_2} + J_3 \quad \text{and} \quad X = \frac{J_3}{J_1+J_2+J_3} \frac{I}{(JT)^{1/2}} \tag{16}$$

In the limit of large I one recovers from (15) the result (9).

For I 0 one obtains from (14)

$$\overline{|L|} = \left[\frac{8}{\pi} \right]^{1/2} \left[\left(\frac{J_1+J_3}{J_1} \right)^{1/2} + \left(\frac{J_2+J_3}{J_1} \right)^{1/2} \right] (JT)^{1/2} \tag{17}$$

The interesting result is that at I = 0 the orbital momentum arises from the excitation of a wriggling mode. In this case the two fragments spin in the same direction while the compound system as a whole rotates in the opposite direction in order to maintain I=0. The excitation of the wriggling mode in HFQ is illustrated in figs.6-7.

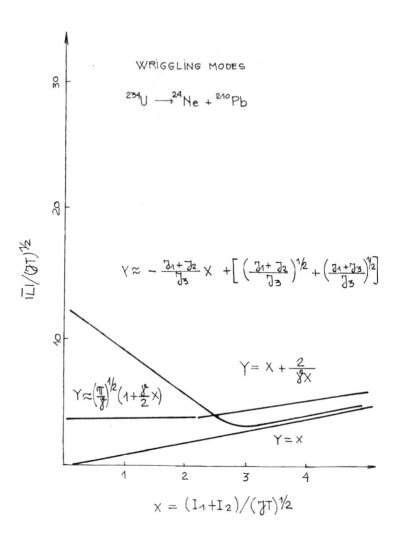

Fig.6. Total orbital momentum of the fragments arising from wriggling as a function of the spin arising from rigid rotation alone plotted in dimensionless form for ^{24}Ne emission from ^{234}U. The upper solid curve shows the result for excitation of the wriggling, while the lower curve corresponds to the coupling of a single wriggling mode to the total fragment spin arising from rigid rotations.

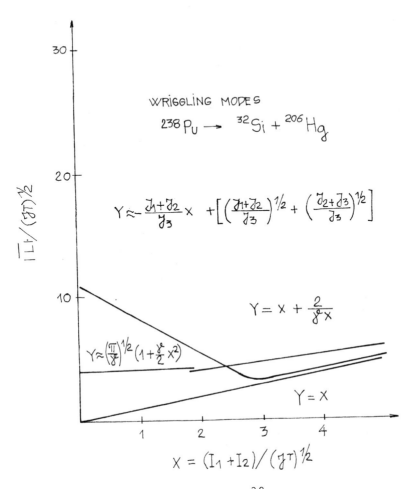

Fig.7. Same as Fig.6 but for ^{32}Si emission from ^{233}Pu.

The collective modes excited during the process may be coupled to the angular momentum arising from relative motion on rigid rotation. For example, the wriggling mode may be coupled to the total spin arising from rigid rotation. If the aligned components due to the rigid rotation are I_1 and I_2 (eqs. 11) and the wriggling one is W the partition function is

$$F = \iint dW \, d\theta \, W \exp(-E/T) \qquad (18)$$

E being a function of I_1, I_2 and W, and θ, the angle between I_1 and W.

The average module of the orbital momentum may be finally approximated by

$$\frac{\overline{|L|}}{(JT)^{1/2}} = X + \frac{2}{X} - \left[\frac{X}{2} + \frac{2}{\gamma X}\right] \exp\left[-\gamma^2 X\right] + \left[\frac{\pi}{\gamma}\right]^{1/2}$$

$$\mathrm{erfc}\left[\gamma^{1/2} \frac{X}{2}\right] \left[1 + \frac{\gamma}{2} X^2\right] \qquad (19)$$

where $\gamma = p \, J \, T$ and $p = \dfrac{4J_1 J_2 + J_2 K_3 + J_1 J_3}{2T J_1 J_2 J_3}$.

The function (19) has the following limiting values

$$\frac{\overline{|L|}}{(JT)^{1/2}} = \left(\frac{\pi}{\gamma}\right)^{1/2} \left(1 + \frac{\gamma}{2} X^2\right) \quad \text{for small } X \qquad (20)$$

$$\frac{\overline{|L|}}{(JT)^{1/2}} = X + \frac{2}{X} \qquad \text{for large } X \qquad (21)$$

The results (9, 17, 20, 21) are illustrated in figs. 6-7).

The wriggling mode generates a random angular momentum in a plane perpendicular to the axis of fragment centers. The wriggling moment and that arising from rigid rotation lead to a fluctuation in the orientation of the total spin, again in the plane perpendicular to the separation axis. The thermal fluctuations, may also lead to fluctuations of the angular momentum projection on the dezintegration axis (tilting).

Now let us analyse the effect of the wriggling mode in decay. The decays which we want to consider are $^{238}Pu \rightarrow {}^{24}Ne + {}^{206}Hg$; $^{238}Pu\ {}^{32}Si + {}^{206}Hg$. If we allow the initial system to envolve to the configuration of two touching inequal spheres ($r_o = 1.22$ fm), we have the quantity J and energy, hence T and J T or $(J\ T)^{1/2}$.

The average orbital momentum generated by the wriggling and the wriggling combined with rigid fragment rotation given by eqs. 17, 18 are illustrated in figs.6-7. As can be seen the excitation of the wriggling causes an important increase in the average orbital momentum (AOM) for light fragment emission. As a rule, the AOM decrease with the increasing mass of the emitted fragment.

The coupling of the wriggling mode to the rigid rotation (eqs.18,19) leads to an important decrease of AOM at small spins and an increase of AOM at large spins.

However, the excitation of these modes causes a modest increase in AOM over the rigid rotational volve, but leads to a sizeable spread in the fragment's angular momenta about the average values. In other words these modes may induce significant spin in the fragments for zero total angular momentum. The induced spins in HFR are around a few units of the Planck constant.

Conclusions

In conclusion, we have investigated the angular momenta associated with a few collective degrees of freedom in HFR.

The excitation of the collective rotational degrees of freedom connected with deviations from the sphericity of the fragments leads to significant corrections in the emission rates. Furthermore, such an excitation appears as an important mechanism for enhancing the HFR. The experimental investigation of the problem is already available with the current detection technique.

The excitation of other elementary modes carrying out the angular momentum such as wriggling tilting, bending and twisting causes a modest increase of the average transferred orbital momentum over the rigid rotational value.

The study of the relevant degrees of freedom in HFR reveals new aspects of HFR and allows a deep understanding of the phenomenon. If is hoped that this beginning of an investigation will prove useful to the study of nuclear reactions dominated by the Coulomb barrier.

Acknowledgements

We grateful acknowledge the pretious help and advice of our colleagues at IFIN-Bucharest and JINR-Dubna. Discussions with V.G.Soloviev, V.M.Shilov, Yu.S.Zamyatnin, S.P.Tretiakova, V.Volkov are greatly appreciated.

REFERENCES

1) D.N.Poenaru, M.Ivascu, A.Sandulescu and W.Greiner, 1984 Phys.Rev. C32, 572

2) Y.I.Shi and W.J.Swiatecki, 1987 Nucl.Phys. A464 205

3) G.A.Pik-Pichak 1986 Sov.J.Nucl.Phys. 44, 923

4) G.Shanmugam and B.Kamalaharam, 1988, Phys.Rev. C38, 1377

5) B.Buck and A.C.Marchant, 1989, J.Phys.G. Nucl.Part. Phys. 15, 615

6) R.Blendowske and H.Walliser, 1988, Phvs.Rev.Lett. 61, 1930

7) M.Ivascu and I.Silisteanu, 1988, Nucl.Phys. A485, 93

8) I.Silisteanu and M.Ivascu, 1989, J.Phys.G, Nucl.Part. Phys. 15, 1405

9) M.Iriondo, D.Jerrestam and R.J.Liotta, Nucl.Phys. A454, 252

10) P.Price Buford, 1989, Nucl.Phys. A502, 41 C and references

11) B.B.Beck, R.R.Bets, J.E.Gindler, B.D.Wilkins, S.Saini, M.B.Tsang, C.K.Gelbke, W.G.Lynch, M.A.Mc Mahan and P.A.Baisden, 1985, Phys.Rev. C32, 195

12) J.J.Kolota, K.E.Rehm, D.G.Kovar, G.S.F.Stephans, H.I. Rosner Kezoe and P.Waitecle, 1984, Phys.Rev. 30, 125

13) R.P.Cristiansen and A.Winther, 1977, Phys.Lett. B65,19

14) P.Price Buford ,private communication

15) F.Baranco, R.A.Brooglia, G.F.Bertch, 1988, Phys.Rev. Lett. 60, 507

16) L.G.Moretto and R.P.Schmitt, 1980, LBL Preprint 8656 Berkeley, California

17) M.Ivascu and I.Silisteanu, 1990, Phys.Elem.Part. and Atomic Nuclei, 21, 1405

18) W.Kutchera, J.Ahmad, S.G.M.Armato et al, Phys.Rev. C 32 (1985) 2036.

TOWARDS A BEAM OF POLARIZED ANTIPROTONS AT LEAR *)

Gerhard Graw

Sektion Physik, Universität München, 8046 Garching, Germany

In this contribution status and progress of our collaboration to obtain a beam of polarized stored antiprotons at LEAR is reported.

1. **Survey**

The polarization of the antiproton beam at LEAR shall be produced with the Filter method[1-3], using in the storage ring a gas target of spin polarized atomic hydrogen in a storage cell as filter target.

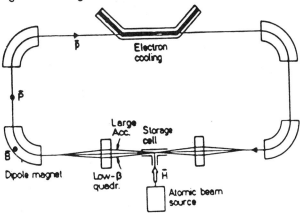

Fig. 1: Scheme of the FILTEX polarization arrangement

For the filter method (FILTEX) it is essential, that beam attenuation will result predominantly from the interaction with these polarized atoms and that furthermore the attenuation (total) cross sections are different for antiprotons with spin orientation parallel or antiparallel with respect to the proton spin

$$\sigma_{\uparrow\uparrow} = \sigma_0 + \sigma_1$$
$$\sigma_{\uparrow\downarrow} = \sigma_0 - \sigma_1.$$

(1)

If in LEAR the spin orientations of the individual antiprotons remain conserved, one of the spin components of the antiproton beam will vanish faster than the other one, thus leading to a definite polarization of the remaining $\bar{p}$ beam.

For 60 MeV as a typical energy of the antiprotons (p = 340 MeV/c)[3] the (unpolarized) attenuation cross section σ_0 is 0.2 barn. If a beam life time

$$\tau_0 = \frac{1}{\sigma_0 n L f} \tag{2}$$

of 10 hours is accepted a target thickness of $nL = 10^{14}/cm^2$ is required.

The $\bar{p}$ polarization will build up linearly with time

$$P_{\bar{p}} = t/\tau_1, \tag{3}$$

the characteristic time

$$\tau_1 = \frac{1}{\sigma_1 n L f} \tag{4}$$

depends only on σ_1, whereas σ_0, which may include effectively other loss mechanisms, determines the beam intensity at the time under consideration.

One of those additional loss processes is $p\bar{p}$ small angle elastic Coulomb scattering. To minimize beam losses, the target has to be placed in the focus of an arrangement of strong quadrupole lenses (low β-section) (Fig. 1) to have most of small angle Coulomb scattering within the acceptance angle of stable storage ring operation. For typical acceptance angles near $\pm 1.0^o$ the remaining Coulomb losses are about 25% of the losses due to nuclear interaction. For lower energies, e.g. at 30 MeV, they increase to about the same magnitude.

An essential point is, that up to now σ_1 is not yet known experimentally for $p\bar{p}$-scattering. In proton proton scattering at 30 MeV the value is large

$$\sigma_1/\sigma_0 = 81 \text{ mb}/105 \text{ mb} = 0.77. \tag{5}$$

Present theoretical estimates for antiproton proton interaction range between $|\sigma_1/\sigma_0|$ of 0.06, obtained by Dover (1986)[4] with the Paris Potential[5] and values of -0.10, -0.13 derived more recently by the Jülich group[6] (1989) from models[7] based on the Bonn Potential[8] treating the annihilation process phenomenologically. The ratio $|\sigma_1/\sigma_0|$ increases with decreasing incident proton energy[6].

If we succeed to obtain a sufficiently polarized $\bar{p}$ beam, the filter target will be used as a polarized target for reaction studies. This offers a unique possibility to study the spin dependence of $p\bar{p}$ interaction in

- elastic scattering,
- charge exchange and in
- specific mesonic annihilation channels,
  in measurements of
- integral and differential spin correlation parameters as well as
- analyzing powers for polarized p's as well as polarized $\bar{p}$'s.

Since the sign of the target spin polarization can be changed rapidly, measurements of high accuracy even for small asymmetries are feasible.

Starting with an initial number $N_0 = 10^{10}$ of stored unpolarized $\bar{p}$'s we expect after $t \approx 2\tau_0$ typically $N = 10^9$ stored polarized $\bar{p}$'s and thus a luminosity of $L = 10^{29}/\text{cm}^2$ sec. With a cross section of 10 mb/sr and detectors with effective solid angles of 0.3 sr typical countrates of 400 events/sec should be observed.

Thus, the spin correlation coefficients can be determined with a sufficient high statistical accuracy even if the beam polarization might be as low as 10%.

2. The p-$\bar{p}$ interaction

The physical motivations of the experiment is to study the $p\bar{p}$-interactions on which we briefly comment in the following.

With the success of LEAR there are now precise data for the total, the elastic and the charge exchange cross sections[10-25]. For elastic scattering at higher momenta (from 497 to 1550 MeV/c at 15 energies) there are in addition data for the analyzing power[27,28]. In the interpretation of these data the emphasis is on models including information about meson exchange. There was pioneering work by Bryan, Phillips[29] (1968), Dover, Richard[30] (1980 and later), the Paris[5] (1982) and Nijmegen[31] (1984) group. I will refer to a very recent study of the Jülich group[6,7] (1989), which is based on the Bonn potential for nucleon nucleon scattering below the π-threshold[8] and defined in terms of meson baryon vertices.

At low energy the $N\bar{N}$ potential contains real and imaginary parts. The real term should be related to the NN - meson exchange (ME) potential by the G-parity transformation. The absorptive part results mainly from annihilation processes and leads to an additional dispersive contribution to the real part of the potential

$$V_{N\bar{N}} = G\,(V_{NN}^{ME}) + \mathrm{Re}\,V^{Ann} + i\mathrm{Im}\,V^{Ann} \tag{6}$$

In the Bonn model different versions of the ME potential have been worked out, the OBE-potential with one boson exchange (first term in Fig. 2) only and a second version (BOX) including in addition direct 2π and $\pi\rho$ exchange and also intermediate Δ-excitation (the second and third terms in Fig. 2.)

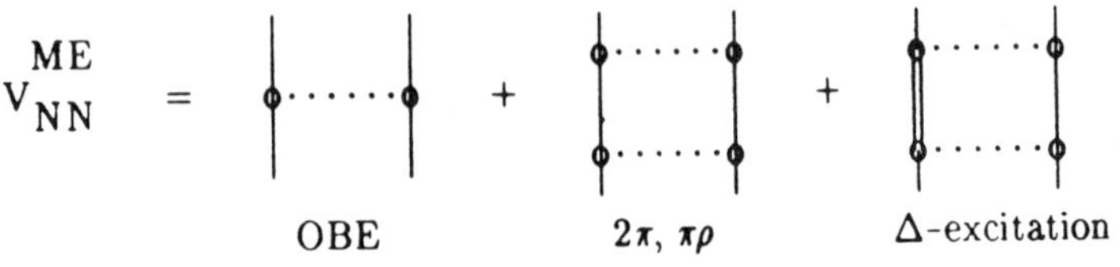

V_{NN}^{ME} =   OBE   + $2\pi, \pi\rho$ + Δ-excitation

Fig. 2: Diagrams included in the calculation of V_{NN}^{ME} of the Bonn-Jülich group

Table 1
List of vertices, taken into account (for more details, see ref. 8)

Vertices	S^π	(T^G)	V_C^{OBE}	V_{LS}^{OBE}	V_{SS}^{OBE}	V_T^{OBE}
<u>NNπ</u>	0^-	(1^-)	PS		$\pm$	$\pm$
NNσ'	0^+	(0^+)	S	-	-	
NNδ	0^+	(1^-)	S	+	-	
<u>NNω</u>	1^-	(0^-)	V	$\pm$	+3	± 2	+
NNρ	1^-	(1^+)	V	+	-3	+2	-
N$\Delta\pi$	0^-	(1^-)				
<u>N$\Delta\rho$</u>	1^-	(1^+)				

The table summarizes the vertices taken into account[6,7,8] (the dominant contributions are underlined), and also the quantum numbers and the spin structure of the corresponding one boson exchange (OBE) potentials. The resulting signs of the $_{NN}^{OBE}$ and $_{N\bar{N}}^{OBE}$ potential terms are indicated. For π, δ, ω exchange they differ because of the negative G-parity. The sign change causes the very strong attraction in $N\bar{N}$[9] and a spin orbit coupling of opposite sign than in NN but somewhat reduced in strength. Part of the vertices are effective and include e.g. correlated two meson exchange (in contrast to direct two meson exchange, treated explicitly in the box diagrams). The parametrization takes care of the G-parity of these terms.

With an annihilation potential parametrized phenomenologically as an optical potential, independent on energy, spin and isospin,

$$V^{Ann} = (U_0 + iW_0)e^{-r^2/2R^2} \tag{7}$$

with $U_0 = -0.6$ GeV, $W_0 = -4.6$ GeV, $R = 0.36$ fm, (in the case of the BOX potential) all data have been reproduced reasonably well.

This is illustrated in figs. 3 to 6, taken from a Jülich report of Hippchen (1989)[7].

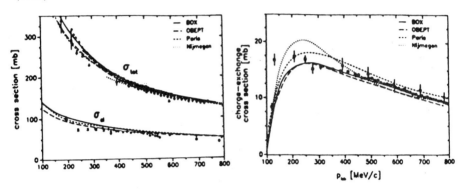

Fig. 3a and b: Energy dependence of σ^{tot}[12-15], σ^{el}[16-19] and σ^{CE}[20,21], with different model calculation (figs. taken form Ref. 7)

In Fig. 3a and 3b the energy dependence of σ_{tot}, σ_{el} and σ^{CE} and calculations with the potentials discussed before are dislayed: BOX: including the exchange of 2 mesons, OBEPT: a pure OBE potential without Box diagrams (but with readusted parameters for V^{Ann}), and results with the Paris and Nijmegen potential. They all lead to a reasonable description of the data.

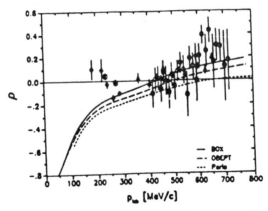

Fig. 4: The energy dependence of the forward scattering amplitude[22-25,7]

In Fig. 4[7] it is shown, that the main behaviour of $\rho = \text{Re } f(0)/\text{Im } f(0)$ is reproduced. However, all potentials fail to describe the positive values at low energy. This anomaly is not yet fully understood[9] but has been discussed tentatively in terms of an additional narrow p wave resonance.

In Fig. 5 a-c angular distributions of elastic scattering, charge exchange and polarisation at selected energies are shown. At most of the other energies a description of comparable quality is obtained. In general all these potentials lead

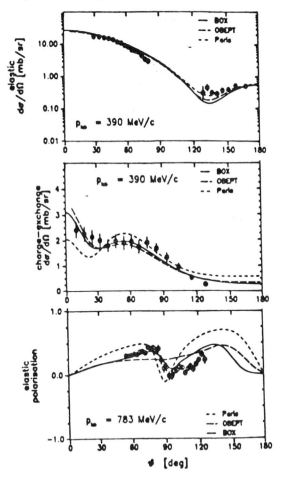

Fig. 5 a-c: Angular distribution of elastic scattering[19], charge exchange[21,26] and analyzing powers[27] for selected energies[7]

to reasonable results In contrast to proton proton scattering in this energy range (e.g. $|A_y\theta| \lesssim 0.04$ for $E_p \leq 52$ MeV[44]) the analyzing powers are large. This results in part from the much larger attraction of the central potential resulting in larger p-wave amplitudes[9]. The A_y data are better reproduced by the more complete BOX version of the Bonn potential than by OBEPT. However, the present data are not sufficient in order to distinguish between the models.

The scattering is dominated by the long range absorption. Haidenbauer et al.[6] (1989) emphasize that the reduction of flux is strongest at distances as large as 1 fm. Thus the heavy mesons contribute only via the very tails of their potentials.

Attempts have been made to calculate V^{Ann} microscopically assuming that quark rearrangement[32] dominates this process. The Jülich group, however, has printed out[6], that also a baryon exchange model with vertices, encluding strangeness production, is able to reproduce an annihilation cross section of the correct order of magnitude and the long ranged behaviour of the annihilation potential.

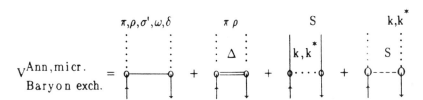

Fig. 6: Diagrams included in the microscopic calculation of the annihilation potential[6]

In Fig. 6 the diagrams included in these calculations are shown. In these calculations, however, the charge exchange cross section is overestimated by a factor of two and the sign for σ_1/σ_0 differs from that obtained in the phenomenological treatment of the absorption.

Apparently there are many interesting and open problems. They are related to the effective treatment of some of the meson exchange diagrams and the mechanism of annihilation. Experiments on the spin correlation are an important tool to distinguish between predictions.

3. TSR Filter experiment set up

Now back to our work.

The conditional approvement of the filter experiment (FILTEX) at LEAR demands to show successful operation of the method for a stored proton beam in the TSR[33] in Heidelberg. With electron cooling and at an average value of the residual gas pressure of 10^{-10} mbar a storage time τ_0 larger than 36 h has been observed for 21 MeV protons[33]. In Indiana the polarization of stored protons has been maintained for several million turns at an energy near an imperfection resonance[34] which is also very encouraging.

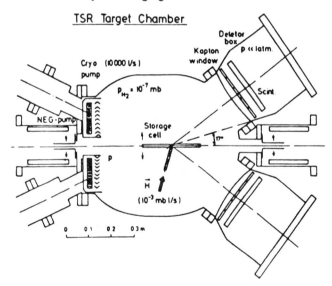

Fig. 7: On scale drawing of part of the components in the target chamber for the test experiment

Our target will be installed in the low β section. In this position the diameter of the electron cooled beam is 1 mm. The diameter of 10 mm for stable orbits in the TSR will be taken as a lower limit for the dimension of the storage cell. The target chamber for the TSR experiment (Fig. 7) is partly completed.

The vacuum chamber and the storage cell with adjustment and opening devices are finished. In Fig. 7 the first of 3 differential pumping stages are shown. They are equipped with non evaporating getter pumps in order to maintain in the ring a vacuum of better than 10^{-9} mbar. The low β focussing can be provided with the presently existing quadrupoles, operated in a special mode. The proton proton scattering in the test experiment will be detected in kinematic coincidence. The efficiency of a symmetric arrangements of (ΔE) scintillator strip detector has been studied in Monte Carlo simulations. For part of the angular ranges the additional detection of the remaining particle energy E_R in a following thicker scintillator provides further information. The effective solid angle of these detectors in left-right and up-down symmetric positions relative to the beam is about 0.3 sr. The polarization created in the circulating beam will be measured from the (differential) spin-spin correlation obtained from the change of the counting rates when the target spin polarization is reversed for a short time interval.

4. Storage Cell

The present target cell design[35] uses teflon coated aluminum walls 0.5 mm thick and cooled to 150 K. The storage cell is 25 cm long (feed tube 10 cm) with an inner diameter of 1 cm, to provide the $1 \cdot 10^{14}$ H/cm² target thickness for an assumed input flux of 10^{17} polarized H-atoms per second.

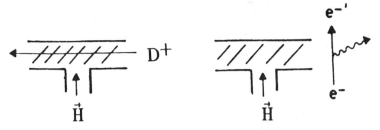

Fig. 8 a and b: Scheme of the Madison and Heidelberg target polarization detection arrangements

The depolarization of the hydrogen atoms in the teflon coated storage cell due to typically 400 wall collisions has been studied both by the Madison group[36], detecting the polarisation of electrons picked up from a beam of unpolarized deuterons (fig. 8a), and by the Heidelberg-Marburg[37] group, detecting the circular polarization of Balmer light emitted by the atoms after inelastic electron excitation (fig. 8b). Both methods use calibrations relative to the polarization P_0 of the incoming hydrogen beam.

At T = 150 K the Madison group observes $P/P_0 = 0.9$ integrating over all target atoms. In Heidelberg the detection of atoms, leaving the cell after typically 400 collisions yields $P/P_0 = 0.7$ in approximate agreement with the Madison results. Fig. 9 shows the temperature dependence recently obtained in Heidelberg. They confirm the fall off at lower temperature, observed in Madison.

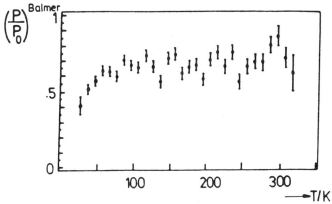

Fig. 9: Polarization of the outgoing hydrogen atoms, normalized to the polarization P_0 of the atomic beam[37]

5. Atomic beam source

Thus it remains the problem to inject into the target cell a flux of 10^{17}/sec polarized hydrogen atoms in one hyper fine state (hfs). A factor of about four has to be gained relative to the atomic beam flux in the best presently available polarized proton sources. A new high flux atomic beam source is developed in a joint effort by the Heidelberg-Marburg-München members of the collaboration in close contact with the Madison group. Most of the atomic beam appartus was provided by München[38,39], beam studies and optimizations of the geometry are done in Heidelberg.

A vertical cut (fig. 10) shows the four step differential pumping device. The inclination of 30° results from geometrical constraints from H beam injection and detector mounting in the target chamber.

The vacuum system using turbomolecular and cryogenical pumps only is designed for a high input flux of $Q_0 = 5$ mb·l/sec of H_2. We maintain $2 \cdot 10^{-3}$, $4 \cdot 10^{-4}$, $1.5 \cdot 10^{-5}$ and $2 \cdot 10^{-6}$ mb at the stages 1 to 4 respectively. The four 6-pole magnets indicated in the figure are not yet installed.

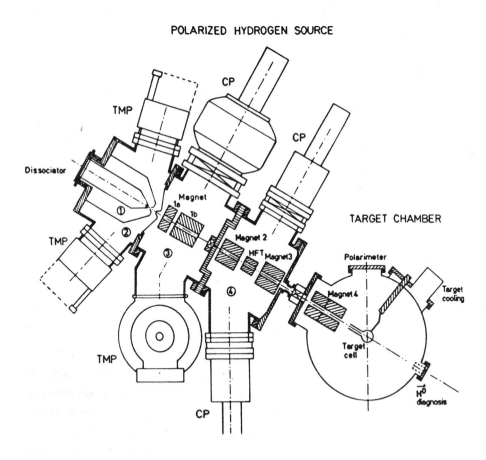

Fig. 10: A schematic of the atomic beam source and the target cell

a)   **Flux measurements**

Fig. 11 shows a measurement of the molecular H_2 flux observed in a compression tube in a position after magnet 3 as function of the gas input[40]. We observe neglectable beam absorption at 300 K and acceptable absorption (up to a factor of 2 in the high flux range) at 50 K.

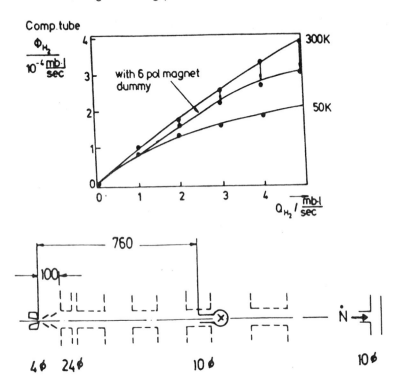

Fig. 11a and b: Compression tube measurements of H_2 beam flux. The geometrical dimensions are given in mm.

To study the beam absorption due to enhanced gas pressure, resulting from beam scattering from the inner walls of the sextupole magnet, geometrical dummies have been used. With the large inner diameter of 24 mm for the magnets and a separation of the first magnet into two short parts 1a and 1b we obtain at high flux conditions a further reduction in the 30% range only.

b. Degree of dissociation

For the degree of dissociation (fig. 12), measured at the same position with a quadrupole mass spectrometer the best values have been obtained[37] with a cooled nozzle of oxydized aluminum of high purity. In this measurement the nozzle temperature was limited to $T \geq 150$ K. High values of dissociation have been observed, ranging from 65% to 40% when increasing the input flux from 2 to 5 mb·l/sec H_2. Later measurements with somewhat improved cooling were found to be hampered by inadequate nozzle material. The optimization of the degree of dissociation is a point of major concern.

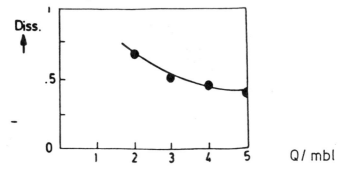

Fig. 12: Degree of dissociation as a function of the input gas flux

c. Beam velocity

Fig. 13 displays results of recent beam velocity measurements[41]. They were done with a 280 Hz rotating chopper in the third chamber. Along a flight

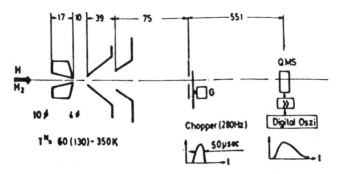

Fig. 13: Time of flight measuring arrangement

path of 55 cm to a quadrupole mass spectrometer with fast amplification system and digital read out the beam bursts spread out due to their velocity distribution. For unfolding, the chopper function (with 50 μsec FWHM width) and a Maxwellian velocity distribution with drift

$$f(v) \, dv = \text{const} \cdot v^2 \exp[-1/2 \, m \, (v - v_D)^2/kT] \, dv. \tag{8}$$

is used.

The solid curve in fig. 14 is a fit to the data points, the dashed one represents the unfolded time distribution.

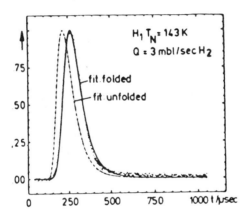

Fig. 14: Typical experimental TOF results and fit curves

For an input flux of $Q = 3$ mb·l/sec H_2 and a nozzle temperature of $T_N = 130$ K we observe for H a drift velocity of $v_D = 1700$ m/sec and a beam temperature T as low as 25 K, corresponding to a Mach number larger than 2.

We further presently try to improve the nozzle cooling. We have to investigate the nozzle geometry, materials and coating to optimize the dissociation and the beam velocity distribution.

e. Stern Gerlach magnets

For the atomic beam focussing with Stern Gerlach sextupole magnets we have decided now to use a system of 5 permanent magnets with $2r_0 = 2.4$ cm inner opening diameter throughout. According to the calculation in Munich a sextupole field

$$B = B_0 \, (r/r_0)^2 \tag{9}$$

with neglectable contributions of higher multipole components and a tip field B_0 of 1.4 Tesla at $r = r_0$ is expected[42].

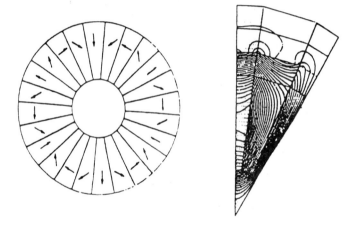

Fig. 15a and b: Distribution of the magnetization direction and of the calculated fieldstrength in a 30° segment.

The design of sextupole magnets using solely permanent magnet material has been discussed by Halbach (1980)[43]. Each of our magnets is made of 24 pieces of Vacodym with the orientation of the magnetization rotated by 60° with respect to the neighbouring piece (fig. 15a). Magnetic field calculations are shown in fig. 15b. For the production of high fields the use of permanent magnets is restricted because of demagnetization effects. In a sextupole the critical area is near the surface of the inner opening where the about 15 KOe field in the opening is for part ot the magnets tangential to the surface and opposite to the direction of the magnetization in the material. For the tangentially magnetized pieces we use Vacodym 400, a material where a magnetization of 1 T is able to withstand an about 20 KOe demagnetizing field. For the other pieces with less critical field orientation we use Vacodym 351 with a higher magnetization of 1.2 Tesla. The demagnetization values are strongly temperature dependent. The glued magnets have to be enclosed in evacuated tightly welded stainless steel vessels to prevent any contact with atomic hydrogen.

The spatial arrangement of the Stern Gerlach magnets is based on beam transport calculations in a ray tracing technique for the appropriate hfs[40]. For each geometry 10 000 trajectories are calculated, emitted from a 4 mm ϕ opening within the 31.4 msr acceptance of the beam collimating system. For a velocity distribution (V_D = 1700 m/sec, T = 25 K), as observed at a 130 K nozzle temperature, we obtain for an optimized system, shown in Fig. 11, that 30% of the particles pass the magnets (2), (3) and (4) and that 18% enter the storage cell opening. Using this magnet arrangement for beams of lower velocity v_D these values increase.

The magnet arrangement is about symmetric with an intermediate focus, where a high field transition is used to select the hfs 1. For polarization reversal in the storage cell we change the sign of a weak magnetic guide field.

6. Expected Intensities

With the presently measured or calculated quantities, we expect a flux N of H atoms in one hfs entering the storage cell entrance

$$N = \frac{1}{4} \cdot \phi_{H_2}^{c \cdot tube} (2 \cdot Diss.) \frac{\Omega^{6-pole}}{\Omega^{c-tube}} \epsilon^{6-pole} \cdot \alpha \qquad (10)$$

and obtain $N = 0.6 \cdot 10^{17}$/sec very close to the design aim. For the H_2 input flux a value of 3 mbℓ/sec and a nozzle temperature of 130 K has been used. We observed in the compression tube (Ω^{c-tube} = 0.13 msr) a flux of $2.5 \cdot 10^{-4}$ mb l/sec H_2. The best value for the dissociation observed was 50%. The reduction factor α due to the residual gas scattered in the sextupole opening is 0.7. The calculated optical sextupole transmission of $\epsilon = 0.18$ is expected to increase further when operation at lower nozzle temperature is achievable. There is also potential to increase the beam dissociation and to improve the supersonic beam expansion.

7. Members of the FILTEX collaboration:

MPI Heidelberg: W. Brückner, E. Jaeschke, D. Krämer, S. Paul, B. Povh, M. Schick, B. Stechert, E. Steffens, K. Zapfe; Marburg: D. Fick, W. Korsch, W. Luck; München: G. Graw, A. Ross, P. Schiemenz; Madison: W. Haeberli, S. Price, T. Wise; GSI: H. Poth; Mainz: T.A. Shibata, T. Walcher; Rutgers: R. Ransome

References

1) H. Döbbeling et al. (1985), Proposal CERN/PSCC/85-80, Nov. 1985 and Addendum 1986 and W. Brückner et al., Memo to the PSCC, Jan. 1989
2) E. Steffens et al. in Physics with Antiprotons at LEAR in the ACOL Eve, ed. by U. Gastaldi et al., pp 245-251, Gif-sur-Yvette, Editions Front.
3) W. Haeberli, Physics at LEAR, with low energy antiprotons. Nucl. Science R.C. Series, ed. by C. Amsler et al., 14 (1988) 195
4) C.B. Dover, in "Low Energy Antimatter", ed. by D.B. Cline, Worl Scientific (1986) 1
5) J. Coté et al., Phys. Rev. Letters 4 (1982) 1319
6) J. Haidenbauer, T. Hippchen, K. Holinde and J. Spaeth, Zeitschrift für Physik, A 334, 467 (1989) and private communication 1989
7) T. Hippchen, KFA Jülich Rep. 494, März 1989
8) R. Machleidt et al., Physics Reports 14 (1987) 2
9) W. Weise, Physics at LEAR with low energy antiprotons. Nucl. Science R.C. Series, ed. by C. Amsler et al., 14 (1988) 287
10) W. Brückner et al., Zeitschrift f. Physik A... (1989)

11) W. Brückner et al., Nucl. Phys. B 8 (1989) 81
12) K. Nakamura et al., Phys. Rev. D 29 (1984) 349
13) A.S. Chongh et al., Phys. Lett. 146 B (1984) 299
14) W. Brückner et al., Phys. Lett. 158 B (1985) 180
15) B.V. Brigg et al., Phys. lett. 194 B (1987) 563
16) D. Spencer and D.N. Edwards, Nucl. Phys. B 19 (1970) 501
17) V. Chalupka et al., Phys. Lett. 61B (1976) 487
18) M. Coupland et al., Phys. lett. 71 B (1977) 460
19) T. Kageyama et al., Phys. Rev. D 35 (1987) 2657
20) R.P. Hamilton et al., Phys. Rev. Lett. 44 (1980) 1179
21) K. Nakamura et al., Phys. Rev. Lett. 53 (1984) 887
22) W. Brückner et al., Phys. Lett. 158 B (1985) 180
23) L. Linssen et al., Nucl. Phys. A 469 (1987) 726
24) H. Iwasaki et al., Nucl. Phys. A 433 (1985) 580
25) V. Ashford et al., Phys. Rev. Lett. 54 (1981) 518
26) W. Brückner et al., Phys. Lett. 169 B (1986) 302
27) R.A. Kunne et al., Phys. Lett. 206 B (1988) 557 and
    Nucl. Phys. B 323 (1989) 1
28) M. Kimura et al., Nuovo Cimento 71 A (1982) 438
29) R.A. Bryan and R.J.N. Phillips, Nucl. Phys, B5 (1968) 201
30) C.B. Dover and J.M. Richard, Phys. Rev. C 21 (1980) 1466, and
    Phys. Rev. C25 (1982) 1952
31) P.H. Timmers et al., Phys.Rev. D 2 (1984) 1928
32) M. Maruyama and T. Ueda, Nucl.Phys. A 364 (1981) 297
    A.M. Green and J.A. Niskanen, Nucl. Phys. A 412 (1984) 448
33) E. Jaeschke et al., Proc. European Part. Acc. Conf., Rome (June 1988)
    G. Bisoffi et al.: The Heidelberg Heavy Ion Cooler Storage Ring, TSR,
    Proc. Nat.Acc. Conf., Chicago (March 1989) and
    D. Habs et al., NIM B 43 (1989) 390
34) A.D. Krisch et al., Phys.Rev.Letters 63 (1989) 1137
35) K. Zapfe, Diss. Univ. of Heidelberg, in preparation
36) W. Haeberli et al. Physics at LEAR, 14 (1988) 195
37) W. Luck, Dissertation Marburg, 1989
38) G. Graw et al., Physics at LEAR, 14,(1988) 221

39) P. Schiemenz, in High Energy Spin Phyiscs, ed. by K.J. Heller, AIP 187 (1988) 1507
40) W. Korsch, Diss. Univ. of Marburg,in preparation
41) M. Schick, Dipl.-Thesis, Univ. of Heidelberg, in preparation
42) A. Ross, Dimplomarbeit, Universität München, August 1989
43) K. Halbach, Nucl.Instr. and Methods 16 (1980) 6
44) K. Imai et al., in Polarization phenomena in Nucl. Physics, AIP, 69 (1981) 141

*) Work supported in part by the BMFT

CONTEMPORARY INTEREST IN COSMIC RAYS
Aspects of the KASCADE Project

H. Rebel

Kernforschungszentrum Karlsruhe
Institut für Kernphysik
P. O. B. 3640
D7500 Karlsruhe / Germany

The lectures give an introduction into the basic facts and problems of modern cosmic ray studies, discussing in particular the phenomena of extensive air showers (EAS) induced by high energy interactions of cosmic particles in the atmosphere. The understanding of EAS appears to be crucial for cosmic ray investigations at energies above ca. 10^{14} eV and has many facets, on the nature of the interaction itself as well as on the origin and propagation of cosmic rays in the universe. The second part the KASCADE project is specifically described.KASCADE (KArlsruhe Shower Core Array DEtector) is an extensive air shower experiment being under construction at the Kernforschungszentrum Karlsruhe (49°N, 8°E, 110 m a.s.l.). The main aim is to obtain information on the primary composition of cosmic rays at energy above 3×10^{14} eV. In this energy range a remarkable bent in the slope of the energies spectrum is observed. Results at lower energies seem to indicate that this is correlated with a change of the chemical composition. Besides other valuable informations the proof of such an effect would have important consequences for our understanding of the origin of cosmic rays. The basic approach of KASCADE is to measure a large number of parameters for each individual shower. These include *electron* and *muon numbers, the lateral distributions* as well as *number, energy,* and *spatial distribution* of *hadrons* in the shower core. Extensive Monte Carlo simulations of air showers are discussed using a specially developed code for hadronic interactions. These simulations of air shower development and of detector response lead to the particular experiment layout. The arrangement will also be able to identify point sources in the quoted energy range.

The presentation of the KASCADE project is based on the results of the KASCADE team.

"Die himmlischen Gestirne machen nicht
Bloß Tag und Nacht, Frühling und Sommer"
Friedrich Schiller, Wallenstein, Die Piccolomini

1. Introduction

Nearly all astronomical knowledge has been gained by the messages which are carried to us by electromagnetic radiation in all wave bands. Since only within the solar system a direct sampling has been possible, the only further direct source of information on the Universe outside our solar system are the cosmic rays (including with this term neutrino radiation and gravity waves, still not detected with certainty). Most abundantly, *primary* cosmic rays are highly energetic charged hadronic particles (nucleons and nuclei). They permeate most of the Galaxy, and possibly the intergalactic space as well. They are continuously bombarding our Earth and produce *secondary* cosmic rays - the full zoo of subnucleonic particles, electrons and photons - by the interactions of the primaries with nuclei of our atmosphere.

Cosmic rays have been discovered by Victor Hess [1, 2] in 1911/12. In his celebrated balloon ascents he showed conclusively that the ionization rate did first fall slowly with increasing altitude, but after a height of about 800 m the rate started again to increase up to the maximum altitude reached (5350 m). It may be of interest to note that Victor Hess reported his first results in 1911 at the *Natur-forscherversammlung in Karlsruhe* [1].

Since that time studies of cosmic rays have shown fascinating aspects. In addition to an intrinsic interest in this still mysterious phenomenon, cosmic rays are unequivocal samples of the Universe, outside the solar system. The subject combines the ingredients from two of the major lines of contempory physics research : *astrophysics* and *elementary particle physics*. Despite a lapse of time of nearly 80 years, the origin of the bulk of cosmic rays and the accelerator mechanisms are still largely a matter of conjecture. Like in the forties when the cosmic accelerators had been the only source of subnucleonic particles, cosmic rays provide the perspective to extend our knowledge on fundamental interactions beyond the energy limits of man-made accelerators.

The current research concerns, in particular, the highest energies of cosmic rays. The questions of interest may be globally classified according to different aspects :

(i) *Information on the nature of the cosmic particles, on the chemical composition of cosmic ray and their energy distribution as observed on the surface on the Earth.*

(ii) *Information on the interactions in the atmosphere, on how the observables reflect the nature of the particles and their elementary structure (say on a partonic level of quark-gluon constituents).*

(iii) *Information on the propagation through the interstellar and intergalactic space, modifying direction, energy and composition of the primary cosmic ray source spectrum.*

(iv) *Information on the sources, on isotropy or lack thereof of the cosmic radiation observed.*
With the latter aspect a new subfield of ultrahigh energy cosmic ray physics has appeared in recent years : Search for point sources which emit γ-rays of the order of 10^{15} eV.
Point sources of optical, radio, X-rays and low energy γ-rays from binary stellar systems are well established [3]. That such radiation can be extended to the 10^{15} eV is relatively recent [4] and stems from the Cygnus X-3 observation of the Kiel group [5]. However, the results are still a matter of debate [6]. Nevertheless these studies have opened a new chapter of gamma-ray astronomy [7].

The background of the present lectures is an introduction to the KASCADE (KArlsruhe Shower Core and Array DEtector) project [8]. It is a larger detector arrangement under construction in Kernforschungszentrum Karlsruhe and addresses experimentally some of these problems, in an energy range which exceeds the largest energies currently available at artificial accelerators (i. e. Fermi Lab Tevatron Collider : $\sqrt{s} = 1.8$ TeV $\hat{=}$ $E_{lab} \simeq 2 \cdot 10^{15}$ eV). Before turning to KASCADE in detail, we survey briefly the present knowledge about cosmic rays in order to orient the novice. KASCADE is based on observation of *extensive air showers,* induced by collisions of the high-energy particles in the atmosphere. Hence, we discuss the features of air showers and how the observables are related to high-energy interaction models and to the elemental composition of the primaries through Monte-Carlo simulations of the rather complex phenomena. Finally an outlook is given on some specifically technical aspects, on the current status and the organization of the project.

2. Present Knowledge on Cosmic Rays

The energy spectrum of primary cosmic rays comprises more than 12 orders of magnitudes (from 10^8 eV to 10^{20} eV) on the energy scale.

In 1970 K. Suga et al [9] have reported on a single event with $4 \cdot 10^{21}$ eV ($= 640\,J$).

The spectrum obeys an overall power law with a change in the slope at around 10^{15} eV (fig. 1). The flux of primary cosmic rays falls from ca one particle/m² · s · MeV at the lowest energies to one particle/km² · century at the highest energies. While the bend ("knee") at 1 PeV is well demonstrated, a great deal of controversy exists about the shape around $5 \cdot 10^{19}$ eV. Some experiments see a cut-off, some a flattened spectrum : an "ankle" and a "toe" [10-13].

A question of rather fundamental nature is the existence or nonexistence of the Greisen cut-off [14] due to the interaction of the cosmic ray particles with the 2.7°K blackbody radiation in the universe. A proton with energies $> 5 \cdot 10^{19}$ eV will interact inelastically with such 10^{-4} eV photons since in the rest frame of the proton, these are γ-rays of ca. 300 MeV, exceeding the photoproduction threshold for pions. Thus the protons could be "cooled" through

$$\gamma + p \to p + \pi^\circ \to p + \gamma + \gamma$$
$$\gamma + p \to n + \pi^+ \to n + \nu_\mu + \mu^+$$

The energy spectrum teaches us that the energy density in ultrahigh energy cosmic ray is rather large, and cosmic rays are significant compared to the energy of the universe. The energy density is ca. 1 eV/cm³, as compared to starlight with 0.6 eV/cm³ and the galactic field with 0.2 eV/cm³.

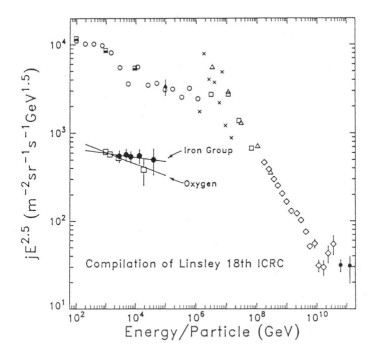

Fig. 1　Differential energy spectrum of the primary cosmic rays between 10^2 and 10^{11} GeV.

The existence of a power law over many decades implies significant restrictions on the accelerating mechanism. In fact, the Fermi statistical acceleration model [15] e.g. by collisions of the charged particles with chaotic magnetic clouds in the interstellar space lead to such a distribution. The "knee" of the spectrum may be associated with a change in the elemental composition. There are experimental indications of an increased proportion of iron and other heavy nuclei in the knee region. The socalled "leaky box" model of the Galaxy [16] allows free diffusion within the galactic volume. When the particles reach the galactic edge, some of them escape in a way dependent on energy and magnetic stiffness. Since the gyromagnetic radius of relativistic particles is proportional to the energy and inversely proportional to the charge, one expects a leakage of protons to start a lower energies, and consequently a depletion of the lighter particles.

Thus, the bend around 10^{15} eV does not necessarily imply a change in the "injection spectrum" of the source. However, the change in the composition is not the only explanation under debate, there may be changes in the accelerator mechanism prevailing in different energy regions or in the hadronic interactions, or some combination of these. A clarification of the presently rather confusing experimental situation is of extreme interest and has far-reaching aspects.

It is the main objective of the KASCADE experiment to obtain information about the chemical composition of the primary cosmic ray particles in the energy range above 10^{14} eV. Below this energy the most direct way to identify the chemical nature of a cosmic ray primary is by measurements above the atmosphere of the earth before the particles have interacted with air nuclei. Such measurements have been performed from high flying balloons and satellites over the last few decades and have yielded results up to energies of a few of TeV [17-19].

Overall the compositions of solar and cosmic ray matter (referred as "low energy composition" (at 100 GeV/nucleon) : 50% protons, 25% α-particles, 13% CNO, 13% Fe (electrons $< 10^{-2}$, $\gamma < 10^{-3}$)) are quite similar. However there are important differences in detail. The overabundance of Li, Be and B in cosmic rays is attributed to spallation reactions of the more abundant C, O, and Fe nuclei.

Though the effect is still controversial, the abundance of Fe in cosmic rays seems to decrease less fast with increasing energy than the total intensity (fig. 5). While at low energies there is approximately 10 times more O than Fe in the primary radiation these two elements are equally abundant near 1 TeV. This result is output of two major experiments, of the measurements of Chicago "Egg" [18] on the space shuttle and of the balloon borne emulsion calorimeters of the Japan - U.S. collaboration known as JACEE [19].

Fig. 6 displays a recent original result from the JACEE collaboration [19] which shows the present knowledge from direct experiments.

However, direct methods are limited in energy due to the steep decline of intensity with energy because detectors on satellites and balloons are limited in weight and hence in size. It is not to be expected that such measurements will be extended by more than a factor of ten beyond the results available today.

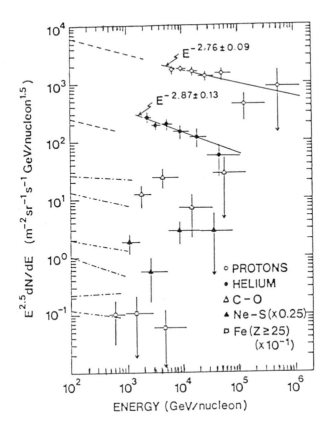

Fig. 2 Energy spectra of the major primary components as determined by the JACEE [19].

Fortunately, when direct methods start to fail due to the low flux, the phenomenon of extensive air showers (discovered in the thirtys [20]) gains importance. Protons or heavy nuclei interact with nitrogen and oxygen nuclei in the atmosphere and generate a shower of charged (and neutral) particles, primarily composed of electrons and photons. Such an air shower is a pancake, hundreds of meters wide and a few meters thick. It moves through the atmosphere nearly with speed of light, with an intensity development reflecting the energy of the primary. Thus the atmosphere takes the role of an amplifier ("multiplier"), and the rare particle on the top of the atmosphere whose energy is distributed to many coherently moving low energy particles becomes detectable by use of an extended array of detectors installed on ground. This is a traditional technique in cosmic ray research. Before considering more details of extensive air showers, fig. 3 concludes these introductory remarks on present days' knowledge and interest in cosmic rays by an overview from standpoint of the techniques and particle rates in various energy ranges. Again let us note that our interest exceeds the energy limits of

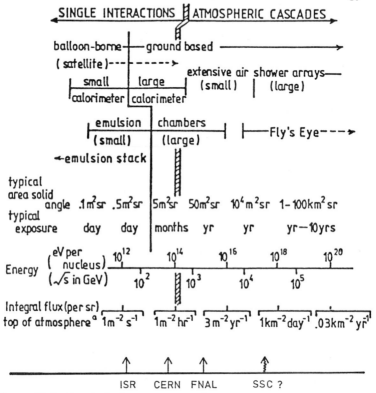

Fig. 3   Detection techniques, rates and energy ranges in cosmic ray studies (after ref. 21).

Earth-bound accelerators. Since some observables of air showers depend on quantities evolving from the specific nature of the elementary interactions and on the nature of the impinging primary particle as well, the issues of interaction and of elemental composition are interrelated. Studies of one ingredient are muddied by the lack of knowledge of the other, and our research proceeds through tentative trials to establish consistency.

3. Characteristics of Extensive Air Showers

A high-energy cosmic ray particle colliding with an air nucleus produces a jet of secondary particles, mostly pions and kaons, and it escapes as an excited baryon with a fraction of the primary energy. The surviving baryon ("leading

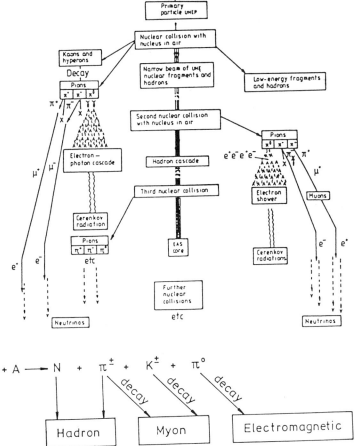

Fig. 4   Schematic diagram of the progression of an extensive air shower [22].

baryon") and the hadrons in the jet undergo further interactions while penetrating deeper into the atmosphere : the hadrons (nucleons, antinucleons, kaons and pions) generate the hadronic cascade via strong interactions and feed the muon component via

$$\pi^\pm \to \mu^\pm + \nu_\mu (\bar{\nu}_\mu)$$

decay. Neutral pions (and η^o's) decay into photon pairs

$$\pi^o \to 2\gamma$$

which subsequently initiate electromagnetic cascades and establish the dominating electron-photon component. Thus, the total number of shower particles will rise with atmospheric depth until the primary energy, continuously pumped from the hadronic backbone to the other components, is consumed. Then the shower will die out, first the hadronic and electromagnetic components, then usually far underground the muonic one; only the neutrinos are left.

Purely electromagnetic showers are determined by the pair-production and radiation length X_0 which are nearly equal at our energy of interest ($X_0 = 37.7 g/cm^2$ in air : $Z = 7.4$; $A = 14.8$). After n halfway attenuation lengths $X_0/\ell n2$, 2^n particles with $E_0/2^n$ are created. The multiplicative process stops when the average energy of the shower particles drops below some critical energy E_c where the dominant energy loss of (relativistic) electrons is by ionization ($dE^{ion} = 2\,MeV/g\,cm^{-2}$) rather than by the bremsstrahlung : $E_c = 84.2\,MeV$. For γ-rays E_c is defined as value when Compton scattering starts to dominate pair production.

The shower reaches its maximum when the average energy of shower particles equals E_c ($N_{max} = E_0/E_c$), that is after $n = \ell n\,(E_0/E_c)/\ell n2$ halfway attenuation lengths $R = X_0/\ell n2$. Hence the depth X_m of the shower maximum has a logarithmic dependence on the incident energy, while N_{max} is proportional to the energy.

Though hadronically initiated showers are a superposition of electromagnetic showers from different interaction points, and X_m is affected by fluctuations of the first hadronic interaction, the proportionality between N_{max} and the energy E_0 of the primary still holds

$$N_{max} = (1.1 - 1.6) E_0 \, (GeV)$$

The longitudinal development of electromagnetic showers can be explicitly calculated by solving the relevant diffusion equations. The evolution of the cascade is described by a formal parameter introduced into the theory : the "age" parameter.The shower maxima deeper are stifted into the atmosphere with increasing energy and there is a lower energy cut-off at sea-level (1033 g/cm$^2 \approx 27$ X$_0$). This feature together with the strong experimental decrease of the primary energy spectrum explains the preference for observations at high mountains.

The "age" parameter enters also into the expression of the particle density dependence on perpendicular distance r from the shower axis

$$\rho_e(r) = N_e/R^2_m \cdot f_e(s, r/R_m)$$

where N_e is the total number of electrons (size) of shower and R_m the Molière multiple scattering unit ($R_m = 79$ m at sea level). The structure function f_e has been derived from electromagnetic cascade theory by Nishimura and Kamata and the full formula for the lateral distribution, which is of course largely determined by Coulomb scattering, is referred as Nishimura-Kamata-Greisen (NKG) formula [23]. The NKG formula approximates the average lateral distribution of the electromagnetic component of hadronic showers globally quite well, though there are deviations for larger r and more adequate empirical expressions are available [23].

With the advent of Monte Carlo techniques on computers electromagnetic extensive air shower calculations can, at least in principle, be made to any degree of reality. Nevertheless the large number of particles requires certain approximations routinely used.

The longitudinal development of the less copious, but equally important muon component differs considerably from the electron-gamma component. This is due to the comparative stability of the muon and the relatively weak interaction with matter so that the muon intensity (total muon number $N_\mu = \int \rho_\mu(r) \, d^2r$) falls only slowly with increasing atmospheric depth. Thus, in principle, the muon component, especially the high energy muons, can provide rather direct information on early stages of the shower (about the primary energy e. g.). Since additionally the Coulomb scattering is of minor importance, the lateral distribution $\rho_\mu(r)$ of the muons reflects the directions of emissions from the charged pions and kaons of the hadronic backbone, i. e. the distribution and the mean value $\bar{p}_T$ of the transverse momenta which determines the corresponding structure function f_μ. The muon component is laterally more extended, and though the number of muons is only a few percent of the electrons near the shower axis, the density $\rho_\mu(r; E_\mu > E_{thres})$ becomes dominant at large distances (of 1 km).

Since the cross section of photoproduction of muons is 2 orders of magnitudes smaller than of hadronic production, extensive air showers initiated by ultra-high energy gamma rays are expected to be "muon-poor". A peculiar result of the Kiel observation [5] of a shower excess from Cygnus X-3 is that the array did not see a significant decrease in the muon content. Other experiments [26, 27] give conflicting results on gamma-like showers.

The observation of extensive air showers comprises the measurements of the *intensity*, the *lateral* and *time distribution* of the three main components by correspondingly large detector arrays (covering usually several km²). The arrival direction is determined by time-delay measurements of the triggering detectors.

Measurements of the arrival times of the hadrons in the core relative to the shower front may be of interest with respect to studies of the mass composition [28, 29]. Differences in the height of the first interaction and the transverse momentum distribution of the secondaries in proton and heavy-ion induced showers are considered to be reflected by different arrival time distributions.

Some newer range of observation techniques is based on atmospheric light emission : *Cerenkov radiation* and *nitrogen scintillation*. The opening angle of Cerenkov light (10^{-4} of the brightness in the night sky) is about 1.5° so that Cerenkov light is highly collimated around the original particle direction. Therefore due to the directionality an extensive air shower must point towards the detectors to be seen.

In case of fluorescence light nitrogen molecules are excited and radiate near - ultra violet photons which can be used to detect the passage for an extensive air shower through the atmosphere. The observation of fluorescence allows interesting studies of the depth X_{max} of the shower maximum. The "Fly's Eye" of the University of Utah [30] is based on such a technique. Coherent *emission of radio waves* (30-60 MHz) from deceleration of slow electrons is of minor importance.

4. Modeling Hadronic Extensive Air Showers

As far as the development of the electromagnetic component is concerned hadronic showers can be considered to be a superposition of individual electromagnetic showers following π° decays from the hadronic backbone. Therefore the development cannot be described with a well-defined shower age, and though globally resembling pure electromagnetic cascades, it is influenced by details of the hadronic interaction and the nature of the primary.

Hadronic cascades develop in subsequent interactions of strongly interacting particles, dependent on the interaction lengths in competition with decay. The probability of a particle i which propagates through the atmosphere to undergo interaction or decay after traveling the distance X [g/cm²] is expressed by

$$P_i(X) = 1 - exp\left\{-X\left(\frac{1}{\lambda_i} + \frac{1}{c\rho(h)\tau_{oi}\gamma_i}\right)\right\}$$

Here λ_i is the interaction mean free path, c the velocity of light, τ_{oi} the mean life time, γ_i the Lorentzfactor (E/m_ic^2). Since the probability for surviving decay is inversely proportional to the geometric path length of the trajectory, represented by $c\rho$ (h), the evaluation of this quantity involves the "atmospheric slant depth" [21], including the actual zenith angle of the trajectory (see ref. [32]).

For particles like muons which practically do not interact and experience only Coulomb scattering and ionization losses, only the decay term in the expression above is of relevance for selecting the decay point. Similarly due to the extremely short life time of the neutral pions ($\tau_0 = 10^{-16}$ s), π° have no chance to interact.

In the energy region of extended air showers the interaction lengths are approximately given by

$$\lambda_{p\text{-}air} = 70 \text{ g/cm}^2 \quad \text{for protons}$$

$$\lambda_{\pi\text{-}air} = 120 \text{ g/cm}^2 \quad \text{for charged pions and kaons}$$

and typically for heavy ions

$$\lambda_{Fe\text{-}air} \approx 15 \text{ g/cm}^2 \quad \text{for Fe ions.}$$

The interaction lengths are inversely proportional to the corresponding (energy dependent) inelastic hadron-air cross sections $\sigma^{inel}_{air}(E)$ through

$$\lambda(E)[g/cm^2] = 2.41 \cdot 10^4 / \sigma^{inel}_{air}(E)$$

The specification of the cross sections and their energy dependence needs more or less bold extrapolations of the accelerator data [33, 34], guided by various theoretical assumptions. From nucleon-nucleon cross sections nucleon-nucleus (A_t) and nucleus (A)-nucleus cross sections have to be deduced, involving Glauber-model type calculations [35].

In a nucleus-nucleus collision event the number ν of nucleons participating in the interactions may be estimated with geometrical arguments [36]. These apply to the incident heavy primary as well as to cluster fragments, since the projectile does not completely dissociate into its constituent nucleons in one collision as assumed by the socalled superposition model. Alpha particles and heavy fragments of the original projectile survive and need a number of interactions before totally reduced to nucleons ("partial fragmentation model" [37]).

The hadronic backbone of the shower in the atmosphere (thickness 1033 g/cm²) is built up by ca. 14 interactions of the leading baryons. The energy degradation is determined by the *inelasticity* K_p, the part of kinetic energy of the particle which is consumed by the production of secondaries. Thus, the energy of the leading particle decreases $E_n = (1 - K_p)^n \cdot E_0$ after n collisions. The inelasticity follows a distribution which is basically determined by the distributions of the longitudinal and transverse momenta as carried by the differential cross sections, and they should strictly derived therefrom. Nevertheless, often shower models adopt an explicit ad-hoc form, for example for nucleons a uniform distribution with an average inelasticity of $\bar{K}_p = 0.5$. Thus, a primary proton loses roughly half of the initial energy in the first interaction.

The development of the hadronic shower is strongly influenced by fluctuations in the position of the first interaction. The depth of the shower maximum proves to be dependent on the product of the inelastic cross section $\sigma_{p\text{-}air}^{inel}$ and the inelasticity K_p. The energy dependence of both quantities may counteract and do obscure each other.

In addition to the interaction lengths λ_i *the inclusive production cross sections* $d^3\sigma_{if}/d^3p$ for the processes

$$i + air \to f + anything,$$

governing the cascade development, have to be specified, for example:

$$\frac{d^3\sigma_{NN}}{d^3p}, \frac{d^3\sigma_{N\pi}}{d^3p}, \frac{d^3\sigma_{\pi\pi}}{d^3p}, \frac{d^3\sigma_{NK}}{d^3p} \ldots$$

In principle, production of strange and charmed particles as well as photoproduction of hadrons (which is essential for γ-initiated showers) have to be taken into account. These cross sections carry the particle production parameters characterizing *multiplicity, elasticity* and *kinematics* of the secondary particles, and specifying:

- the type and the relative production probabilities of the secondaries f
- the average number ñ and the distribution of the number of secondaries
- the energy, the longitudinal and transverse momentum distributions of the secondaries which in turn determine the inelasticity $K_i = \Sigma_\mu E_\mu^f / (E_i - m_i c^2)$

Any production model which is used as generator for Monte Carlo simulations of extensive air showers has to provide a consistent set of prescriptions for calculating these parameters and their energy dependence (empirically adjusted to experimental data and continued to higher energies).

For illustration tab. 1 gives a typical set of parameters, used in the time-honoured standard model based on the socalled CKP formula of Cocconi, Koesters and Perkins [38]. It has been fitted to accelerator data in the region of tens of GeV. High-energy nucleons are assumed to lose on average 50% of their energy in each collision, only pions are admitted as secondary particles,

Tab. 1 "Preferred" parameters of the CKP model [38].

Nuclear mean free path	$\lambda_p = 80 \text{ g cm}^{-2}$
Nucleon-air nucleus coefficient of inelasticity	$P(K_p)dK_p = \{(1+\alpha)^2 K_p^\alpha \ln K_p\}dK_p$ with $\langle K_p \rangle = 0.5$
Pion mean free path	$\lambda_\pi = 120 \text{ g cm}^{-2}$
Pion-air nucleus coefficient of inelasticity	$K_\pi = 1.0$
Total number of produced secondaries (pions) in nucleon-air nucleus collisions	$n_\pm = 3.75 \, E_p^{1/4} - 1.0$
Distribution in energy of secondary particles (after G Cocconi et al 1961 unpublished)	$f(\varepsilon_\pi)d\varepsilon_\pi \times \{\frac{1}{T}e^{-\varepsilon_\pi/T} + \frac{1}{u}e^{-\varepsilon_\pi/u}\}d\varepsilon_\pi$ where T and u are the mean energies of the particles moving forwards and backwards in the CMS.
Distribution in transverse momentum among produced secondaries	$f(p_t)dp_t \times (\frac{p_t}{p_o})^{3/2} e^{-p_t/p_o} dp_t$

produced with equal numbers of π^+, π^- and π^0, and with explicitly given (exponential) distributions in energy and in transverse momentum. The $E_p^{1/4}$-energy variation of the average multiplicity ñ (which is probably to strong with respect to recent data) originates from a statistical model of Landau [39].

The extrapolations of the differential cross sections of inclusive particle production and related distributions are discussed with the question to which extent such distributions "scale" i. e. whether they are independent from the energy, provided they are written in terms of adequate variables. For a consideration of the scaling behaviour of the Lorentz invariant cross sections of the form

$$E \, d^3\sigma / d^3p = f(p_L, p_T, \sqrt{s})$$

which a priori appear to be functions of the c.m. energy $\sqrt{s}$ and the longitudinal and transverse momenta, Feynman's momentum scaling variable

$$x = p_L / p_L^{max} = 2p_L^{c.m.} / \sqrt{s},$$

i. e. the fraction of the maximum possible momentum, is introduced. The conjecture of Feynman [40] says that in the limit of highest energies

$$f(x, p_T, s) \to f(x, p_T)$$

without any further specification of the explicit form of f(x). Feynman scaling proves to be a powerful procedure to scale up the cross sections as experimentally determined at highest available accelerator energies to the ultrahigh energies of extensive air showers.

A natural consequence evolving from Feynman scaling is the energy dependence of the average multiplicity

$$n = \frac{1}{\sigma^{inel}} \int \frac{d^3\sigma}{d^3p} d^3p = \int f(x, p_T) \frac{d^3p}{E}$$

which follows a logarithmic energy dependence

$$\bar{n} \propto \ln\sqrt{s}$$

A current debate of vital interest for the interpretation of cosmic ray data concerns the violation of Feynman scaling (see ref. 34). Experimental results [41] give evidence for a scaling violation in the small x (pionization) region (i. e. at large angles and small longitudinal momenta). The situation in the forward (diffractive dissociation) region (large x) is unclear as this region is usually difficult to explore by collider experiments. Wdowczyk and Wolfendale [42] have put forward a procedure with strong scaling violation, essentially replacing Feynman's hypothesis by

$$\frac{E}{\sigma^{inel}} \frac{d^3\sigma}{d^3p} = \kappa \cdot (s/s_0)^a f(x \cdot (s/s_0)^a, p_T)$$

i. e. substituting

$$x \to x \cdot (s/s_0)^a$$

where $s_0^{1/2}$ is some reference energy with respect to which scaling violation is exhibited, and a an empirical parameter characterizing the degree of scaling violation (a = 0 : Feynman scaling). Additionally, an inelasticity factor κ (s/s_0) allows reducing the fraction of the c. m. energy to go into charged particle production. The fit of a (= 0.18) to the data leads to a moderate rise ($\bar{n} \propto (s/s_0)^{0.09}$) of the average multiplicity.

When considering the characteristics of the momentum distributions a useful quantity is the rapidity variable y (see ref. 43).

The variable has two useful properties

- for $p_T^2 \geq m$

  $y \simeq \eta \equiv -\ln(\tan(\theta/2))$ ("pseudorapidity")

  with

  $\tan\theta = p_T / p_L$

  The pseudorapidity is of particular interest since mass and momentum of the particle (often not well known) do not enter.

- a rapidity η interval $\Delta y = y_2 - y_1$ is invariant under Lorentz transformations along the incident beam direction, hence the multiplicity density dn/dy too.

In some sense as alternatives to prescriptions (like scaling) which describe the general characteristics of the collisions by empirical expressions with parameter values adjusted to accelerator data, but without broad theoretical "a priori" basis, more modern approaches intend to deduce the features of the production cross sections, in particular the rapidity and transverse momentum distributions of the secondaries within a microscopic model, i. e. from a description of the processes from viewpoint of the interacting partons. Such an approach which presently necessarily includes a lot of phenomenological ingredients, is the *dual multichain parton model*. In the formulation of J. N. Capdevielle [44] as a Monte Carlo generator [45] it guidelines the extrapolation to ultrahigh energies and is the basis of the CORSIKA code : COsmic Ray SImulation for KASCADE (tab. 2)

The model is based on ideas of Capella and Tran Than Vanh [46]. Multiple production is described by chains or coloured strings, stretched between elementary constituents of the colliding hadrons.

The interacting valence quarks of projectile and target rearrange by gluon exchange the colour structure of the system. As a consequence, constituents of the projectile and target (a fast quark and slow diquark e. g.) form a colour singlet string with partons of large relative momenta. Due to the confinement *the stretched chains start to fragment (spontaneous $q\bar{q}$ -production) in order to consume the energy within the string. We recognize a target string (T) and a projectile string (P), which are the only chains in pp collisions. In multiple collision processes in a nucleus, sea constituents are excited too, and mediate nucleon-A interactions. While in the intermediate step the projectile diquark remains inert, chains with the sea quarks of the projectile are formed.*

The distribution of the rapidity of the secondaries can be obtained by folding the respective structure (internal quark momentum distribution) and fragmentation functions. The limitation to the target and projectile strings leads to a two-component structure of dn/dyd^2p_T whose shape can be parametrized by Gaussians [47], two of which represent the fragmentation of the colliding hadrons, a third one parametrizes the contribution from multiple scattering in the target nucleus

In recent years negative binominal distributions [48]

$$P_{NB}(n, \bar{n}, k) = \binom{n + k - 1}{n} \left(\frac{\bar{n}/k}{1 + \bar{n}/k}\right)^n \left(\frac{1}{(1 + \bar{n}/k)^k}\right)$$

depending on two adjustable parameters ñ and k have been found to provide very good fit formula for the multiplicity distributions. The dispersion of P_{NB} accounts for the widening of experimental multiplicity distributions as observed with increasing energy. The origin of the success of this class of distributions in

$$D^2 = n + n^2/k$$

describing experimental findings has not yet completely clarified [48]. The parameters of P_{NB} follow empirical energy dependencies.

For any multiple production (nondiffractive) collision the generator plays at dice, defines the actual value of the number n of the secondaries, generates a set of the n individual rapidities (longitudinal momenta), and in the same time a sample of transverse momenta. The rapidity distribution is parametrized as a sum of Gaussians. The underlying transverse momentum distribution is the "QCD" inspired form proposed by Hagedorn [49]. The "event" is rejected if conditions of conservation of energy and momenta are not satisfied.

Tab. 2 *Features of the hadronic multiproduction model used in CORSIKA for $\sqrt{s} > 10\,GeV$.*

A. NN-distributions

Longitudinal momentum : Rapidity distribution dn/dy

DUAL PARTON MODEL - PARAMETRISATION

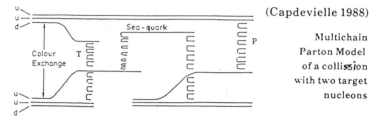

(Capdevielle 1988)

Multichain Parton Model of a collission with two target nucleons

Transverse momentum :

$$\frac{d(dn/dy)}{2\pi p_T \cdot dp_T} = A\left(\frac{p_0}{p_T + p_0}\right)^m$$

(Hagedorn 1983)

$$\bar{p}_T = 2p_0/(m-3)$$

Average multiplicity $\bar{n}(s)$

$$\bar{n} = A + B \ln s + C \ln^2 s \quad \text{or}$$
$$= D + E s^F \qquad \text{(Alner et al 1986)}$$

Multiplicity distribution

$$P_{NB}(n, \bar{n}, k) = \text{Negative Binomial Distribution}$$

(UA5 Collaboration)

$$k^{-1} = a + b \ln \sqrt{s} \qquad \text{(Giovanini and Howe 1986)}$$

The model reproduces the recent results from ISR, SPS, Fermilab and p$\bar{p}$ - collider

B. Extensions to N-air and Nucleus-air distributions

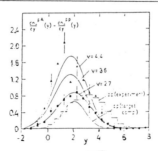

In case of a diffractive excitation the cross section is assumed to be inversely proportional to the squared mass of the excited system [see ref. 45].

The sketched procedure is embedded in the architecture of a program developed by P. Grieder [50] and actually also accounting for the hadronic interactions at lower energies [32].

5. Signatures of the Chemical Composition

The appearance of an extensive air shower at the level of observation far down in the atmosphere is critically influenced by the first interactions at the beginning of the cascade, through the attenuation of the hadrons as determined by the absorption cross sections and the inelasticity in the interactions.

A shower initiated by a heavy ion of mass A ("A shower") of the energy E_0 is approximately borne out by the superposition of the effects of A leading nucleons with E_0/A. This simple *superposition hypothesis* may not be completely correct, in particular as in the case of heavy primaries a nuclear-fragmentation pattern may additionally show up (apart from specific nucleus-nucleus collision effects), but it makes plausible that statistical fluctuations of the individual interaction processes affect the development of a heavy-ion induced shower less than in case of a proton initiated one. Thus, we conjecture that the fluctuations of most observable shower parameters tend to decrease with increasing A [51]. Whether this feature will serve as an experimental source of information depends on the question to which extent intrinsic shower fluctuations can be observed distinctly separated from experimental uncertainties and fluctuations of instrumental origin.

There have been some attempts analysing the shape of the fluctuation distribution of various ground parameters of extensive air showers, in particular of the muon component [52-54], but with less convincing success and conflicting results.

It is intuitively evident that the observation of very early stages of the shower development would be most informative [55]. Direct observations of the longitudinal development of the atmospheric showers are usually restricted to

Fly's Eye [30] and experiments based on the detection of Cerenkov light (Cerenkov pulse width technique [56]). However, the distribution of the point of the first interaction, which would allow most unambiguous interpretation is not accessible due too little light intensity. One measures, instead, the distribution of the depth X_{max} of the shower maximum, which has been shown to display the interaction length of the primary approximately through simple proportionality [57]. Nevertheless quantitative conclusions remain dependent on the properties of the hadronic interactions. Various refinements of the X_{max}-technique [58] including a fluctuation method [59] are discussed in literature [see tab. 3].

Following the current extrapolations of experimental results the average number of secondary particles ñ exhibits approximately a logarithmic increase with energy, in any case slower than with $E^{1/2}$. Hence A nucleons of energy E_0/A produce more secondary particles than one nucleon of energy E_0. Since most secondaries are pions, a fraction of which decays into muons, a larger muon content is expected for heavy ion induced showers. This is the basis of the N_μ - N_e correlation method and its variants [60, 61]. However, we note that quantitative conclusions are affected by the detailed form of the energy dependence of the average multiplicity. Here the "scaling violation" debate meets the problem of chemical composition of cosmic rays.

The inelasticity specifying the relative energy losses of the leading particles appears to be nearly independent from the collision energy. Therefore the secondaries of a proton induced shower will have in average a higher energy, reflected by a harder energy spectrum of hadrons in the shower core [62].

Magnitude and distribution of the transverse momenta affect the path length of the particle travel and hence the lateral distributions. The slow energy dependence of the average transverse momenta implies that showers from light primaries are less spread out and more penetrating into atmosphere than heavy ion showers. The large spread-out and the lower energy of the hadrons from heavy primaries increase the number of hadrons in the "pancake" which are timely delayed (in the order 20-100 ns) with respect to the front [28, 29]. The delay rates are considered to be sensitive to the composition.

Tab. 3 Signatures of the chemical composition

SIGNATURES OF CHEMICAL COMPOSITION

(1) <u>Weak energy dependence of the average multiplicity ñ</u>

Muon content N_μ/N_e at fixed primary energy

$\tilde{n}_p(E) \propto \ln E$

$\tilde{n}_{Fe}(E) \propto A \ln(E/A)$ (superposition hypothesis)

(2) <u>Approximate constancy of the average inelasticity</u>

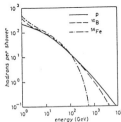

Energy spectra of shower hadrons ($E_0 = 10^{15}$ eV)

(3) <u>Approximate constancy of the average transverse momentum</u>

Delay of relative arrival time of hadrons

Flatter lateral distributions

(4) <u>Inelastic cross section differences</u>

Shift in the point of first interaction reflected by the distribution of the shower maximum depth : X_{max} technique of "Fly's Eye"

(5) <u>Shape of the fluctuation distributions</u>

These effects have been extensively studied by realistic air shower simulation calculations done with the program CORSIKA on the basis of the layout [8] of KASCADE. We illustrate the results - pars pro toto - by the N_μ-N_e correlation (Fig. 5). In order to get a realistic picture the steeply falling energy spectrum ($E^{-2.7}$) was folded in. Generally, we see the muon enrichment of Fe induced shower, but we recognize also the role of the fluctuations, particularly conspicuous for p-showers (reaching even the region $N_\mu/N_e = 0.001$ where photon induced showers are located). More results are found in ref. 8.

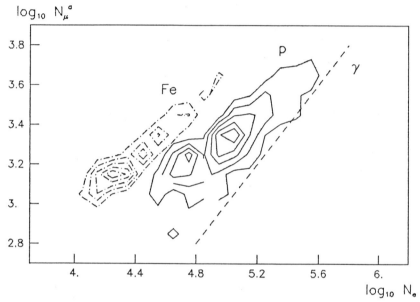

Fig. 5 :   The N_μ-N_e correlation of p and Fe induced air showers with primary energies of $2.5 \cdot 10^{14} - 2.8 \cdot 10^{15}$ eV.

6. The Karlsruhe Cosmic Ray Project KASCADE

The basic approach of KASCADE is to measure a large number of characteristic parameters for each individual shower. Fig. 22 displays schematically the detector field which covers an area of 200 x 250 m² size with 316 detector stations (> 3 m²) and with a central hadron calorimeter. It samples the intensity of the electromagnetic and muonic components for a determination of the *size, lateral distribution, core position* and *arrival direction* (by the method of relative arrival times) of the showers. Details are given in ref. 8.

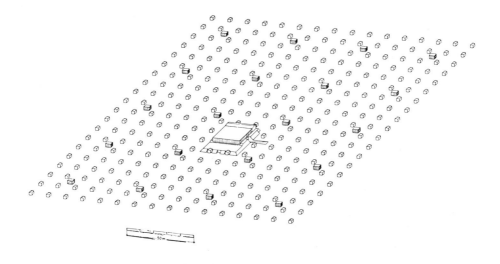

Fig. 6   Schematic view of the KASCADE detector field.

Each station houses 4 scintillation detectors of 0.78 m^2 each for measuring the electron density and a fifth scintillation detector of 3.2 m^2 for muon detection. The latter is shielded by 20 radiation lengths of lead and iron to absorb the electromagnetic component. This results in a muon threshold of ca. 300 MeV. The electron detectors are covered by 5 mm of lead to convert gamma rays for a better timing of the shower front.

The central detector is a hadron calorimeter of 10 interaction lengths thickness. It will be made of iron and concrete with 7 layers of ionization chambers filled with liquid tetramethylsilane. The electrodes of the ionization chambers are divided into pads of 25 x 25 cm^2 which are read out independently, so a spatial resolution of 25 cm is obtained in both horizontal coordinates. The spacing of the layers is chosen in such a way as to obtain an energy resolution of about 35 % independent of energy. An additional layer of scintillators is introduced to provide a fast hadron trigger. Beneath the central calorimeter muons above 2

GeV will be detected and localized by means of multiwire proportional chambers.

The total sensitive area for muons is 1300 m² (i. e. 3.2 % of the array area). The corresponding number for electrons is 1000 m² (2.5 %).

There will be two types of registered events, those triggering the array and a subsample of these (0.8 %) in which the shower core hits the central detector. For a maximum zenith angle of 30° the expected rates are ca. 5000 and 40 d⁻¹, respectively, for a primary energy above 10^{15} eV.

7. Concluding Remarks

These lectures tried to guide your interest to a specific aspect of space which is a giant hall with an unknown number of irregularly distributed accelerators, injecting beams with unknown angular distributions, energy ranges and kinds of particles. The type of questions put in modern cosmic ray research has been illustrated with the scientific motivation and the experimental layout of the Karlsruhe project KASCADE. This detector arrangement, expected to be in operation in 1993, is rather unique. It is fair to say that it will be one of the most effective and powerful instruments for cosmic ray research in the energy range around the "knee" as it determines coherently many parameters of the dominant components (electromagnetic, muonic and hadronic as well) of each individual shower observed.

The experiment achieves high accuracy since the degree of sampling is comparatively high and the fluctuations are governed by the intrinsic fluctuations of the shower development rather than by the limited detector area.

Due to the large sensitivity of the arrangement it is well able to compete in the search of ultrahigh energy radiation from *cosmic point sources*, in addition to studies of a various interesting but experimentally unclear effects like the Tien Shan effect [63] (events with anomalously large penetration in the calorimeter) or the identification of *diffuse gamma rays* [64] produced by cosmic rays in the interstellar medium.

Acknowledgement

The main aim of these lectures is to familiarize a "low-energy" nuclear physics audience with the dominant features and ideas of current cosmic ray research, illustrated with the specific motivation of the KASCADE experiment. The presentation of KASCADE is based on the results of a collaboration of the authors' team of ref. 8. I'm particular grateful to H. J. Gils, P. K. Grieder, T. Thouw and G. Schatz for many useful discussions on the subject, and Dipl. Phys. H. J. Mathes and cand. phys. J. Horzel for carefully reading the manuscript.

References

1. V. Hess, Physikal. Zeitschrift 12 (1911) 998
2. V. Hess, Physikal. Zeitschrift 13 (1912) 1084
3. M. S. Longair, High Energy Astrophysics, Cambridge University Press, Cambridge 1983, p. 219-235
4. A. A. Watson, Nature 326 (1987) 541
5. M. Samorski and W. Stamm, Astrophys. J. 268 (1983) L 17
6. G. Chardin and G. Gerbier, Astron. Astrophys. 210 (1989) 52
7. P. Sokolski, Introduction to Ultrahigh Energy Cosmic Ray Physics, Chapt. 14, p. 181 - Frontiers in Physics 76, Addison-Wesley Publishing Comp., Inc. (1989)
   P. V. Ramana Murthy and A. W. Wolfendale, Gamma-ray astronomy, Cambridge University Press, Cambridge 1986
8. P. Doll et al, KfK-Report 4686 (Juni 1990)
9. K. Suga et al, Acta Phys. Hung. 29 (1970) 423
10. P. Sokolsky, Introduction to Ultrahigh Energy Cosmic Ray Physics, Chapt. 7, p. 84, Frontiers in Physics 76, Addison-Wesley Publishing Comp., Inc. (1989)
    A. W. Wolfendale, Rep. Progr. Phys. 47 (1984) 692
    M. M. Winn, J. Ulrichs, L. S. Peak, C. B. A. McCusker and L. Horton, J. Phys. G : Nucl. Phys. 12 (1986) 633
11. G. B. Khristiansen, Proc. 19th ICRC 1985, La Jolla, 9, 487
12. R. M. Baltrusaitis et al, Phys. Rev. L. 54 (1985) 1875
13. M. Teshima, Proc. 20th ICRC, Moscow 1987, 1, 404
14. K.Greisen, Phys. Rev. L. 16 (1966) 748
    G. T. Zatsepin and V. A. Kuzmin, JETP Lett. 4 (1966) 78
15. E. Fermi, Phys. Rev. 75 (1949) 1169
    M. S. Longair, High Energy Astrophysics, Cambridge University Press 1981, p. 377
16. J. Ormes and P. Freier, Astrophys. J. 222 (1978) 471
17. W. R. Binns et al, Nucl. Inst. Meth. 185 (1981) 415
18. S. Swordy et al, Nucl. Inst. Meth. 193 (1982) 591
19. T. X. Burnett et al, Astrophys. Journ. 349 (1990) L25
20. D. Skobelzyn, Ann. Phys. Leipzig 43 (1927) 354
    P. Auger, R. Maze and T. Grivet-Mayer, Compt. Rend. Hebd. Seanc. Acad. Sci. 206 (1938) 1721
    W. Kohlhörster, I. Matthes and E. Webber, Naturw. 26 (1938) 576
21. A. W. Wolfendale, Rep. Progr. Phys. 47 (1984) 655
22. G. D. Rochester and K. E. Turner, Contemp. Phys. 22 (1981) 425

23. K. Kamata and J. Nishimura, Prog. Theoret. Phys. **6** (1958) 93, Supp.
    K. Greisen, Progr. Cosmic Ray Phys. **3** (1956) 1
24. A. M. Hillas and J. Lapikens, Proc. 15th ICRC 1977, Plovdov (Bulgaria), **8**, 460
    D. J. van der Welt, J. Phys. G : Nucl. Phys. **14** (1988) 105
25. W. R. Nelson et al, Stanford Report SLAC - **265** (1985)
26. T. Kufune et al, Astrophys. J. **301** (1986) 230
27. R. Morse, in : Very High Energy Gamma-Ray Astronomy, ed. K. E. Turner, Reidel Publishing Company, Dordrecht, Boston, Lancaster and Tokyo 1987, p. 197
28. H. T. Freudenreich et al, Phys. Rev. **D41** (1990) 2732
29. J. A. Goodman et al, Phys. Rev. L. **42** (1979) 854 - Phys. Rev. **D26** (1982) 1043
30. R. M. Baltrusaitis et al., Nucl. Instr. Meth. **A240** (1985) 410
31. P. Sokolsky, Introduction to Ultrahigh Energy Cosmic Ray Physics, Chapt. **15**, p. 198, Frontiers in Physics **76**, Addison-Wesley Publishing Comp. Inc. (1989)
32. P. K. F. Grieder, unpublished report 1970, Physikal. Institut, University of Bern (Switzerland) and Institute of Nuclear Study, University of Tokyo (Japan) - Nuovo Cim. **7** (1977) 1
33. J. G. Rushbrooke, CERN-Report EP/85-178 (1985) - Proc. Int. Europhysics Conf. on High Energy Physics, Bari, Italy, 1985
34. Ch. Geich-Gimbel, Report Bonn-HE-87-30; Int. J. Mod. Phys. **A4** (1984) No. 4
35. T. K. Gaisser, U. P. Sukhatme and G. B. Yodh, Phys. Rev. **D36** (1987) 1350
    Cheuk-Yin Weng and Zhong-Dao Lu, Phys. Rev. **D39** (1989) 2606
36. G. Schatz, Int. Report, Kernforschungszentrum Karlsruhe 1989 (unpublished)
    M. K. Hegab, M. T. Hussein and N. M. Hassan, Z. Phys. **A336** (1990) 345
37. H. E. Dixon, K. E. Turner and C. J. Waddington, Proc. Ray Soc. **339A** (1974) 157
38. G. Cocconi, L. G. Koesters and D. H. Perkins, LBL-HEP Seminars 28 (1961) Part 2, UCID 1444
39. L. D. Landau, Isv. Akad. Nauk. SSSR, Ser. Fiz. **17** (1953) 51
40. R. P. Feynman, Phys. Rev. Lett. **23** (1969) 1415
41. K. Alpgard et al, Phys. Lett. **121B** (1983) 209
42. J. Wdowczyk and A. W. Wolfendale, Nature **306** (1983) 24; J. Phys. G : Nucl. Phys. **13** (1987) 411
43. Particle data group, Phys. Rev. Lett. **204** (1988) 92
44. J. N. Capdevielle, Proc. Int. Symp. on Ultrahigh Energy Interactions, Beijiing, **7** (1986) 23
    J. N. Capdevielle and S. Zardan, Proc. 20th ICRC Moscow 1987, **5**, 160
45. J. N. Capdevielle, J. Phys. G : Nucl. Phys. **15** (1989) 909
46. J. N. Capella and Tran Thanh Vanh, Phys. Lett. **B93** (1980) 146 - Acta Phys. Pol **B14** (1983) 459
47. A. Klar and J. Hüfner, Phys. Rev. **D31** (1985) 491
48. A. Giovannini and L. van Howe, Z. Phys. **C30** (1986) 391
49. R. Hagedorn, Nuov. Cim. Suppl. **3** (1965) 147
50. P. K. F. Grieder, Shower Simulation Program and Model Options, Report Physikal. Inst. Universität Bern 1980, unpublished - private communication (1988)

51. J. F. de Beer et al, J. Phys. A : Gener. Phys. 1 (1968) 72
52. J. W. Ellert et al, J. Phys. G : Nucl. Phys. 2 (1976) 978
53. J. Procureur, J. Phys. G : Nucl. Phys. 9 (1983) 835
54. S. I. Nikolsky, J. N. Stamenov and S. Z. Ushev, Sov. Phys. JETP **60** (1984) 18
55. G. Thornton and R.Clay, Phys. Rev. L. 43 (1979) 1622
56. Y. A. Fomin and G. B. Khristiansen, Sov. J. Nucl. Phys. 14 (1972) 360
57. R. W. Ellsworth et al, Phys. Rev. D**26** (1982) 336
58. J. Linsley, Proc. 15th ICRC, Plovdin (Bulgaria) 1977 **12**, 90
    A. A. Andam et al, Phys. Rev. D**26** (1982) 23
59. M. N. Dyakanov et al, Proc. 20th IRCR, Moscow 1987, **6**, p. 147
60. B. S. Acharya et al, Proc. 13th ICRC 1983, Bangalore, 4, 8
61. T. Cheung and P. K. MacKeown, J. Phys. G : Nucl. Phys. **13** (1987) 687
62. P. K. F. Grieder, Nuov. Cim. **84A** (1984) 285
63. A. D. Serdyukov et al, Proc. 20th ICRC 1987, Moscow, **6**, 356
64. P. V. Ramana Murthy and A. W. Wolfendale, Gamma-ray astronomy, Cambridge University Press, Cambridge 1986, Chapt. 4

CONCLUDING REMARKS

V.G.Soloviev

Joint Institute for Nuclear Research, Dubna, USSR

Usually, at conferences and symposiums the basic assumptions and the obtained results are expounded. At schools, there is a possibility to discuss in more detail a nuclear model or an experimental setup and data processing as well as the obtained results. It is also possible to elucidate a new theoretical trend or experimental method. I can say that all these specific features are demonstrated at the present school.

An atomic nucleus is a very complex system of many particles. It is well known how difficult is to solve the problem of three and especially four bodies and how difficult is to measure experimentally their characteristics. In studying the nuclear structure it is necessary to understand and describe each nucleus, i.e. the system of 16, 100, 235 and 238 particles and to find out the specific features of each nucleus, how one nucleus differs from another. It is also difficult to describe interactions between nuclei.

In studying the structure of atomic nuclei the models used to describe the characteristics of excited states become more complicated. As soon as we pass to a new region for any nuclear parameter we get nontrivial and very often unexpected results. No matter what region is that : a new region in excitation energies, in angular momenta, in energy resolution, in masses of colliding nuclei and others. The lectures at this school give the examples of such unexpected results.

Now let us dwell upon high-spin states. About 25 years ago high-spin states were of interest from the point of view of answering the question : "At what angular momenta the transition from a superfluid to a normal state proceeds ?". In studying high-spin states nobody expected backbending, specific change of the shape, and superdeformed states to which the lectures by Prof. W. Nazarewicz were devoted. Therefore, the recent discovery of identical γ - ray sequences both in ^{151}Tb and ^{152}Dy and in ^{150}Gd and ^{151}Tb is completely unexpected. The structure of these identical superdeformed bands was discussed in the lecture by Prof. W. Nazarewicz in terms of the strong-coupling approach.

Superdeformed states were not predicted theoretically though in calculations by Pashkevich, Strutinsky and others in the dependence of the potential energy on the quadrupole deformation at large angular momenta there were observed local minima. In those years people couldn't believe that the shell correction method and the description of an average field would be so good. The discovery of superdeformed states means that the mean field approximation is good and the Saxon-Woods potential can be used to describe and average nuclear field at large excitation energies, large angular momenta and large deformation parameters. I would like to emphasize that a possible separation of an average field of an atomic nucleus is not a mathematical procedure but a reflection of its fundamental properties. Just because of an average field there is such a variety of nuclear properties.

Another unexpected discovery was the cluster radiactivity. It is known that a large enhancement of favoured α decays is only partially explained by pairing. For experimental observation of the cluster radiactivity one needs enhancement many orders of magnitude larger than for the α decay. Therefore, many physicists are sceptical about a possibility of observing the cluster radiactivity. Therefore, the paper by A. Sandulescu, D. Poenaru and W. Greiner published in "Particles and Nuclei" was absolutely unexpected. It was stated that in very heavy nuclei there may be a new type of decay which can be interpreted as a fission extremely asymmetric in masses or emission of a heavy cluster. This paper can be considered as theoretical predictions of rare decay modes. The observation of natural radioactive emission of ^{14}C from ^{223}Ra by J.H. Rose and G.A. Jones opened a vast research with many exciting possibilities. A further observation of the cluster radiactivity was made in the Institute of Atomic Energy in Moscow and the Joint Institute for Nuclear Research at Dubna.

The Rumanian theoreticians made a great contribution to the study of the cluster radiactivity which was reflected in the lectures of this school, and expecially, in the lectures by Prof. A. Sandulescu "Clusters as solitons on the nuclear structure", by Prof. D. Poenaru "Cluster radiactivity in various regions of parent nuclei" and by Prof. I. Silisteanu "Microscopic approach to cluster radiactivities".

In studying the nuclear structure, apart from a detailed investigation of the properties of several nuclei it is necessary to expand the region of nuclei studied. Therefore, the study of nuclei far from stability is an important section in nuclear physics. Here also there were some unexpected things. Transitional nuclei are very rare among stable isotopes, their properties were earlier considered as exceptions from general rules. Most of the nuclei far from stability are transitional. Much attention was paid to the study of the properties of transitional nuclei at the school. Prof. P. von Brentano demonstrated the trixiality and gamma softness in the Ba, Xe region. Prof. F. Poughenon demonstrated recent progresses in the study of neutron deficient light nuclei. The lectures by Prof. J. Gizon were devoted to the study of Ba and Ce isotopes; and the lectures by Prof. C.J. Gross, to light isotopes of Mo and Zr. The calculation of levels within the IBM of light isotopes of Ba was presented by Dr. D. Bucurescu.

Atomic nuclei are of great interest due to the variety of their properties and also as a tool for studying the elementary particle interactions. One of the interesting problems is the study of parity nonconservation in nuclear reactions which has been the theme of the lectures by Prof. Dumitrescu. "Grand unification and double beta decay" was the lecture of Prof. A. Faessler in which he demonstrated the importance of studying double beta decay of answering the question whether neutrino is a Dirac or a Majorana particle.

Of great interest is the problem of how strong are the changes in characteristics of nucleons inside a nucleus in comparison with free nucleons. In this connection, it is interesting to get information on renormalisation of the constant of axial-vector weak interaction G_A in nuclei. For this purpose one needs experimental data on Gamow-Teller beta decay of proton-rich nuclei and their comparison with the results of calculations. This was the theme of the lectures by Prof. O. Klepper. The importance of inclusion of particle-particle interactions in calculating β^+ decays has been pointed out by Prof. A. Faessler.

Mixed-symmetry states have been observed in even-even nuclei spanning the mass range A = 50 to 240. In vibrational and γ-soft nuclei the lowest energy mixed-symmetry state is 2^+, observed by W.D. Hamilton e.a. while in rotational nuclei is 1^+ of the M1 scissors mode, observed by A. Richter e.a. Many theoretical efforts have been devoted to understanding this new type of collective excitation and most of them used IBM-2. Microscopic description within the RPA of 1^+ states in deformed nuclei, made by A. Faessler and R. Nojarov, has been represented by Prof. A. Faessler, Prof. P. von Brentano in his lectures gave an account of the experiments on fragmentation of 1^+ scissors mode in the reaction (γ, γ') with the energy resolution of 4 keV. Prof. D. Warner represented the data on excited 1^+ states in the (t,α) reaction which have been obtained in

Daresbury. These experiments allow one to approach the explanation of the microscopic structure of 1^+ states.

Some lectures were devoted to the experimental study of the structure of low-lying states. Prof. P. Brandolini expounded the data on measurement of g factors for several excited states in ^{140,144}Sm and ^{142}Gd. In his lectures Prof. D. Warner gave the experimental data on excitation of states of deformed nuclei in the (t,α) and (^{12}C, ^{11}C) reactions. Prof. G. Graw expounded interesting experimental data on double quadrupole excitation in ^{112}C via (d,d') and (p,p') and on direct and multipole excitations in ^{96}Zr from 22 MeV (d,d') reactions. Experiments have been performed at the U-120 cyclotron in Bucharest for studying intruder states in Sb nuclei and were reported by Dr. M. Ionescu-Bujor.

The systematisation of experimental data and an empirical technique to extract p-n interaction energies for specific protons and neutrons in a wide variety of nuclei was described in lectures by Prof. R. Casten. The lecture by Prof. H. Rebel was devoted to "New interest in cosmic rays. Aspects of the KASCADE Project".

In the lectures by Profs. J.Gizon, G. Graw, D. Warner and K. Klepper the newcoming and acting accelerators and detectors for studying the nuclear structure were described. Such detectors as NORDBALL, EUROGAM, GAMMA-SPHERE and other new accelerators with high energy resolution, electron accelerator CEBAF and the polarized antiprotons at LEAR allow one to assert the New Era in Nuclear Spectroscopy. This is the nearest future of nuclear physics for 10 - 15 years.

I think that as concerns the nuclear structure and mechanisms of nuclear reactions the first stage of investigations has been performed. We have a detailed information on collective rotational and vibrational low-lying states and giant resonances, high-spin states as well as one and two-quasiparticle states, on the fragmentation of deep hole and high-lying particle states. Giant resonances constructed on excited states, the decay of giant resonances, fragmentation of two-phonon states and so on are now being studied. But this is only a small part of information about an atomic nucleus as a system of many particles. For instance, we have experimental data only on 10^{-6} part of the normalisation of the wave functions of neutron resonances. We still have no answer to the questions: "What is the structure of neutron resonances? Are there statistical regularities or have their wave functions large many-quasiparticle components?" Undoubtedly, there is a vast field of activities in studying the structure of an atomic nucleus. This work is for many decades. The study of the nuclear structure will lead to many brilliant and unexpected results. It will demonstrate once more what properties of a many-particle system cannot be exhibited in interactions of elementary particles. I call nuclear physicists to face the future with optimism and wish all the participants of the School successes in nuclear physics development.

I express my most appreciative thanks to the Organizing Committee for splendid organisation of this interesting school.